MW01351400

Aruba Certified Mobility Associate
OFFICIAL CERTIFICATION STUDY GUIDE
(EXAM HPE6-A42)

First Edition

Steven M. Sowell

HPE Press
660 4th Street, #802
San Francisco, CA 94107

Aruba Certified Mobility Associate
Official Certification Study Guide
(Exam HPE6-A42)
Steven M. Sowell

Published by:

Hewlett Packard Enterprise Press
660 4th Street, #802
San Francisco, CA 94107

ISBN: 978-1-942741-72-5

Printed in Mexico

WARNING AND DISCLAIMER
This book provides information about the topics covered in the Aruba Certified Mobility Associate certification exam (HPE6-A42). Every effort has been made to make this book as complete and as accurate as possible, but no warranty or fitness is implied.

The information is provided on an "as is" basis. The author and Hewlett Packard Enterprise Press shall have neither liability nor responsibility to any person or entity with respect to any loss or damages arising from the information contained in this book or from the use of the discs or programs that may accompany it.

The opinions expressed in this book belong to the author and are not necessarily those of Hewlett Packard Enterprise Press.

TRADEMARK ACKNOWLEDGEMENTS
All third-party trademarks contained herein are the property of their respective owners.

GOVERNMENT AND EDUCATION SALES
This publisher offers discounts on this book when ordered in quantity for bulk purchases, which may include electronic versions. For more information, please contact U.S. Government and Education Sales 1-855-447-2665 or email sales@hpepressbooks.com.

Feedback Information

At HPE Press, our goal is to create in-depth reference books of the best quality and value. Each book is crafted with care and precision, undergoing rigorous development that involves the expertise of members from the professional technical community.

Readers' feedback is a continuation of the process. If you have any comments regarding how we could improve the quality of this book, or otherwise alter it to better suit your needs, you can contact us through email at hpepress@epac.com. Please make sure to include the book title and ISBN in your message.

We appreciate your feedback.

Publisher: Hewlett Packard Enterprise Press

HPE Aruba Contributors: Don McCracken, Kevin Zhu, Kimberly Graves, Venu Dhanraj Puduchery, Fardin Rahim Raoufi, Leo Banville

HPE Press Program Manager: Michael Bishop

About the Author

Steven Sowell has been an IT professional and entrepreneur since 1984, with a focus on WLAN technologies for the past 16 years. He has sold, designed, and deployed WLANs covering millions of square feet, for sporting arenas, hospitals, offices, and outdoor areas. Steven has been a courseware developer, editor, and author for fifteen years, and an industry-certified instructor and engineer since 1994. He has managed and groomed teams of technology instructors and engineers to reach the highest levels of certification, and to exceed customer expectations.

Introduction

This book is based on the Implementing Aruba Wireless course. It will help you to prepare to pass the Aruba Certified Mobility Associate (ACMA) certification exam (HPE6-A42). The material in this book will also help you to understand general wireless technologies, and to understand, deploy, and configure HPE Aruba-based WLAN systems.

Certification and Learning

Hewlett Packard Enterprise Partner Ready Certification and Learning provides end-to-end continuous learning programs and professional certifications that can help you open doors and succeed in the idea economy. We provide continuous learning activities and job-role based learning plans to help you keep pace with the demands of the dynamic, fast paced IT industry; professional sales and technical training and certifications to give you the critical skills needed to design, manage and implement the most sought-after IT disciplines; and training to help you navigate and seize opportunities within the top IT transformation areas that enable business advantage today.

As a Partner Ready Certification and Learning certified member, your skills, knowledge, and real-world experience are recognized and valued in the marketplace. To continue your professional and career growth, you have access to our large HPE community of world-class IT professionals, trend-makers and decision-makers. Share ideas, best practices, business insights, and challenges as you gain professional connections globally.

To learn more about HPE Partner Ready Certification and Learning certifications and continuous learning programs, please visit http://certification-learning.hpe.com

Audience

This book is designed for presales solution architects involved in supporting the sale of HPE Aruba WLAN solutions. It is also designed for engineers that need to learn how to deploy and configure these solutions.

Assumed Knowledge

This is an entry level book. As such, you do not need much prerequisite knowledge. It is assumed that you have a very basic understanding of general networking concepts, such as Ethernet, routing, and switching technology and an interest in learning about the HPE Aruba product line—physical and virtual—to help you understand customers' business issues and to propose appropriate solutions. However, you do not need expert-level knowledge in these areas to comfortably learn the material contained in this book.

Minimum Qualifications

There are no prerequisite qualifications for this certification.

Relevant Certifications

After you pass the exam, your achievement may be applicable toward more than one certification. To determine which certifications can be credited with this achievement, log in to The Learning Center and view the certifications listed on the exam's More Details tab. You might be on your way to achieving additional certifications.

Preparing for Exam HPE6-A42

This self-study guide does not guarantee that you will have all the knowledge you need to pass the exam. It is expected that you will also draw on real-world experience and would benefit from completing the hands-on lab activities provided in the instructor-led training. To pass the certification exam, you should get as much hands-on experience as possible.

Recommended HPE Training

Recommended training to prepare for each exam is accessible from the exam's page in The Learning Center. See the exam attachment, "Supporting courses," to view and register for the courses.

Obtain Hands-on Experience

You are not required to take the recommended, supported courses, and completion of training does not guarantee that you will pass the exams. Hewlett Packard Enterprise strongly recommends a combination of training, thorough review of courseware and additional study references, and sufficient on-the-job experience prior to taking an exam.

Exam Registration

To register for an exam, go to http://certification-learning.hpe.com/tr/certification/learn_more_about_exams.html

CONTENTS

1 WLAN Fundamentals and RF Basics

LEARNING OBJECTIVES

- ✓ Before you begin learning about Aruba-specific implementations, you will learn about fundamental WLAN technologies and RF concepts. We begin with a discussion of WLAN organizations. You will understand Radio Frequency Bands and channels, and the standards used to regulate them. These standards ensure interoperability between devices.
- ✓ You will then learn about radio frequency coverage and interference. Your understanding of these concepts is vital to both the successful implementation and diagnosis of WLAN systems. You will learn about antenna technology and how various options can help ensure you get the right coverage in various deployment scenarios.
- ✓ How do you know if a particular implementation is successful? Of course, you will seek end user feedback and measure the performance of the network yourself. For this endeavor, you must know what metrics to explore. Toward this end, you will learn about RF power, signal strength, and how it is measured.
- ✓ We end with an introduction to WLAN mobility concepts. You will learn about one of the great advantages of WLAN systems—the ability for users to roam about while remaining connected.

WLAN Organizations

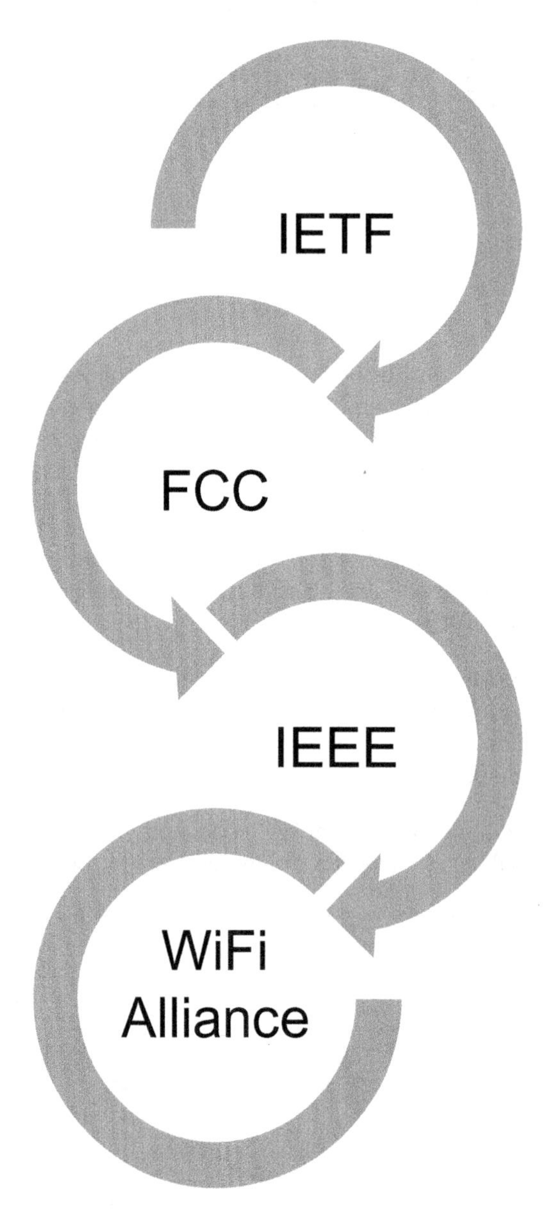

Figure 1-1 WLAN organizations

There are several organizations that develop WLAN standards (Figure 1-1). These regulatory bodies restrict how RF technology is deployed. Organizations involved in these regulations include the following:

- Internet Engineering Task Force (IETF)—International standards
- Federal Communications Commission (FCC)—Local regulatory domain for RF spectrum (United States)
- The Institute of Electrical and Electronic Engineers (IEEE)—802.11 standards
- The Wi-Fi Alliance—WLAN standards

The IETF is an international standards organization for the Internet.

The FCC places certain restrictions on power and channel usage in the US. Other countries have similar entities. For example, channel and power restrictions in Europe are controlled by the European Telecommunications Standards Institute (ETSI). In Japan, it is Nippon Telegraph (NTT).

Within the unlicensed spectrum bounds defined by the FCC, the IEEE creates the 802.11 standards. In other words, the FCC controls which frequencies in the spectrum we can use and at what power settings we can use them. The IEEE committee standardizes how data is transmitted over those frequencies.

802.11 channels, as defined in the standard, must fall within the FCC unlicensed ranges. Over the years, new technology has been introduced to improve the speed and reliability of WLANs. In response, the IEEE continues to amend standards and to create new standards.

The Wi-Fi Alliance further defines standards within the bounds set by the 802.11 standard. The Wi-Fi Alliance promotes interoperability between the vendors of WLAN equipment.

RF bands and channels

2.4 GHz ISM band and channels

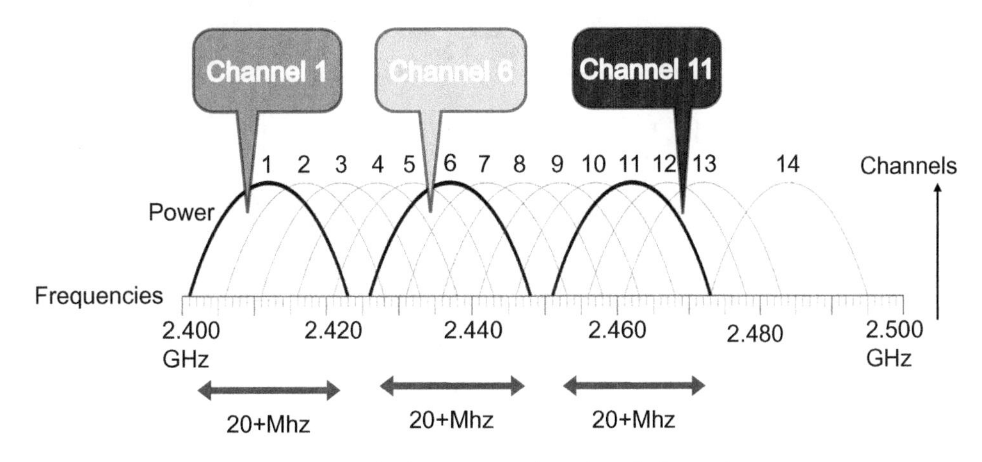

Figure 1-2 2.4 GHz ISM band and channels

The 2.4 GHz ISM (Industrial, Scientific and Medical) Band is used by 802.11, 802.11b, 802.11g, and 802.11n. Using so-called "spread-spectrum technology" data is encoded and spread across a spectrum of frequencies, a bit over 20 MHz wide. However, the channels are only 5 MHz apart, so adjacent channels overlap. When two physically nearby APs use overlapping channels, they can hear each other's RF signals. This results in interference, which can severely degrade performance. This specific type of interference is often called "co-channel interference."

Think about having a business meeting, when multiple separate conversations occur in the same room. If two people talk loudly, it distracts others. Parts of the conversation are missed and meaning is lost. And so it is with WLANs. Due to the overlap, some or all of the transmission may be lost.

In Figure 1-2, you can see that channel 1 overlaps with channels 2–5. There is enough of a gap between channels 1 and 6, and so they do not interfere with each other. Similarly, there is a gap or space between channels 6 and 11, and so there is no co-channel interference between them.

So while there are 11 channels available in the US (13 in Europe, 14 in Japan), only channels 1, 6, and 11 are actually used. This is because they are the channels that do not interfere with each other. Other channels such as 2 and 8 may not interfere with one another because there is enough bandwidth between them, but the only channel plan that can take advantage of three usable channels in the same RF space are 1, 6, and 11.

5 GHz U-NII bands and channels

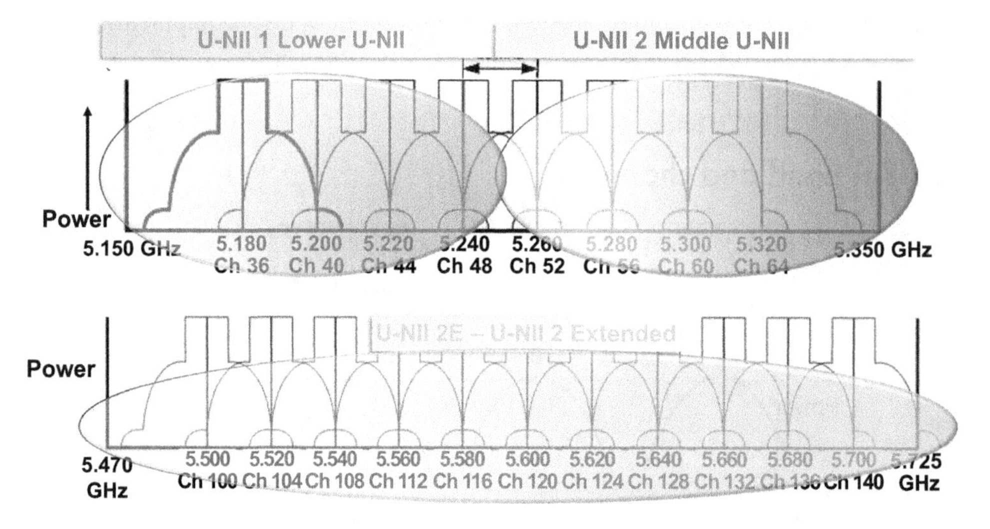

Figure 1-3 5 GHz U-NII 1, 2, 2E bands and channels

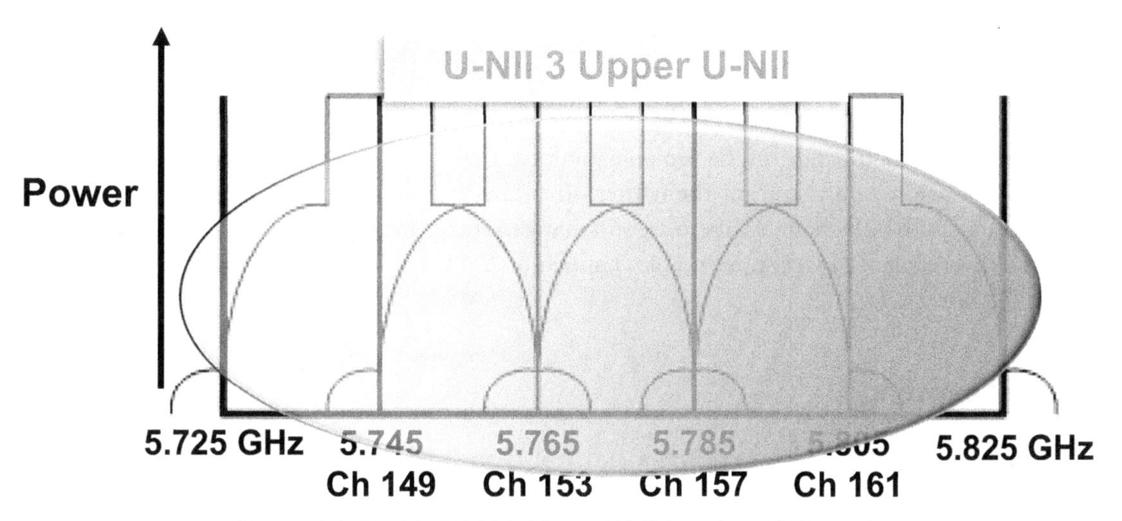

Figure 1-4 5 GHz U-NII 3 Upper U-NII bands and channels

802.11a, 802.11n, and 802.11ac use the 5 GHz Unlicensed National Information Infrastructure (U-NII) bands, as defined by the FCC. Like the 2.4 GHz band, actual channel availability can vary by country. Figures 1-3 and 1-4 show the four sets of U-NII Bands:

- U-NII 1, also known as Lower U-NII
- U-NII 2, also known as Middle U-NII
- U-NII 2E, also known as U-NII 2 Extended
- U-NII 3, also known as Upper U-NII

Originally, U-NII bands 1–3 were defined for use by 802.11a. Each of these three bands was broken into four useable channels. Look at Figure 1-3 and notice that the frequencies used in the U-NII 1 and 2 bands are contiguous. There is a gap between the U-NII 2 and 3 bands that was previously not available for unlicensed use in WLANs. The frequencies in this gap are used for aircraft and military radar systems.

As discussed later, engineers figured out how to open up this space for WLAN use, while avoiding interference with these radar systems. Soon thereafter, the U-NII 2E band was made available, adding an additional 11 channels in the 5 GHz range. (Again, depending on the country).

Originally, the Lower U-NII could be used for indoor use only, the Middle U-NII for indoor or outdoor use, and the Upper U-NII for outdoor use only. This meant that you were likely to have 8 U-NII channels that were useable for indoor WLANs. Only channels that are directly next to each other have any risk of interfering. They are usually at low enough power that there is minimal or no interference. Still, it is typically considered a best practice to avoid using adjacent channels for physically adjacent Access Points.

So, unlike the 2.4 GHz spectrum, all 5 GHz channels are usable, without risk of co-channel interference. However, many client devices lack support for certain 5 GHz channels. This is especially true of the U-NII 2e channels.

The U-NII 2 extended bands can be problematic for actual real-world use, as many NICs either do not support them, or only scan for them after other channels are scanned. However, even without the use of the U-NII 2E channels, there are still many more nonoverlapping channels in 5 GHz than 2.4 GHz. If we can get by with three usable channels on 2.4 GHz, surely eight nonoverlapping channels are fine for 5 GHz.

Channel bonding

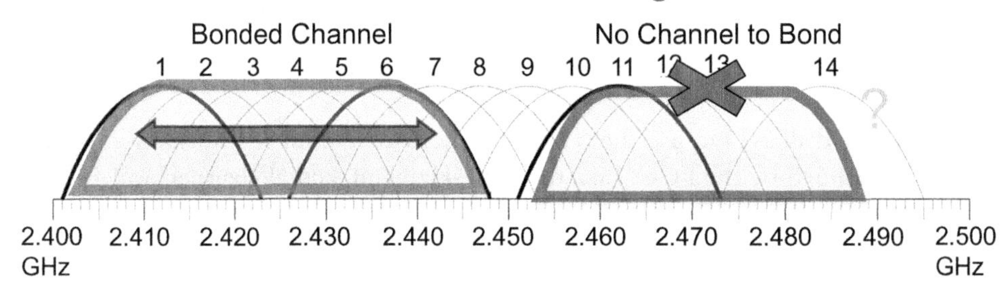

Figure 1-5 2.4 GHz channel bonding

One of the performance enhancements available in 802.11n and 802.11ac is channel bonding. Channel bonding technically does not bond two channels together (Figure 1-5). It is actually using the frequency ranges of those two channels and treating them as a single, much wider channel. This is similar to the concept of modem bonding employed years ago to achieve a higher throughput on dial-up links. Another analogy is that of "link aggregation," which bonds two or more wired Ethernet links into a single, logical link.

By having a wider channel, more information can be transmitted at a time. 802.11n is also referred to as High Throughput or HT. HT20 indicates 802.11n is deployed without channel bonding, using a standard 20 MHz wide channel. HT40 indicates 802.11n is deployed with channel bonding, using a 40 MHz wide bonded channel.

Bonding channels in 2.4 GHz is not a viable solution. With only three usable channels, bonding would effectively leave only one usable channel.

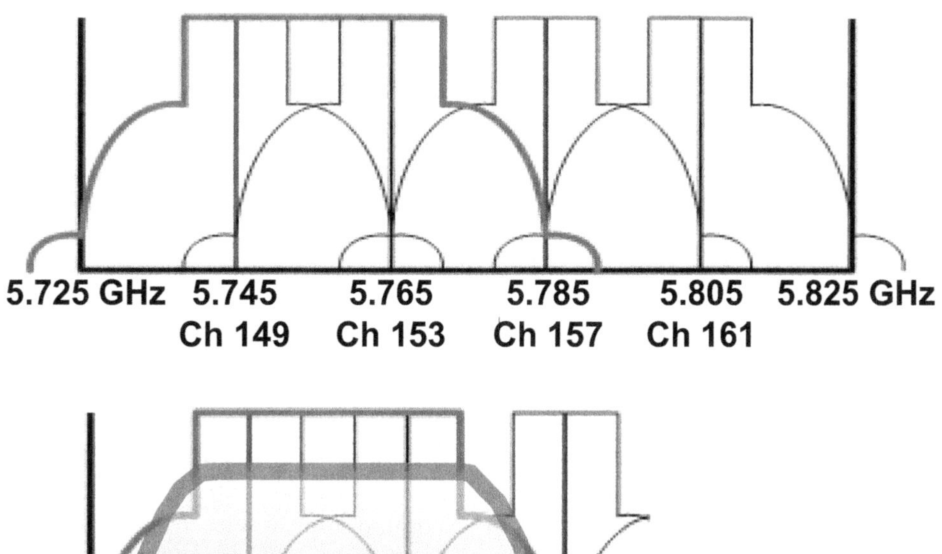

Figure 1-6 5 GHz channel bonding

Bonding is really only effective in 5 GHz for enterprise class WLANs (Figure 1-6). So, for other than home use, bonding is really a 5 GHz discussion. This becomes more evident with 80.11ac, which is purely a 5 GHz technology.

802.11ac 5 GHz allowed channels in FCC countries

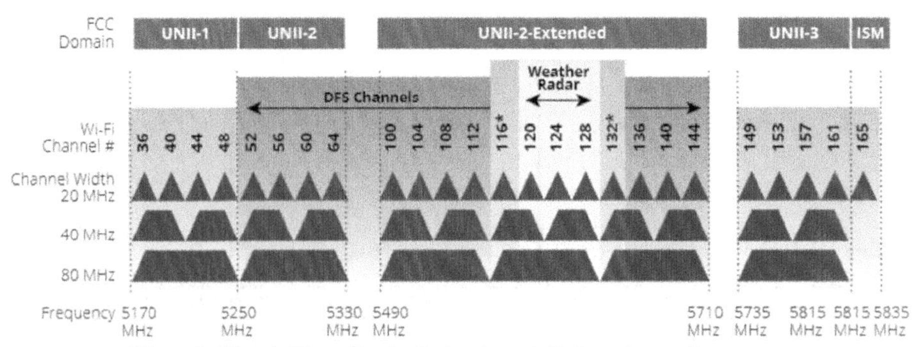

*Channels 116 and 132 are Doppler Radar channels that may be used in some cases.

Figure 1-7 802.11ac 5GHz Allowed Channels in FCC Countries

Figure 1-7 shows channels defined for 5 GHz bands (US regulations) and how the 802.11ac standard can use these bands. As previously described, each 5 GHz channel is 20 MHz wide. The 802.11a standard can leverage each separate channel for communications. With bandwidth limited to a 20 MHz-wide spectrum, (and due to limitations in encoding methods) 802.11a was limited to a maximum data rate of 54 Mbps.

The newer 802.11n standard supports both the 2.4 GHz and 5 GHz spectrum. The 802.11n standard introduced a new concept called "Channel Bonding." This allows you to "bond" two adjacent, 20 MHz-wide channels into a single 40+ MHz-wide path. Users now have over twice the bandwidth. Along with superior encoding methods, maximum theoretical data rates reach 600 Mbps.

The 802.11ac standard further capitalized on this channel-bonding capability, enabling you to bond up to FOUR channels (802.11ac Wave 1) or even EIGHT channels (802.11ac Wave 2), for very high potential bandwidths. However, there is a point of diminishing returns—if you convert twelve 20 MHz channels into three 80 MHz channels, you increase co-channel interference. For this and other reasons, most engineers agree that either 20 MHz or 40 MHz channel width is currently optimal, overall, for a corporate WLAN deployment.

As previously mentioned, 2.4 GHz channel bonding is not feasible in a corporate, enterprise-class network, because there are only three nonoverlapping channels.

The following list shows channels defined for 5 GHz bands (US regulations) and lists the number of 20, 40, 80, and 160 MHz channels available: (channel 144 is now allowed in the US for one additional 20 MHz, one 40 MHz, and one 80 MHz channel). Please refer to Figure 1-7 while reviewing the information below.

US U-NII I and U-NII II bands:

- U-NII I: 5150-5250 MHz (indoors only)
- U-NII 2: 5250-5350 MHz
- 8x 20 MHz channels

- 4x 40 MHz channels
- 2x 80 MHz channels
- 1x 160 MHz channel
- U-NII II requires DFS (& TPC if over 500mW/27dBm EIRP)

US intermediate band (U-NII 2 extended):

- 5450-5725 MHz
- 12x 20 MHz channels
- 6x 40 MHz channels
- 3x 80 MHz channels
- 1x 160 MHz channel
- Requires DFS (& TPC if over 500mW /27dBm EIRP)
- 5600-5650 MHz is used by weather radars and is temporarily not available in the US.

US U-NII 3/ISM band

- 5725-5825 MHz
- 5x 20 MHz channels
- 2x 40 MHz channels
- 1x 80 MHz channel
- Slightly different rules apply for channel 165 in ISM

Learning check

1. Which channels are considered nonoverlapping and will not cause RF interference when co-located?
   a. 1,6,11
   b. 2,4
   c. 1,2,3
   d. 40,48
   e. 6,7,8

2. Channel bonding is a best practice technology for combining bandwidth of _______ channels?
   a. 5 GHz
   b. 2.4 GHz

Answers to learning check

1. Which channels are considered nonoverlapping and will not cause RF interference when co-located?

   a. 1,6,11

   b. 2,4

   c. 1,2,3

   d. 40,48

   e. 6,7,8

2. Channel bonding is a best practice technology for combining bandwidth of _______ channels?

   c. **5 GHz**

   d. 2.4 GHz

802.11 standards and amendments

Compare 802.11a/b/g/n/ac data standards

Table 1-1 provides a comparison of the 802.11a/b/g/n/ac data standards.

Table 1-1 802.11 a/b/g/n/ac data standards

IEEE Standard	Transmission Speed	Frequency and Band	Comment
802.11 (1997)	1, 2 Mbps	2.4 GHz ISM	Original standard. Rarely used anymore. FHSS and DSSS.
802.11b (1999)	1, 2, 5.5, 11 Mbps	2.4 GHz ISM	First standard to gain consumer popularity. Backward compatible with 802.11 DSSS.
802.11a (1999)	6, 9, 12, 18, 24, 36, 48, 54 Mbps	5 GHz U-NII	Slowly gained popularity due to less interference in the 5 GHz frequency range. OFDM.
802.11g (2003)	1, 2, 5.5, 6, 9, 11, 12, 18, 24, 36, 48, 54 Mbps	2.4 GHz ISM	Popular standard, quickly being replaced by 802.11n. Backward compatible with 802.11 DSSS and 802.11b. OFDM.

Table 1-1 802.11 a/b/g/n/ac data standards (Continued)

IEEE Standard	Transmission Speed	Frequency and Band	Comment
802.11n (2009)	6.5–600 Mbps. 300 Mbps is commonly supported with some devices going to 450 Mbps.	2.4 GHz ISM and 5 GHz U-NII	Quickly becoming the standard for both home and enterprise use. Offers high performance along with backward compatibility.
802.11ac (2013)	600 Mbps–6.93 Gbps theoretically, but most WLAN clients operate at a lower speed.	5 GHz U-NII	Wider channel support (up to 160 MHz). Improved Modulation (256-QAM). Up to eight spatial streams. Backwards compatible with 802.11a/g/n. Multi-User MIMO with 802.11ac wave 2

Table 1-1 summarizes a few of the 802.11 standards that are commonly discussed among WLAN experts. As far as the IEEE is concerned, there is only one current standard. It is denoted by IEEE 802.11 followed by the date that it was published. IEEE 802.11-2012 is the only version currently in publication. The standard is updated by means of amendments. Amendments are created by task groups (TG). Both the TG and their finished documents are denoted by 802.11 followed by a non-capitalized letter, for example, 802.11a or 802.11b. The amendments are then incorporated into the standard periodically.

The 802.11 standard was originally ratified in 1997. It was capable of transmitting at 1 and 2 Mbps using either Frequency Hopping Spread Spectrum (FHSS) or Direct Sequence Spread Spectrum (DSSS). The 802.11 standard used the 2.4 GHz spectrum, as recently discussed. This band has been referred to as the Industrial Scientific and Medical (ISM) band. Most 802.11 installations used FHSS. The original 802.11 standard is rarely used anymore. However, the FHSS technology is still used for Bluetooth, although in an improved mechanism.

As an interesting side note, the actress Hedy Lamarr was instrumental in developing this technology. She is an interesting character in history, and worth a quick Internet search.

The 802.11b and 802.11a amendments were both ratified at the same time in 1999. 802.11b products shipped immediately, but 802.11a products did not ship until almost a year later. 802.11b provided an upgrade to 802.11. 802.11b uses DSSS to support transmission speeds of up to 11 Mbps. It provided backward compatibility with 802.11 DSSS and was the first WLAN technology to gain widespread consumer acceptance. This was mainly due to decreased equipment prices.

Like 802.11 and 802.11b, 802.11g operates in the 2.4 GHz spectrum, and so is downward compatible with 802.11b. However, instead of DSSS, 802.11g uses a modulation technique called Orthogonal Frequency Division Multiplexing (OFDM). This is the same modulation technique used by 802.11a, and so both 802.11g and 802.11a support the same theoretical data rates, up to 54 Mbps. Remember, as compared to 2.4 GHz, the 5 GHz spectrum has less co-channel interference (more channels) and fewer non-802.11 sources of interference.

The 802.11n amendment increased the data rate to approximately 300 Mbps and theoretically 600 Mbps maximum. 802.11n added the term High Throughput (HT), indicating speeds up to 300 Mbps for WLANs.

The latest amendment to the 802.11 standard is 802.11ac, which only operates in the 5 GHz range. 802.11ac features theoretical data rates up to 6.93 Gbps. Speeds from 150 Mbps and up are also referred to as Very High Throughput or VHT.

802.11h—DFS and TPC

Earlier, we learned that certain radar systems use portions of the 5 GHz spectrum. Originally, that portion of spectrum was "carved out" by governing bodies, like the FCC, and not available for unlicensed use for WLANs. Specifically, this relates to the formerly unused portion between the U-NII 2 band and the U-NII 3 band.

You may recall that at some point, engineers figured out how to let WLAN systems operate in this spectrum. The IEEE codified these techniques into the 802.11h standard, released in 2003. Soon after, the formerly forbidden spectrum was released as the U-NII 2 extended band.

The two enabling features of 802.11h are called Dynamic Frequency Selection (DFS) and Transmit Power Control (TPC).

With DFS, Access Points constantly listen for RF energy that is likely sourced by some radars or similar systems. If this is detected, the AP must move to a new channel. DFS finds an appropriate, non-interfering channel, and informs all clients that it is moving to this channel. The AP automatically restarts on the new channel, and the clients automatically reassociate to the AP.

TPC offers another level of protection from interference. When devices communicate, they negotiate toward the minimum power level that still allows reliable communications. The objective is to maintain effective data transfer for the WLAN, while minimizing the potential to interfere with other devices.

Vendors are not required to support DFS. However, if they choose not to, they must block the use of the U-NII-2 and U-NII-2E bands.

802.11e—Wireless Quality of Service

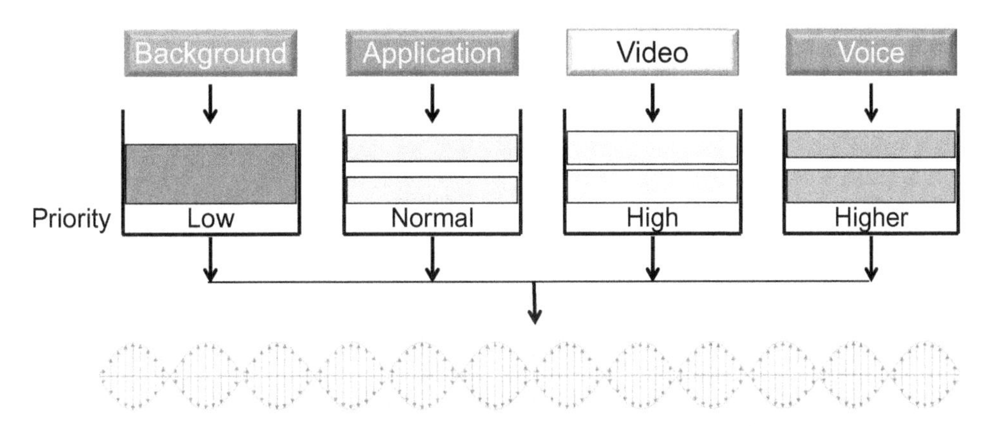

Figure 1-8 802.11e—Wireless Quality of Service

The IEEE's 802.11e standard defines a Quality of Service (QoS) mechanism to define appropriate levels of service for specific applications. Wi-Fi Multimedia (WMM) is a subset of the 802.11e standard, which has become more commonly supported on end systems. As typically deployed, specific application traffic is placed into one of four queues. In Figure 1-8, the top row depicts traffic that is waiting to be transmitted over the WLAN.

This traffic is categorized and organized into queues. There is a special queue reserved for Voice traffic, and another queue for Video traffic. Other business-critical application traffic is placed in the Best Effort queue. Finally, there is a queue labeled Background that may be used for "scavenger-class" traffic—for applications that are not business critical.

Each of the four queues receive a certain priority, or level of service. Since voice and video traffics are more sensitive to delay, they receive higher levels of priority. High network latency can cause voice or video frames to be dropped.

The Best Effort queue will receive what is considered a "normal" priority, while the Background queue will receive the lowest priority.

As shown at the bottom of Figure 1-8, the data from these queues is modulated onto a 2.4 GHz or 5 GHz carrier wave and transmitted as a wireless signal. The highest priority traffic will tend to be sent first, while the low-priority traffic will have to wait longer for service.

802.11i Security

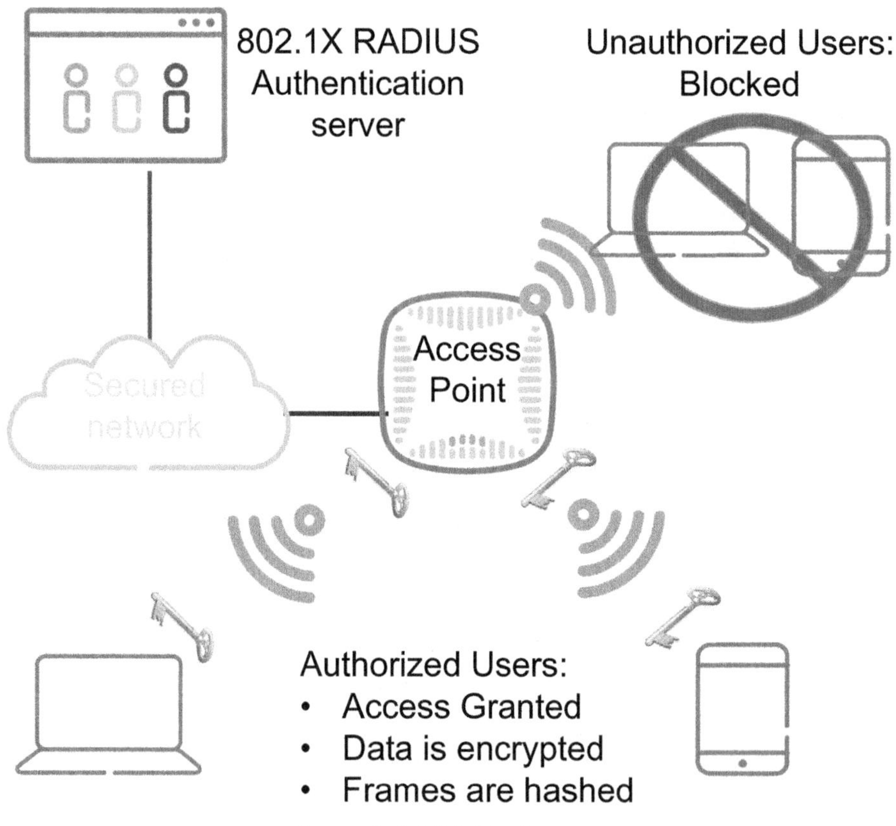

Figure 1-9 802.11i Security

802.11i is an IEEE security standard that supports authentication, encryption, and hashing.

- **Authentication** ensures only valid users can connect. This can be accomplished using a passphrase, which is a preshared ASCII text string. Another option is to assign unique usernames, passwords, and/or certificates to each user. These credentials are authenticated against a back-end RADIUS server. Of course, users with incorrect credentials are blocked from accessing the network, as shown in Figure 1-9.
- **Encryption** prevents attackers from seeing packet data. This is based on the Advanced Encryption Standard (AES) algorithm and uses CCMP (see note).
- **Hashing** algorithms offer protection against man-in-the middle attacks, or anti-replay. This detects if an attacker captures a packet, modifies the data, and then resends it. Hashing is also based on the AES algorithm.

 **Note for CCMP**

The actual protocol is (are you ready?) the Counter Mode with Cipher Block Chaining-Message Authentication Code Protocol. We shorten this first to Counter Mode CBC-MAC Protocol, and then on down to simply CCMP.

The IEEE's 802.11i standard is implemented by the Wi-Fi alliance as Wi-Fi Protected Access version 2 (WPAv2). In other words, WPAv2 uses the same AES-based protocols as 802.11i, for both encryption and hashing.

The Wi-Fi alliance's WPAv1 standard uses the same RC4 encryption algorithm used by the outdated and insecure Wired-Equivalent Privacy (WEP). However, WPAv1 vastly improves upon the security by adding an encryption "wrapper" called the Temporal Key Integrity Protocol (TKIP). Unlike WEP, WPAv1 also provides hashing, based on the "Michael" algorithm in which a Message Integrity Code (MIC) is created for each frame. WPAv1's algorithms can also be used to support legacy devices in an 802.11i network.

Wi-Fi alliance certifications

The Wi-Fi alliance provides standards to help ensure interoperability between APs and clients from different vendors. Table 1-2 provides an overview of the Wi-Fi Alliance certifications.

Table 1-2 Wi-Fi alliance certifications

Wi-Fi Alliance Security	Authentication Mechanism	Encryption	Hashing
WPA-Personal	Preshared Key	RC4-TKIP	Michael-MIC
WPA-Enterprise	802.1x/EAP RADIUS Server	RC4-TKIP	Michael-MIC
WPA2-Personal	Preshared Key	AES-CCMP TKIP (optional)	AES-CCMP Michael (optional)
WPA2-Enterprise	802.1X/EAP RADIUS Server	AES-CCMP TKIP (optional)	AES-CCMP Michael (optional)

Study Table 1-2 and notice that WPA-Personal and WPA-Enterprise use the same *encryption* protocol and algorithm for security, but use different *authentication*. WPA-Personal uses a simple Preshared Key (PSK) method of authentication. This is a much simpler authentication mechanism than WPA-Enterprise, which uses EAP/RADIUS authentication to a RADIUS server. The details of WPA/WPA2 and EAP/RADIUS Authentication will be covered more in a later module. This slide is just to introduce the 802.11i standard.

WMM is a Wireless Multimedia standard to provide QoS and prioritize Voice and Video traffic.

Other 802.11 Standards

The following list describe few more important 802.11 standards.

- **802.11k (Radio measurement standard)** improves a WLAN Client's search for nearby APs, which could be potential roaming targets. It does this by creating an optimized list of channels. When the signal strength of the current AP weakens, your device scans for target APs from this list.
- **802.11v (Improved transition between access points)** enables smooth client transition between access points, using a technique called Basic Service Set (BSS) transition-management. The network's control layer provides endpoints with the load information for nearby access points. Supporting clients take this information into account when deciding among possible roam targets. Support for this protocol continues to increase, but some clients may not yet support it.
- **802.11r (Roaming standard)** facilitates client roaming between APs on the same network. 802.11r uses a feature called Fast Basic Service Set Transition (FT) to authenticate more quickly. FT works with both PSK and 802.1X/EAP authentication methods.
- **802.11ax (Higher throughput data standard)** is a new WLAN data standard due to be ratified in 2019. It is designed to improve overall spectral efficiency. It is predicted to support data rates of up to approximately 10 Gigabits/second.

"Roaming" is the ability of an end-user device to seamlessly move from one AP to another. The first three standards in the list above can serve all to improve the roaming capability of tablets, smartphones, and handheld WLAN devices. Often these personal devices do not roam very well on their own, and roaming is very important for enterprise-class WLAN deployments.

Note

More information about roaming can be found in the Aruba document "Optimizing Aruba WLANs for Roaming Devices" and "RF and Roaming Optimization for Aruba 802.11ac Networks."

802.11 Frame Types

The following describes the three 802.11 frame types and the most important subtypes associated with each frame type.

Management Frames are used to establish, control, and maintain client connections (Figure 1-10).

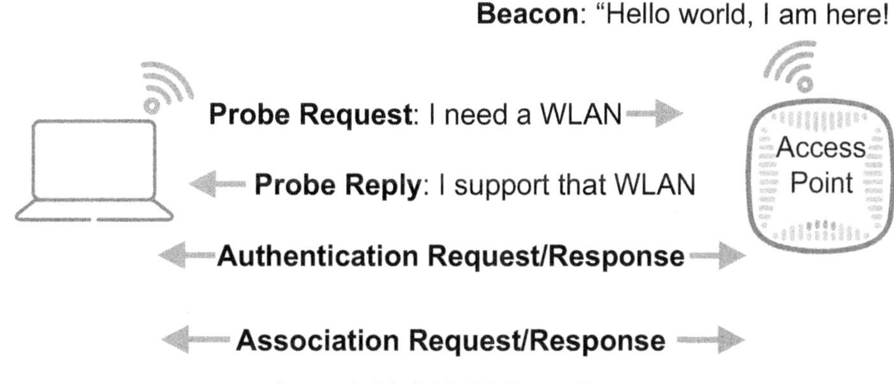

Figure 1-10 802.11 Frame Types

- APs broadcast periodic Beacon frames—"Here I am," along with basic information about the WLAN. This facilitates an endpoint's *passive* discovery of APs and WLANs.
- Clients send probe request frames to facilitate *active* discovery of APs and WLANs.
- APs send a Probe Reply in response to endpoint Probe Request. This contains essentially the same information as a Beacon frame. Clients use the information in Beacon and Probe frames to build a list of available networks.
- Authentication and deauthentication frames act as a handshake mechanism for the initial client connection request.
- Association frames are used to complete a client's 802.11 connection. Of course, disassociation frames are used to disconnect clients.

Control Frames are used to control access to the channel and to acknowledge receipt of frames.

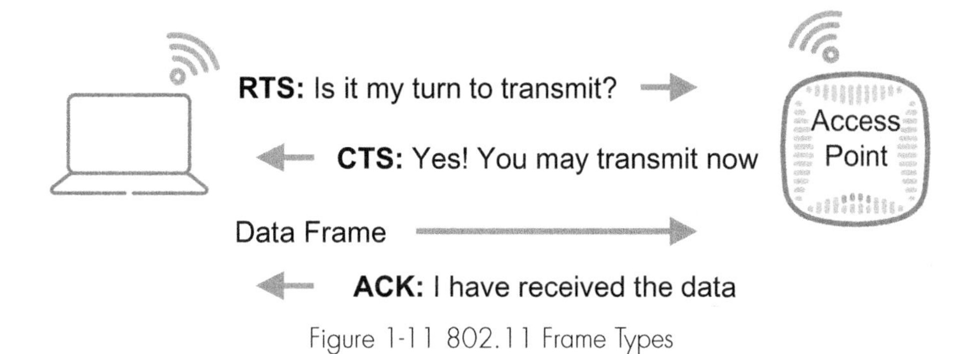

Figure 1-11 802.11 Frame Types

- A device transmits a Request-To-Send (RTS) frame to gain exclusive rights to an AP's channel. The AP confirms this request with a Clear-to-Send (CTS) frame (Figure 1-11).
- Each frame received by a device must be acknowledged, and this is done with an ACK frame. A Block ACK frame is used to acknowledge successful receipt of a series of frames. This is more efficient that acknowledging each individual frame with an ACK.

Data frames are used to transmit actual data over the WLAN, and on to some destination, such as a web site or corporate application. The interaction of Data and ACK frames is already shown in the previous figure.

- Data frames send data, as described above
- QoS data frames transmit data using a QoS method based on 802.11e, as recently discussed.

Learning check

3. Which 802.11 standards support data speeds up to 54 Mbps?
   a. 802.11a
   b. 802.11b
   c. 802.11g
   d. 802.11n
   e. 802.11ac

4. Which 802.11 standards and amendments operate at 5 GHz?
   a. 802.11a
   b. 802.11b
   c. 802.11ac
   d. 802.11g
   e. 802.11n

5. Which 802.11 standard defines security mechanisms?
   a. 802.11a
   b. 802.11i
   c. 802.11e
   d. 802.11k
   e. 802.11v

Answers to Learning check

3. Which 802.11 standards support data speeds up to 54 Mbps?
    a. **802.11a**
    b. 802.11b
    c. **802.11g**
    d. 802.11n
    e. 802.11ac

4. Which 802.11 standards and amendments operate at 5 GHz?
    a. **802.11a**
    b. 802.11b
    c. **802.11ac**
    d. 802.11g
    e. **802.11n**

5. Which 802.11 standard defines security mechanisms?
    a. 802.11a
    b. **802.11i**
    c. 802.11e
    d. 802.11k
    e. 802.11v

RF Basics

RF coverage versus capacity

Wireless network design involves a tradeoff between maximizing AP coverage area, versus ensuring good performance and capacity, given multiple users per AP. The first model for designing is based upon purely providing RF coverage in a given area. Meanwhile, the second approach takes into account the number of users and their application bandwidth and speed requirements. Both approaches are concerned with providing 100% coverage for all areas. However, the capacity-based design adds additional requirements.

Coverage

One approach to WLAN design is to maximize the square footage that each AP can cover. This results in relatively few APs spread farther apart. To ensure coverage, each AP operates at high power settings. The motivation is to save money by using fewer APs.

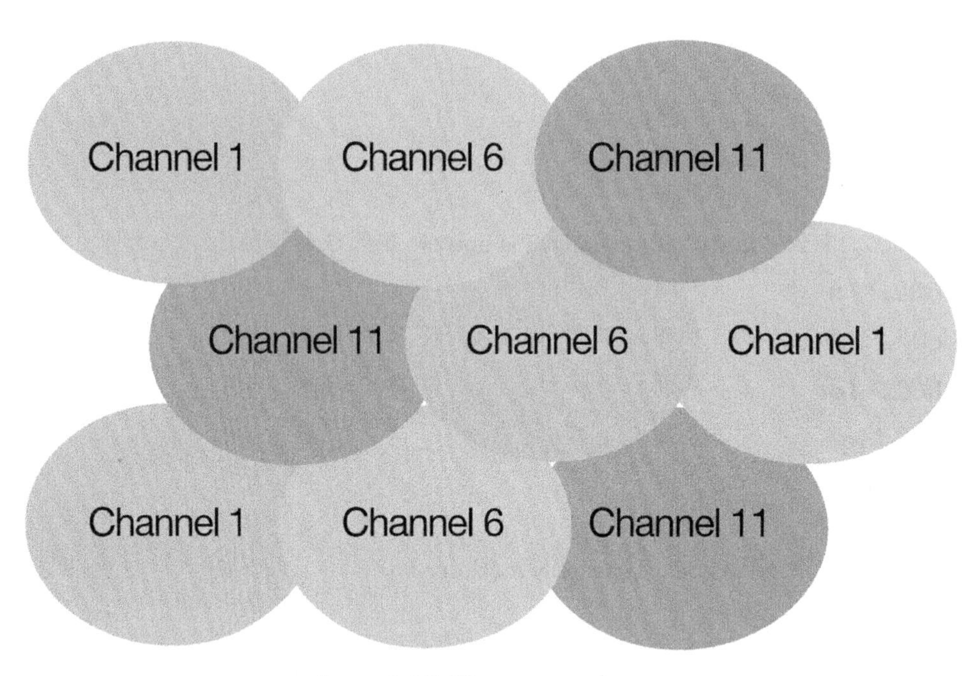

Figure 1-12 RF coverage design

Capacity

Another approach is to design the WLAN with an eye on expected number of users and bandwidth requirements for their applications. Using this approach, APs are more closely spaced, and so can operate at a lower power setting to achieve proper coverage. The motivation is to ensure that all users connect at high data rates, so they will experience superior throughput.

Also, since each AP covers a smaller area, there will be fewer users per AP. Compare and contrast the Figure 1-12 for coverage design versus Figure 1-13 for a capacity design. Both design options cover the same amount of floor space in an office.

In a coverage design, you might have, say 40 users per AP. Using a Capacity design for the same area you may only have 15 or 20 users per AP. This adds yet another performance improvement.

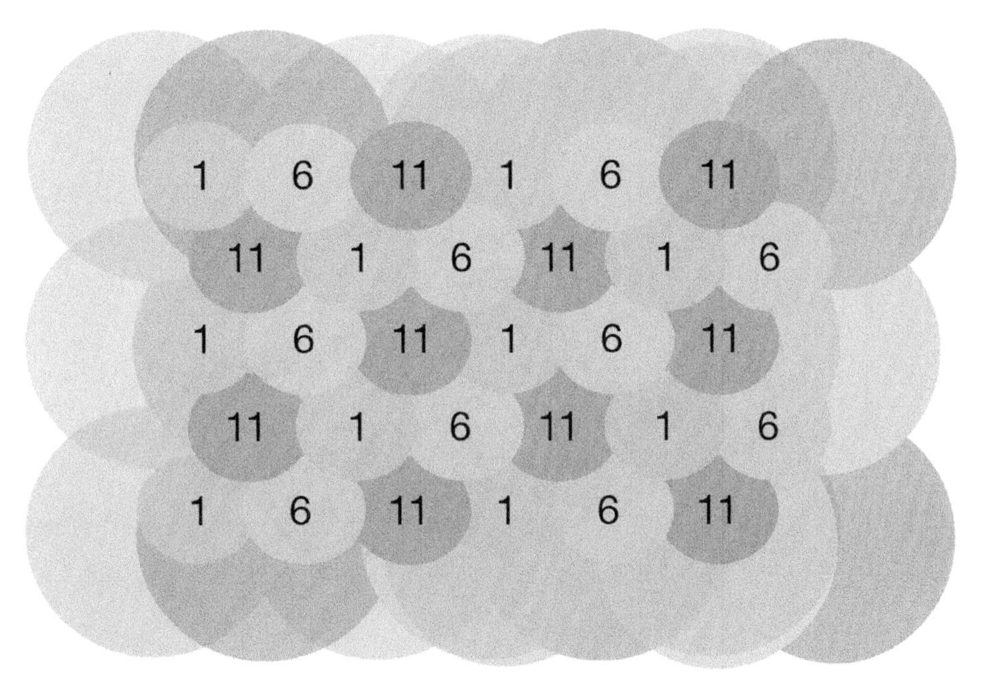

Figure 1-13 RF capacity design

Another advantage to this design is resiliency. Should an AP fail, surrounding APs can increase their power to "fill in the gap." Also, if new walls or other obstacles are erected, this design can more easily adapt.

Which is best?

Larger cells might be appropriate for a low-density deployment, such as a big open area with only a few users that use something like a simple handheld scanner, which is low bandwidth. The coverage design saves money, since you purchase fewer APs.

However, typical office spaces would be better served with smaller cells that included fewer users in each cell. Yes, this means you need to purchase more APs to cover the same area. However, if you try to save money by purchasing few APs, your users will likely be plagued by poor performance, less reliable service, and poor roaming characteristics.

Channel availability

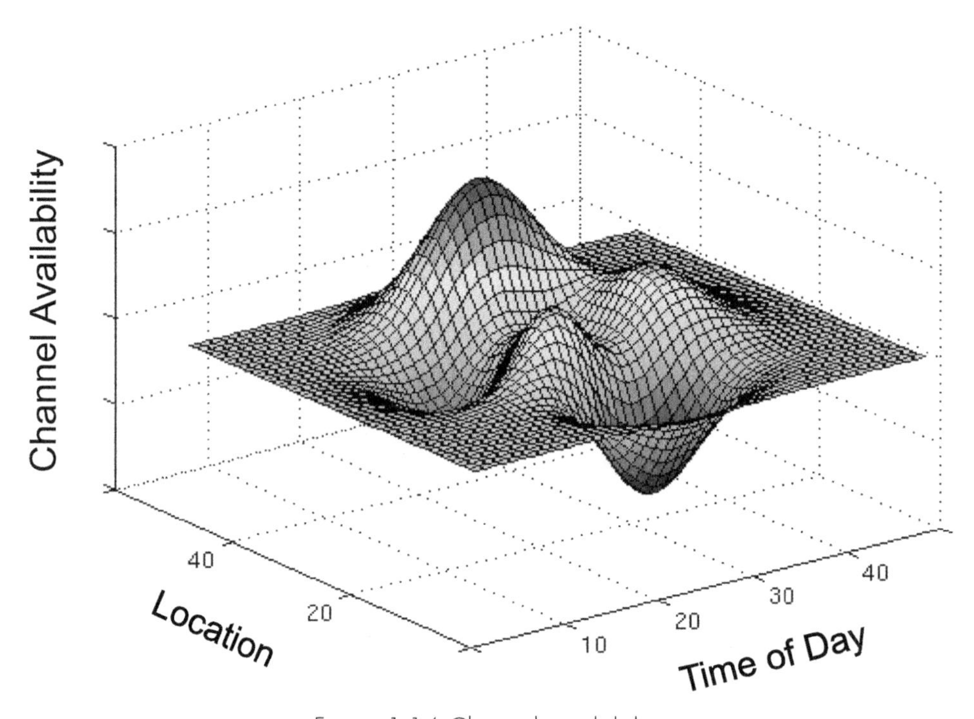

Figure 1-14 Channel availability

One challenge to designing and troubleshooting wireless networks is the ever-changing nature of RF environments. RF Interference can come from other 802.11 devices, in the form of co-channel interference. You will also encounter non-802.11 RF transmitters, such as microwave ovens and Bluetooth devices.

Both sources of interference can cause signal degradation, poor performance, and dropped connections. Interference is both inevitable and unpredictable. It varies by device (microwave and cordless phone), usage pattern (time variance), and location (local emissions regulations and construction).

In Figure 1-14, darker shades or "hot" colors like red and yellow represent high levels of radio energy, at a specific location, during a specific time of day. This energy could be emanating from some interferer, such as a microwave oven. Lighter shades or "cool" colors like blue represent very low levels of radio energy.

The RF environment must be managed to ensure reliable Wi-Fi operation. By the end of this course, you will understand several factors that cause interference and get some ideas for dealing with them.

As previously mentioned, 5 GHz is nearly always "cleaner air." With more available channels, there is less co-channel interference. Also, fewer non-802.11 interferers use the 5 GHz spectrum, as compared to 2.4 GHz. Most of the interfering devices you will encounter, such as those mentioned here, operate in the 2.4 GHz spectrum.

RF Problems

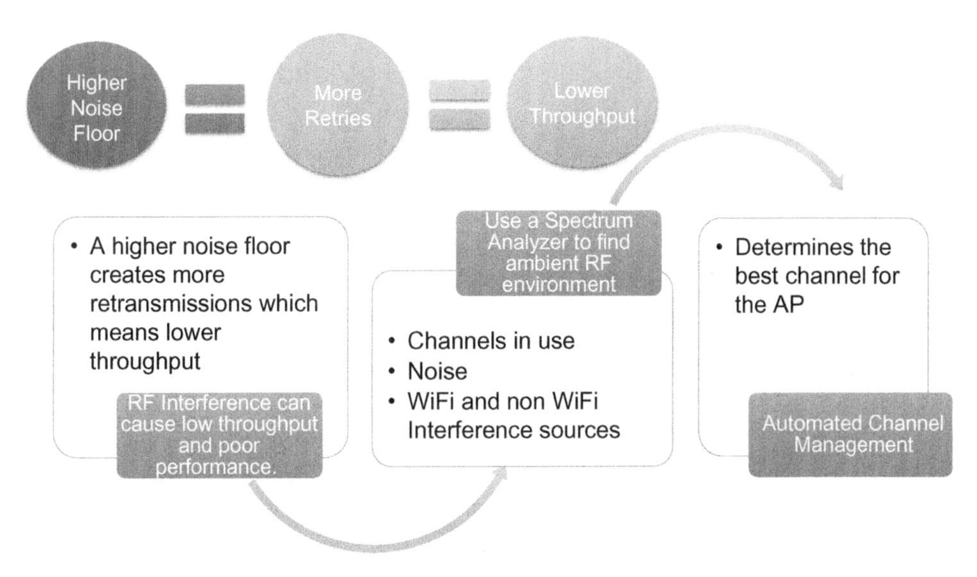

Figure 1-15 RF problems

A common WLAN problem is slow throughput. For example, a WLAN client may be connected to an AP at a 54 Mbps data rate (Figure 1-15). However, during file transfer, actual throughput may be closer to 5 Mbps. A common cause of this is a noisy RF environment. This raises the "noise floor"—the general "background noise" present in any RF environment. This elevated noise floor can drown out desired transmissions and cause APs or clients to retransmit data more often.

Imagine walking down the street, talking to a friend, when a noisy truck passes by. Although your friend is very close to you, and speaking clearly, you may miss some of the conversation. This is because the truck is louder than that of your friend's voice. "What?" you say, and your friend must repeat himself, like endpoints may need to retransmit garbled data frames. Frequent retransmissions waste time, which results in lower throughput.

The term "co-channel interference" is used to describe the condition when your own APs interfere with other AP signals on the same (or overlapping) channel. Meanwhile, non-802.11 sources of interference is often referred to as "noise." Co-channel interference can be mitigated by following best practices for network design and deployment and by using automated channel management. Noise can often be mitigated by finding and eliminating or reducing the source. If it is not feasible to eliminate the source, you can mitigate the interference by situating APs strategically away from the source. You can also tune controller/AP settings to avoid the interference and/or make them less sensitive to it.

HPE Aruba controllers have automated AP channel and power management features. This forms a very effective defense against RF Noise. The Controller can automatically and dynamically change AP channel and power settings to optimize performance.

RF WLAN interferers

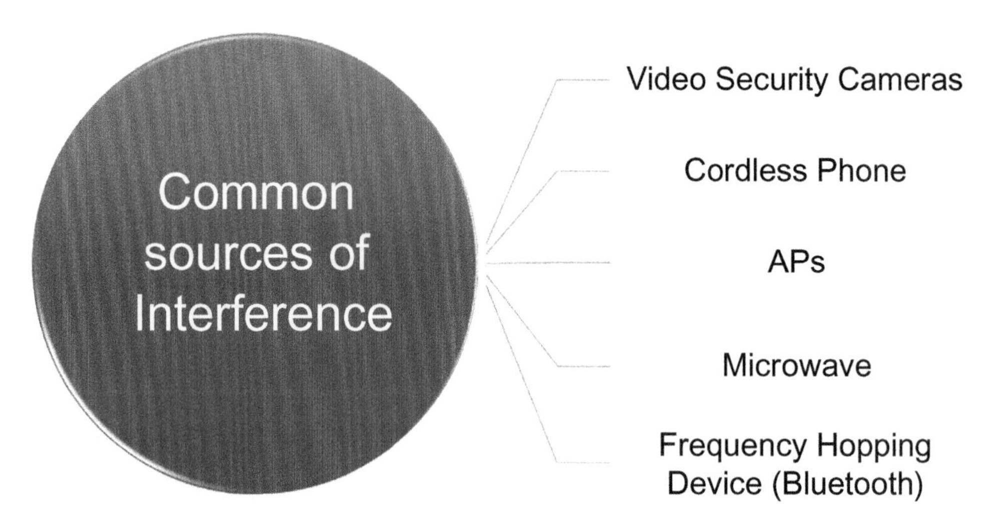

Figure 1-16 RF WLAN interferers

Some of the most common sources of interference in a WLAN are video cameras, cordless phones, third party APs, Microwave Ovens, and Bluetooth or other Frequency Hopping Devices. Most of these interferers operate in the 2.4 GHz spectrum, with fewer interferers typically found at 5 GHz. A spectrum analyzer is an effective tool, used to identify and locate sources of interference (Figure 1-16).

Two elements of interfering devices to be aware of are "duty cycle" and "decibels" (dB). Duty cycle describes how often an interferer is active, over a given time period. Decibels provide a scale to measure the strength of the signal.

Suppose you are standing near a busy highway and a loud truck passes by. The loudness of the truck noise is measured in dB. RF Power or signal strength measurements are also based on dB. Now, you speak to your friend for two minutes, and a very slow-moving truck crawls along next to you, for the entire two minutes. That is a duty cycle of 100%. The interference was active for 100% of the time. However, if a fast plane flies overhead in 30 seconds, that interference has only a 25% duty cycle.

Think about an actual WLAN deployment. There are often many Bluetooth devices, which operate in the 2.4 GHz range. However, Bluetooth typically has a very low duty cycle, and so this may not be as much of a concern as first imagined. Although this is not always the case, Bluetooth devices typically have little real-world impact on WLANs.

Meanwhile, there may only be a few microwave ovens in your facility, but while they are in use, the interference profile has a 100% duty cycle. This can be an egregious source of interferes. Non-802.11

based security cameras are also of high duty cycle and often have a big impact on the WLAN—especially when operating at high dB levels.

Aruba spectrum analysis

An Aruba spectrum analyzer will help you to visualize the RF environment. Any Aruba AP can be configured to act as a spectrum analyzer. This configuration will be covered in a later module.

The channel utilization chart

The channel utilization chart reveals both 802.11 and non-802.11 RF energy. This can include microwave ovens, RF transmitters, DSSS cordless phones, for example. In Figure 1-17, there is a column for each potential channel in the 2.4 GHz spectrum. The height of each column indicates the strength of the signal.

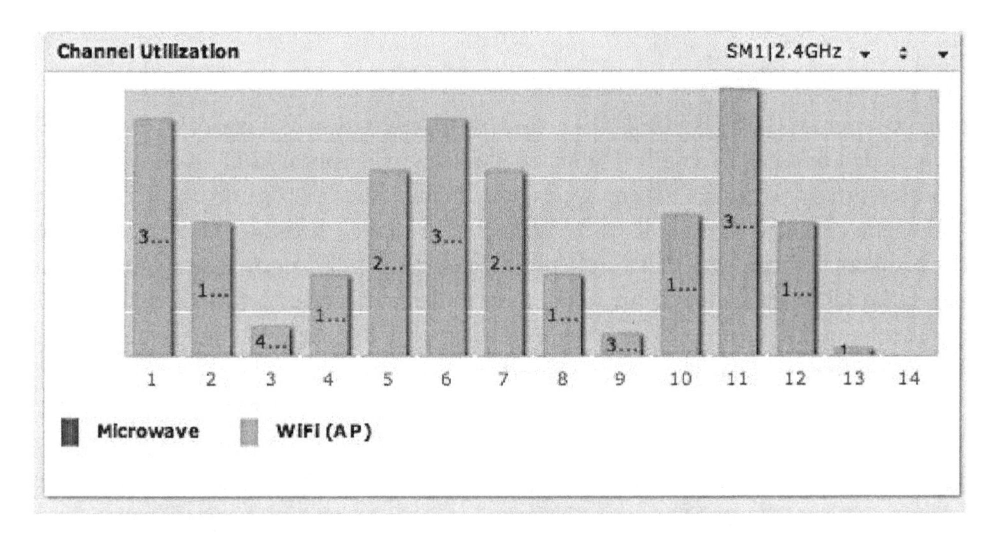

Figure 1-17 Aruba spectrum analysis: Channel utilization chart

The Active Devices chart

The Active Devices chart (Figure 1-18) shows APs or other RF transmitters. It can indicate associated RF dB levels, duty cycle, and affected channel. This particular pie chart indicates that nearly all of the RF energy seen is from APs. This is good news!

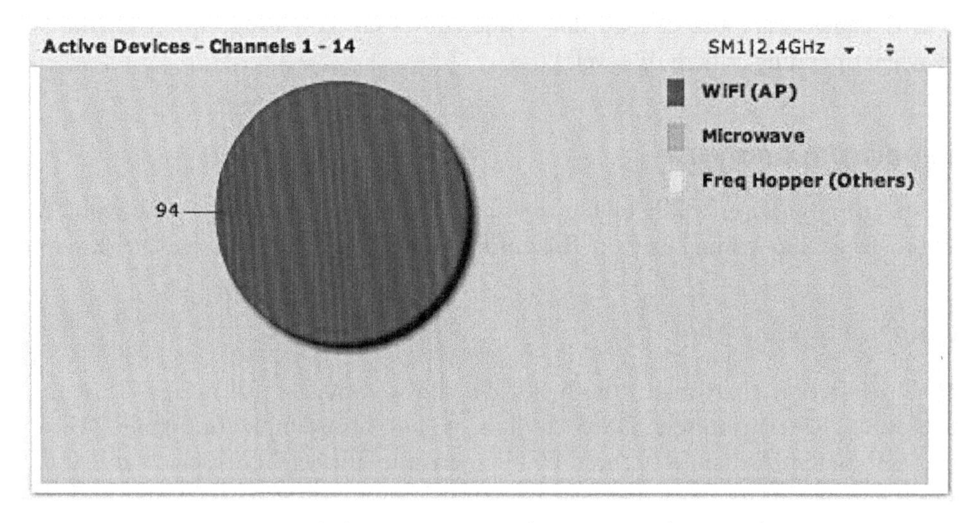

Figure 1-18 Aruba spectrum analysis: Active devices chart

The Swept Spectrogram chart

The Swept Spectrogram chart (Figure 1-19) is sometimes referred to as a waterfall chart. It shows RF energy, color coded by signal strength. Darker shades (that will appear as red or orange in the actual chart) indicate hotter or stronger RF energy. Lighter shades (such as blue or purple) indicate cooler, weaker RF energy. Look at the small white line or "tick mark" directly above the channel 1 label (ch1). The band above it consists mostly of lighter colors such as green or blue. However, the darker shade near to the adjacent tick mark just to the right indicate that there is quite a bit of red, or strong RF energy, for channel 2. We know that channel 2 overlaps with channel 1, and so channel 1 is probably experiencing the negative effects of this interference.

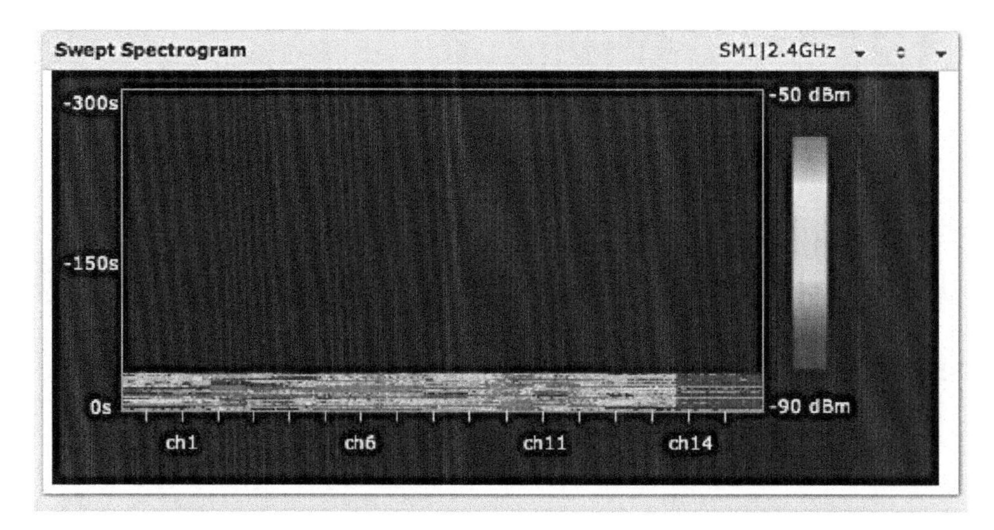

Figure 1-19 Aruba spectrum analysis: Swept spectrogram chart

Real-time Fast-Fourier Transform (FFT) chart

The FFT chart draws a "max hold line." This is the highest dB value received over a period of time. Looking at this chart, you can see the strength of the signal over each channel. With experience, you may also be able to interpret what type of device is transmitting, by analyzing the overall shape of the waveform. This type of experience is far less necessary, since the Aruba Spectrum analyzer (Figure 1-20) will automatically detect several types of devices/categories:

- 802.11 Devices
- Access Points
- Non-802.11 Devices
- Microwave—2.4 GHz only
- Bluetooth—2.4 GHz only
- Fixed Freq (Others)
- Fixed Freq (Cordless Phone)
- Fixed Freq (Video)
- Fixed Freq (Audio)
- Freq Hopper (Others)
- Freq Hopper (Cordless Network)
- Freq Hopper (Cordless Base)
- Freq Hopper (Xbox)—2.4 GHz only
- Microwave (Inverter)—2.4 GHz only
- Generic Interferer

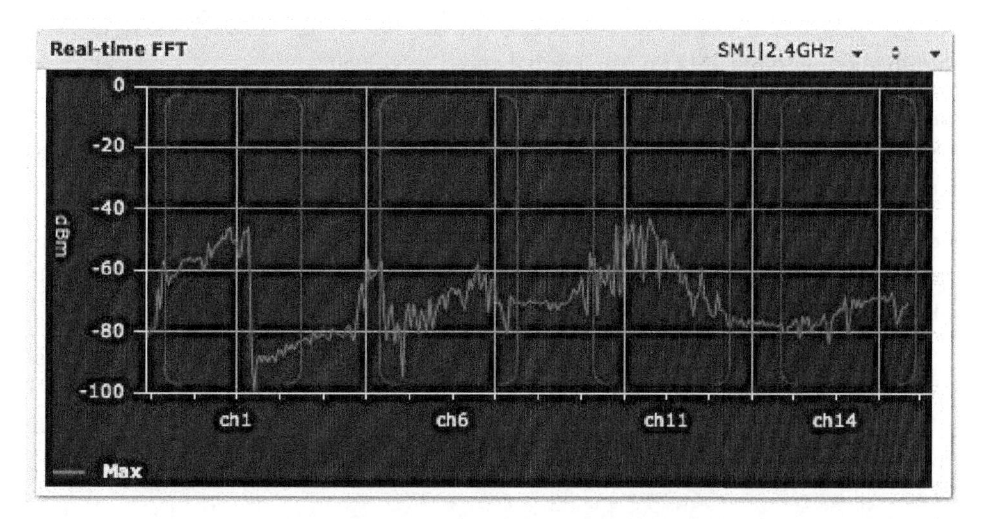

Figure 1-20 Aruba Spectrum analysis: Real-time FFT chart

Antenna technology

Antenna types

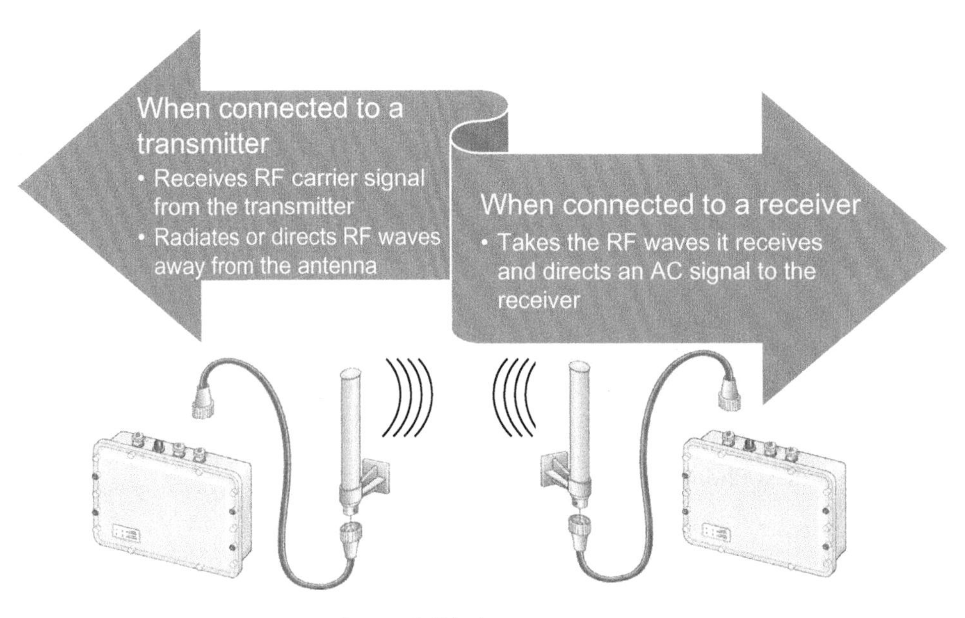

Figure 1-21 Antenna types

The antenna can perform two functions (Figure 1-21). Some devices, like an AM/FM radio only *receive* RF signals. Other devices only *transmit* RF signals. An example of this is an AM/FM radio station broadcasting unit. It can either transmit or receive, but not both.

Wireless APs and endpoints both transmit and receive, and so the antenna serves both functions. When the device transmits data, the antenna receives an oscillating carrier signal from the transmitter and radiates or directs the RF waves outward from the antenna. When the device receives data, the antenna receives the RF signal and directs an oscillating carrier to the receiver.

Two common antenna types are omnidirectional and directional:

- **Omnidirectional**—"Omni" means "all" or "in all ways or places," and so an omnidirectional antenna radiates energy in all directions. Often referred to as simply an "omni" antenna, they radiate energy in a kind of oval or a "squashed sphere" shape.
- **Directional**—A directional antenna focuses more energy in a single direction, resulting in less energy in all other directions. Such an antenna may also be referred to as a "sectional" or "sector" antenna, which again describes its radiation pattern.

Reading radiation patterns horizontal plane (azimuth)

AP-320 Antenna Pattern Plots

Horizontal planes (top view, AP facing forward)

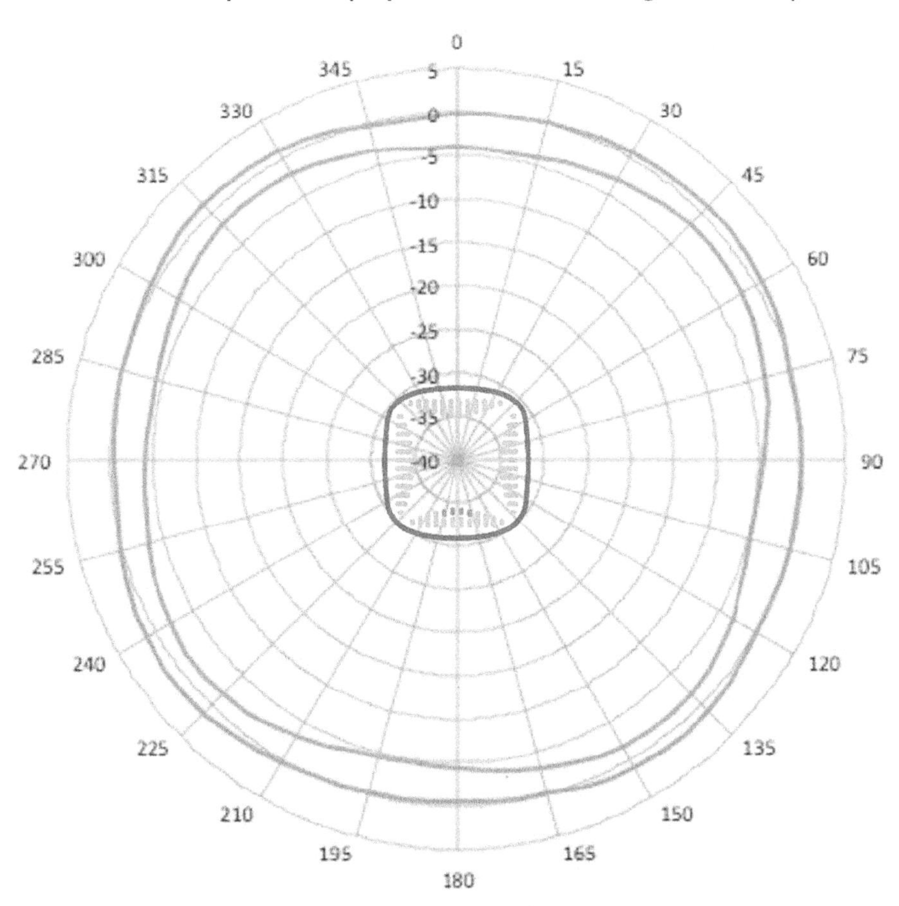

Inside pattern: 2.4GHz WiFiAverage Azimuth

Outside pattern: 2.4GHz WiFiAverage Downtilt 30

Figure 1-22 H-plane 2.4 GHz Wi-Fi (antennas 1, 2, 3, 4)

Most vendors provide antenna radiation patterns—how the antenna shapes the RF signal that emanates from an AP. Figure 1-22 shows an overhead view of an antenna radiation pattern, as if you were hovering over the AP, looking down upon it. This is called the horizontal coverage, or H-plane. It can also be referred to as the azimuth.

Reading Radiation Patterns Vertical Plane (elevation)

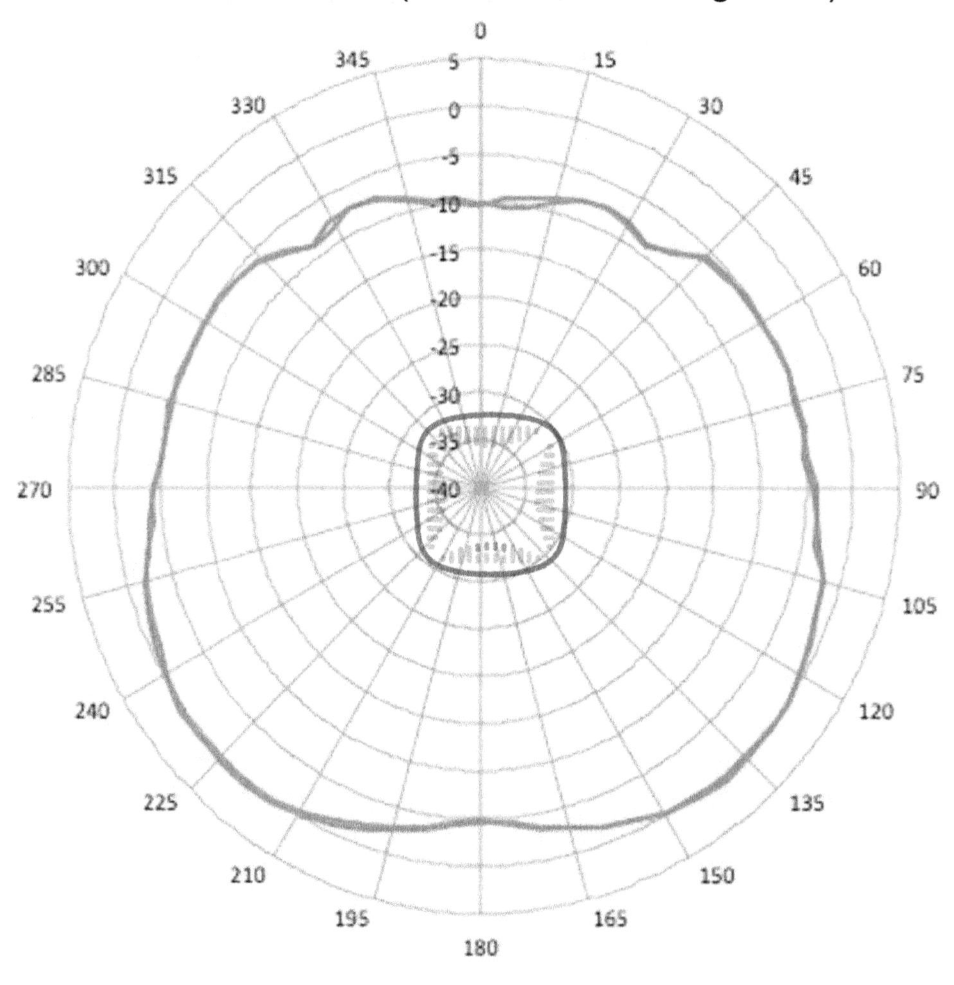

Figure 1-23 E-plane 2.4 GHz Wi-Fi (antennas 1, 2, 3, 4)

Figure 1-23 shows a vertical, side view of the antenna's radiation pattern. This is called the E-Plane (E for Elevation).

Together, the azimuth and elevation charts give you a three-dimensional idea of the intended coverage area. Remember, physical obstructions and RF interference may change the actual radiation pattern in any given space. Of course, this affects where clients can receive good RF signals from the AP.

Passive E-plane antenna gain

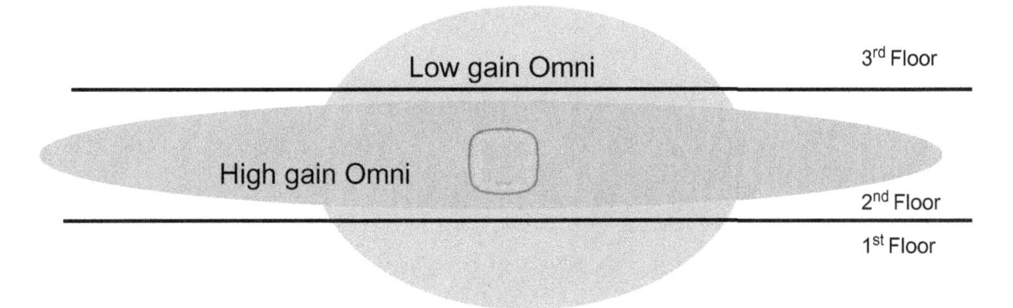

Figure 1-24 Side view—Vertical (E-plane) coverage

Antennas do not draw power from the AP or any other source, and so do not "add power" to the signal. This is why we refer to them as "passive devices." While the antenna does not increase the overall power of the signal, it can *shape* how that signal radiates into the environment.

In Figure 1-24, the more rounded shape represents the E-plane of a low-gain, Omni antenna. Remember, omni means that the antenna radiates energy in all directions—360 degrees around the antenna. Low-gain means that the antenna does not focus much of the energy in any given direction. In other words, it transmits about the same amount of energy in all directions. This low-gain omni is analogous to a fairly round, slightly flattened ball or balloon.

Now, imagine you hold the balloon, with one hand on top, the other on bottom, and squash it by gently pressing your hands closer together. What happens to the balloon? The horizontal dimension increases, at the expense of the vertical dimension.

This is like a high-gain omnidirectional antenna, as represented by the oblong shape in Figure 1-24. Such an antenna is designed to direct more power in the horizontal direction to cover more floor space. This means that less energy radiates in the E-plane, and so less vertical space is covered.

 **Note**

Humans are not perfect, and so cannot create an antenna with zero gain—one that radiates an exact equal amount of energy in all directions (a perfect sphere of energy). However, we can imagine a theoretical "zero gain" antenna, which is called an isotropic antenna. We can then compare the gain of actual, real-world antennas against this.

We measure the gain of antennas in units called dBi—decibels of gain over an Isotropic antenna. The lowest-gain antenna typically created are the little "rubber ducky" or dipole antennas often seen included with APs with external antennas. These have a gain of 2 dBi–2.2 dBi.

Passive H-Plane Antenna Gain

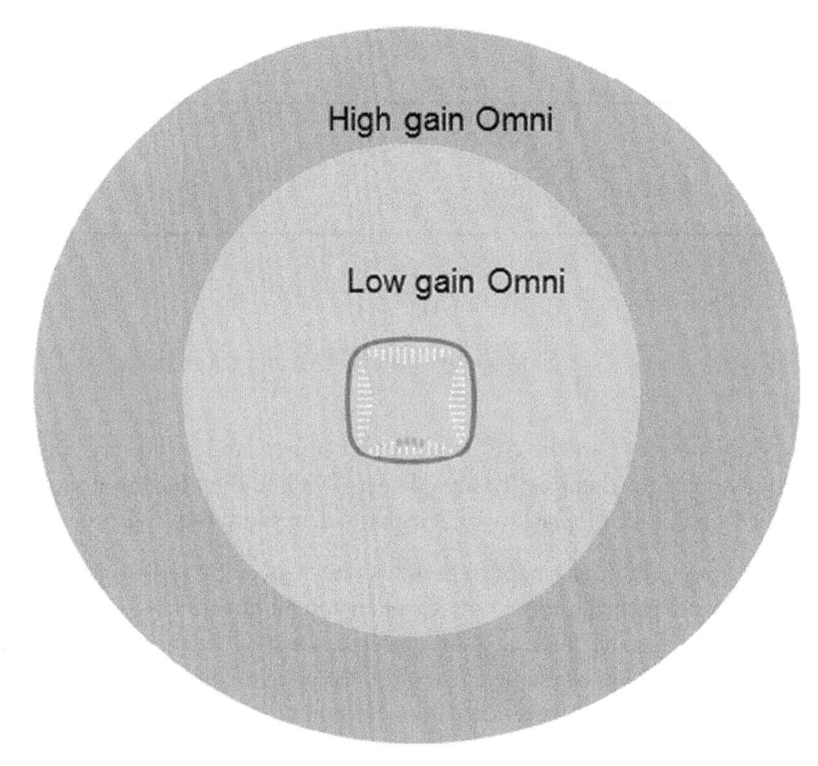

Figure 1-25 Top view—horizontal (H-plane) coverage

Figure 1-25 shows the H-plane coverage of the low-gain antenna (the smaller circle), and the high-gain antenna (the larger circle). Remember, they are both Omni antennas, and so have a 360-degree, circular coverage pattern on the floor of an open room or a field. However, the high-gain omni covers a larger area when viewed from above.

AP/Antenna mounting options

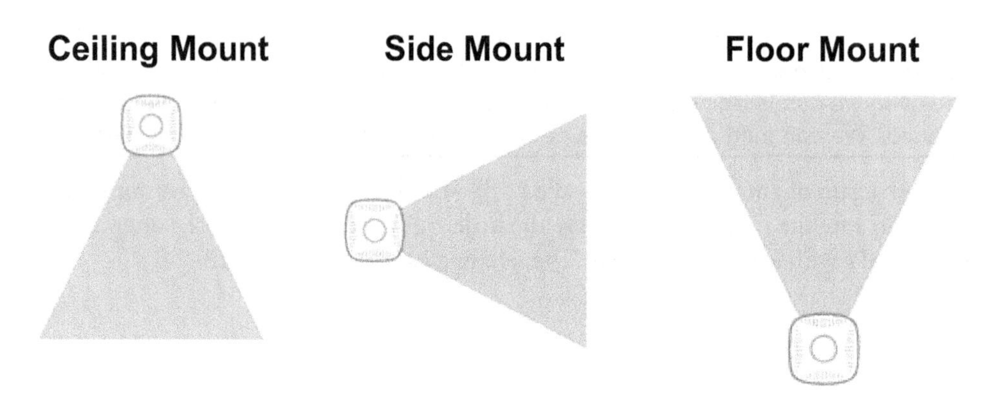

Figure 1-26 AP/Antenna mounting options

Antennas, or APs with integrated antennas can be mounted in one of three ways (Figure 1-26).

- **Ceiling mount**—This is a very common mounting method for the typical indoor, medium-density deployment, such as a typical carpeted office space or hospital. The APs are mounted flat, or parallel to the ceiling. You can mount them below the ceiling for easier, quicker installation, and easier maintenance (you can easily see AP status lights). However, some people might object to APs being visible to passersby, and so are willing to spend a bit more time and money to hide the APs above the ceiling.
- **Side mount**—APs are mounted to walls, beams, columns, or other structural supports that exist in the space to be covered. This is a far less common method of mounting APs—humans absorb radio energy, and so variations in user density can have large effects on coverage. Still, when other methods are less suitable for some reason, this can be a viable alternative.
- **Floor mount**—For this option, APs are mounted in, under, or just above the floor of the coverage area. This is not a very common method—occasionally used in very high-density deployments, such as a sports arena. Since the floor, carpet, chairs, and humans absorb a lot of energy, each AP's coverage area is a very small "pico cell."

Single Input Single Output (SISO)

Figure 1-27 Single Input Single Output (SISO)

Legacy WLANs (based on 802.11a/b/g) use Single-Input Single-Output (SISO) radio technologies, where only one antenna transmits or receives at a time. One device transmits a signal over *one* antenna (Figure 1-27). Other devices receive this signal on *both* antennas and send it to the radio for processing. The radio chooses the signal with the best reception and discards the other signal; so effectively only one stream of data is used for each transmission. This concept is known as antenna diversity. This "diversity" is used to mitigate an RF problem called "multipath distortion."

Multipath distortion occurs when signals propagate along multiple paths. Perhaps a signal travels directly from an endpoint to the AP's left-side antenna. This is good. However, a portion of the signal bounces off something metallic, like a filing cabinet, and arrives at the same antenna a few microseconds later. These two signals are slightly out of phase and distort each other. This is called multipath distortion and can result in frame retransmissions and low data throughput.

The AP detects this multipath distortion on its left-side antenna, and so instead uses its other antenna to receive the signal. After all, if two signals converge at the left-side antenna, they are therefore NOT converging anywhere else—including the right-side antenna.

This is very similar to our use of two ears. The sounds you hear in both ears are sent to your brain (like an AP radio) for processing. This two-eared approach improves hearing because one ear may receive a superior sound signal than the other. Also, you can detect the direction of the signal's source, turn toward it, and speak directly toward them. Thus, they receive your voice with better clarity. Likewise, antenna diversity uses the antenna that received the clearest signal to transmit back.

Multipath propagation scenario

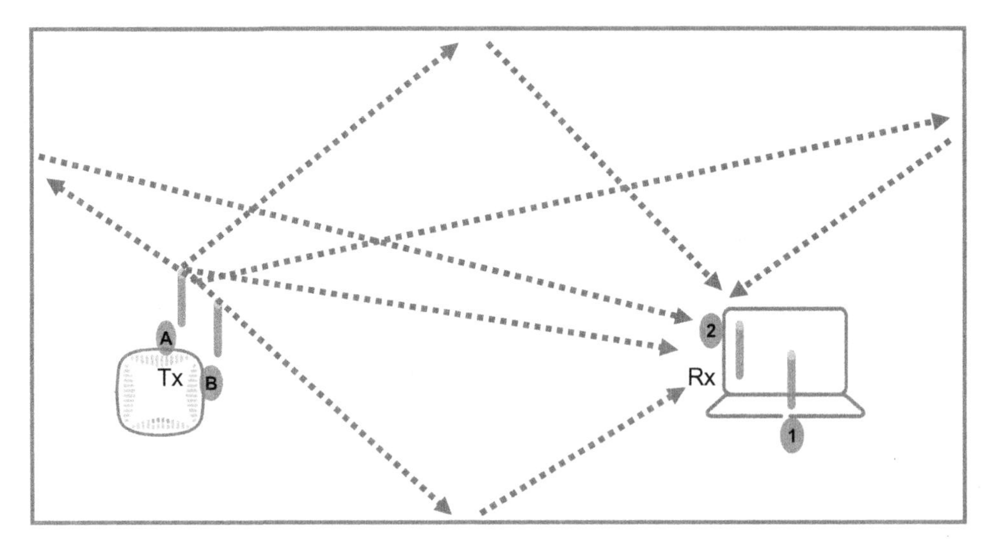

Figure 1-28 Multipath propagation scenario

Figure 1-28 shows the challenge of multipath propagation, as previously introduced. In this scenario, when the AP transmits, one of its antennas propagate RF energy. Some of this energy travels a relatively straight-line path to both of the endpoint's antennas.

However, some of this RF energy bounces off of some metallic object, such as iron pipes in the ceiling, a metal desk, or large piece of machinery. These metallic surfaces are represented by the box surrounding Figure 1-28. Some of this reflected energy just happened to be angled to converge at the endpoint's antenna #2.

This reflected energy took a longer path, and so arrives at the antenna a few microseconds later than the RF signal with a straight-line path. In this case, this multipath *propagation* has caused multipath *distortion*—a degradation of signal received by antenna #2. Of course, the endpoint can mitigate this distortion by dropping the signal from antenna 2, and using the signal from antenna #1.

Note that many deployments have little or no problem with this issue. Most items in your deployment may be made of nonreflective materials—carpet, upholstered and wood furniture, drop ceilings, and so forth. In this case, there is less multipath propagation, since most signals are *absorbed*, instead of *reflected*. WLAN deployments with a lot of heavy machinery or other metal objects may have more challenges with this issue.

As you gain experience, you will be on the lookout for this issue. During initial deployments, you might walk the site, looking for highly reflective environments. If you are doing remote troubleshooting, you cannot see the deployment site. If a remote site is having performance issues, you might ask about the environment. Is it highly reflective, with many metallic objects? Or is it mostly absorptive, with mostly fibrous materials.

Multiple Input Multiple Output (MIMO)

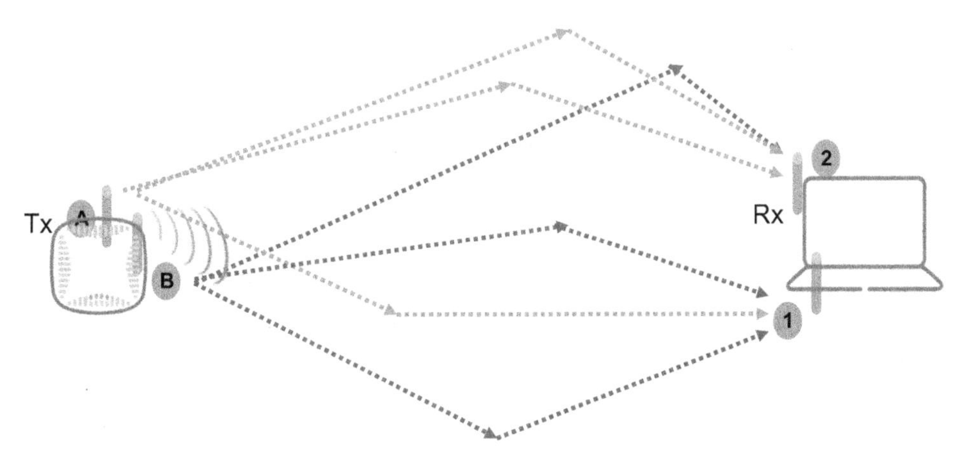

Figure 1-29 Multiple Input Multiple Output (MIMO)

The legacy 802.11a/b/g SISO challenge of multipath propagation is turned into a benefit with 802.11n/ac Multiple Input Multiple Output (MIMO) technology (Figure 1-29).

802.11n and 802.11ac use MIMO antenna technologies to transmit and receive multiple data streams via multiple antennas, at the same time. These so-called "spatial streams" provide significantly faster throughput, as compared to legacy SISO technology.

When you purchase an 802.11n or 802.11ac-based device, the number of antennas it can use to transmit and receive are specified, using an "N by M" matrix, where N is the number of transmit (Tx) antennas and M is the number of receive (Rx) antennas.

For example, some devices may be sold as a "2 x 2" MIMO—it can use two transmit and two receive antennas. Some inexpensive devices may by 1 x 1—only 1 transmit and 1 receive. Obviously, this is not really MIMO, as there is only 1 antenna. The maximum "N by M" matrix is 4 x 4 for 802.11n and 8 x 8 for 802.11ac.

Another feature introduced in 802.11ac Phase 2 is called Multi User-MIMO (MU-MIMO). This is a great departure from all previous WLAN technologies in which one-and-only-one device may transmit at a time (within each AP's area of coverage). As the name implies, MU-MIMO allows different users' data to traverse diverse spatial streams, bouncing along through the air across up to eight paths simultaneously. Thus, several Wi-Fi clients will be able to share this larger pool of streams and antennas.

Learning check

6. What technology enables APs to send different users' data along separate spatial streams?
   a. MIMO
   b. SISO
   c. Multi-User MIMO
   d. Multi-User SISO

7. What is the top view or horizontal view of an AP radiation pattern?
   a. Azimuth
   b. Elevation
   c. Gain
   d. E-Plane

Answers to learning check

6. What technology enables APs to send different users' data along separate spatial streams?
   a. MIMO
   b. SISO
   c. **Multi-User MIMO**
   d. Multi-User SISO

7. What is the top view or horizontal view of an AP radiation pattern?
   a. **Azimuth**
   b. Elevation
   c. Gain
   d. E-Plane

RF transmit power

RF Power

Most vendors specify an AP's RF power at its antenna connector. This RF power can be specified in one of two different units of measure—Milliwatts or dBm (decibels relative to one Milliwatts).

0 dBm is equal to 1 Milliwatts. As dBm numbers become negative, Milliwatts become fractional numbers. mW represents the data linearly, dBm represents the data logarithmically.

RF power dictates the size of the APs coverage area, or how far the signal will travel. This affects the quality of the signal received by endpoints, and therefore data rate.

dBm versus Milliwatts

There are several concepts related to RF power, signal strength, and measurement. To properly design a WLAN, it is important that you understand these concepts.

The power output of a transmitter, and the strength of a received signal can be measured in Milliwatts, or in dBm. They are simply two different scales that can be used to measure the same data. This is similar to how a weight can be measured in either kilograms or pounds.

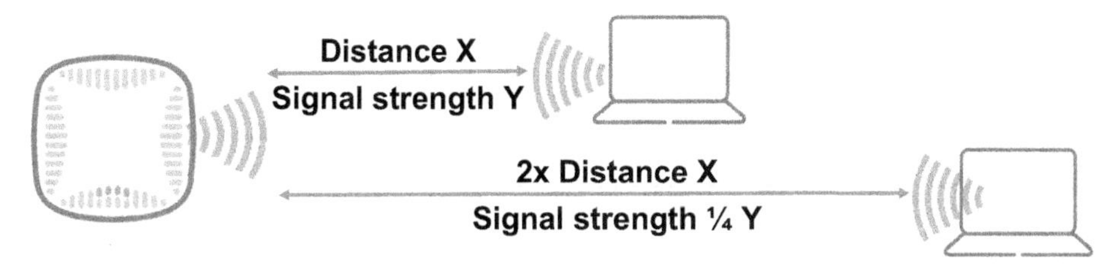

Figure 1-30 dBm versus Milliwatts

RF power does not work in a linear fashion. It is logarithmic (Figure 1-30). For example, if you stand some distance "X" away from an AP, your signal strength will be some value, Y. If you stand twice as far away, your signal will not be half as strong, it will be one-fourth as strong.

Milliwatts can be used to measure RF power. However, it is a linear scale, being used to measure a logarithmic phenomenon. At a close distance, the client's signal strength might be .00001 mw. At some farther distance, the signal strength could be .0000001 mw. The numbers can get cumbersome.

Meanwhile, dBm is a logarithmic scale, used to measure a logarithmic phenomenon. dBm may seem a bit foreign at first, but it is easier to use.

Suppose an AP sends 100 mW (equal to 20 dBm) of energy to its antenna. This energy is spread across the entire three-dimensional coverage area of the AP. Also, the energy wanes due to Free Space Path Loss, as discussed later.

This means that the endpoint's antenna receives only a small fraction of the total energy radiated by the antenna. A user standing very close to the antenna may receive a strong signal, say –40 dBm… or .0000989 mW. This is a very good, strong signal. Somebody farther away may only receive a single strength of –80 dBm, or .0000000099 mW—a very, very weak, unusable signal.

Most WLAN/RF professionals find it easier to say "My signal strength is –80 dBm" as opposed to "My signal strength is .0000000099 mW."

dBm and mW relationships

If you understand certain relationships between dBm and mW, you will be a more effective WLAN engineer. The two main ways these measurement scales can be perceived involves the "rule of 10s" and the "rule of 3s." A 3 dBm increase is equal to double the power. A 10 dBm increase is equal to 10 times the power. This rule is also inversely true for a 3 dBm decrease or a 10 dBm decrease. A 3 dBm decrease is equal to half of the power. A 10 dBm decrease is equal to 1/10th the power.

Let us look at dBm as relates to typical AP transmit power. This is most appropriately indicated in the top portion of Table 1-3, from the 0 dBm row, up to the +20 dBm row. When setting or measuring AP transmit power, you should understand the "Rule of Threes." Each –3 dBm shift represents a 50% reduction in power. If you are not used to thinking "logarithmically," this is not intuitive. Again, look at the top portion of Table 1-3, which shows some potential AP power output in dBm, and the corresponding output in mW.

Table 1-3: AP power ouput examples

dBm	mW	
+20	100	
+17	50	
+14	25	
+11	12.5	
+8	6.25	
+5	3.125	
0	1	
dBm	**mW**	**Endpoint experience**
–50	.00001	Excellent
–60	.000001	Very good
–70	.0000001	Usable
–80	.00000001	Unacceptable
–90	.000000001	

Do you see the 3 dBm pattern now? Do you see how each 3-dBm reduction results in 50% less power in the mW column?

Please understand that each end-user device only receives a small portion of an AP's transmitted energy. Say you configure an AP to transmit at 14 dBm (25 mw). Remember, an omni antenna radiates energy in a semispherical pattern, so the 25 mw of power is spread evenly within this large balloon-like shape. A client only receives that tiny portion of the signal that hits its antenna, a very small portion of the total AP power.

The dBm "Rule of Tens" is very appropriate here. This is indicated in the bottom portion of Table 1-3, from the –50 row down to the –90 row. If you are very close to an AP, your endpoint may show a signal strength of –50 dBm, which equals .00001 mw. This is excellent signal strength! A bit farther away, and you may get a –60 dBm signal—still very good. A signal strength of –70 dBm is not exciting, but usable, at .0000001 mw. Can you see that each 10 dBm shift down drops the power by 1/10th? This is the rule of tens.

A common target for WLAN quality is around –65 dBm to –67 dBm. When doing a site survey, we set a test AP to 50% power, at a particular location. Using site survey software, we walk around taking measurements to draw the –67 dBm outline for that AP, onto a floorplan. We then place the next AP so its –67 dBm ring overlaps the previous by 15%–20%.

Another key value is between –90 dBm and –100 dBm. This is considered a typical "noise floor" and represents the typical background noise in most environments.

Signal to Noise Ratio (SNR)

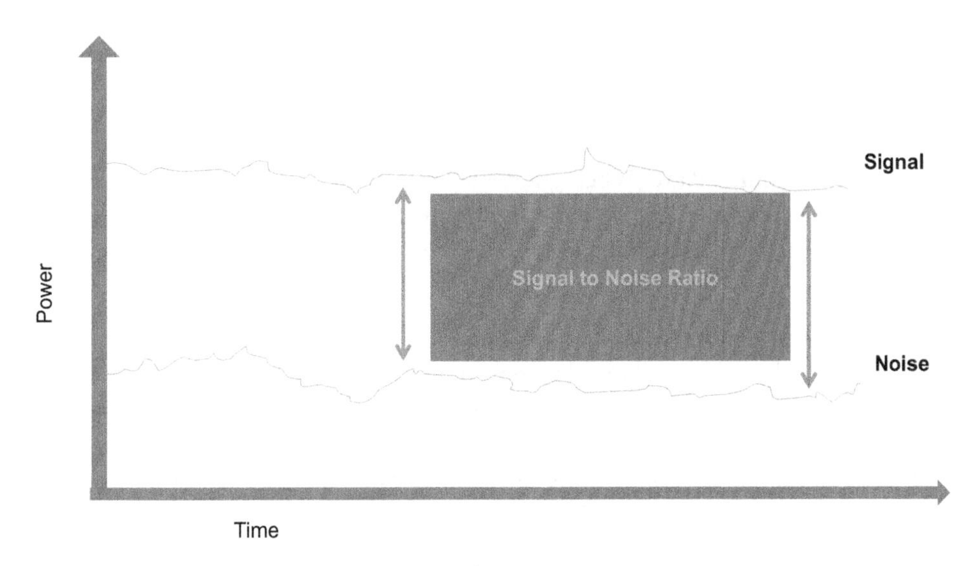

Figure 1-31 Signal to Noise Ratio (SNR)

The Signal to Noise ratio (SNR) is a measurement of the power of the WLAN signal compared to the background noise of the RF environment. SNR is a very important value and is used to evaluate the quality of the signal (Figure 1-31).

Each WLAN radio must be able to receive signals that are significantly stronger than the ambient RF noise level. This means they can discern the difference between signal and noise. Many WLAN client radios need at least 10 dB–20 dB difference between signal and noise level.

Effective or Equivalent Isotropic Radiated Power (EIRP)

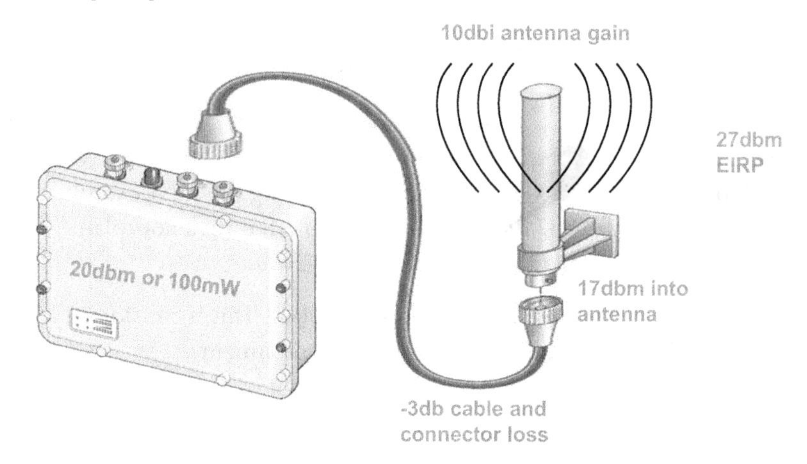

Figure 1-32 Effective or Equivalent Isotropic Radiated Power (EIRP)

EIRP can stand for either Effective Isotropic Radiated Power (EIRP) or Equivalent Isotropic Radiated Power, depending on which book you read (Figure 1-32). Regardless, we simply say "EIRP" and think, "Effectively, how much Power is being Radiated into the atmosphere, Isotropically (equally in all directions)." This is defined as the current, actual output of an antenna system including the RF transmitter, any amplifiers or attenuators plus the antenna gain. The sum of this gives the total RF output power.

Each country's authority sets their own legal limits on maximum EIRP, above which, you are breaking the law. We will rarely even care about this concept in most indoor installations, since we are typically doing indoor installations using Aruba APs with integrated antennas. Manufacturers ensure that these products are legal at maximum power. EIRP is mostly relevant when doing outdoor deployments, especially long-distance "bridge shots" using high-gain antennas. (In other words, using the RF system as a WAN, instead of a LAN.)

Let us see how it works.

Calculating EIRP

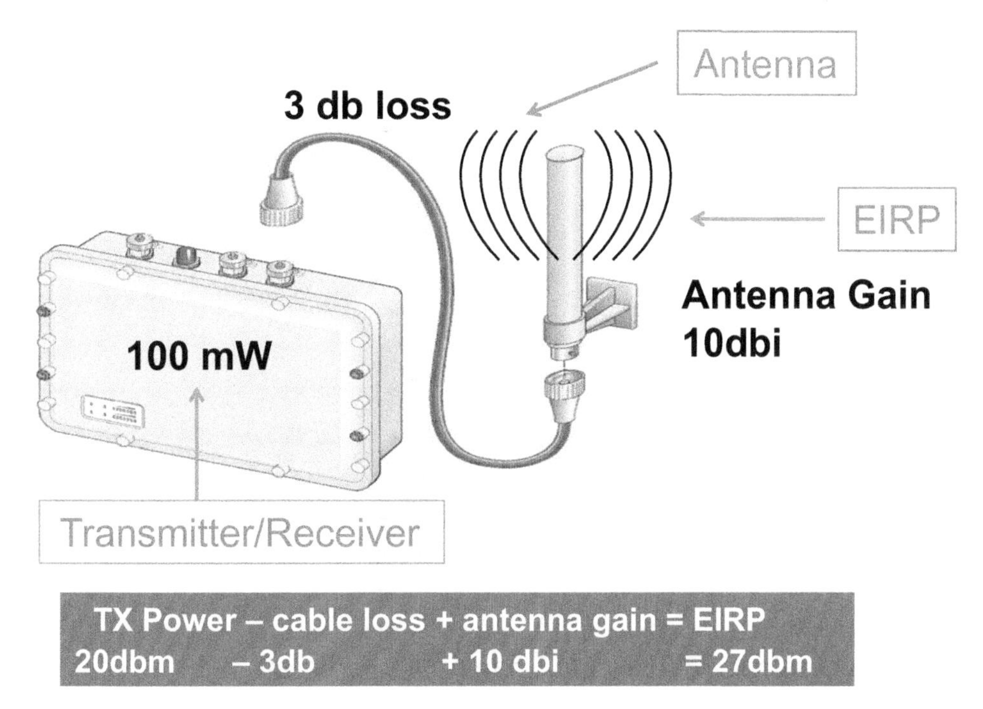

Figure 1-33 Calculating EIRP

Say you have set an AP's transmit power to 20 dBm (100 mW) (Figure 1-33). You have connected around 50 ft–75 ft of cable to the AP, with associated connectors. The resistance of the cable and connectors creates 3 dBm of loss, so we are down to 17 dBm by the time the signal reaches the antenna. We get this number by subtracting the 3 dBm of the cable loss from the 20 dBm of the transmitter.

This antenna has 10 dBi of gain, added to the 17 dBm and arriving at the antenna gives us an EIRP of 27 dBm. This is well within the FCC's 36 dBm maximum limit for EIRP. Other country's limits may vary.

WLAN mobility

WLAN logical configuration

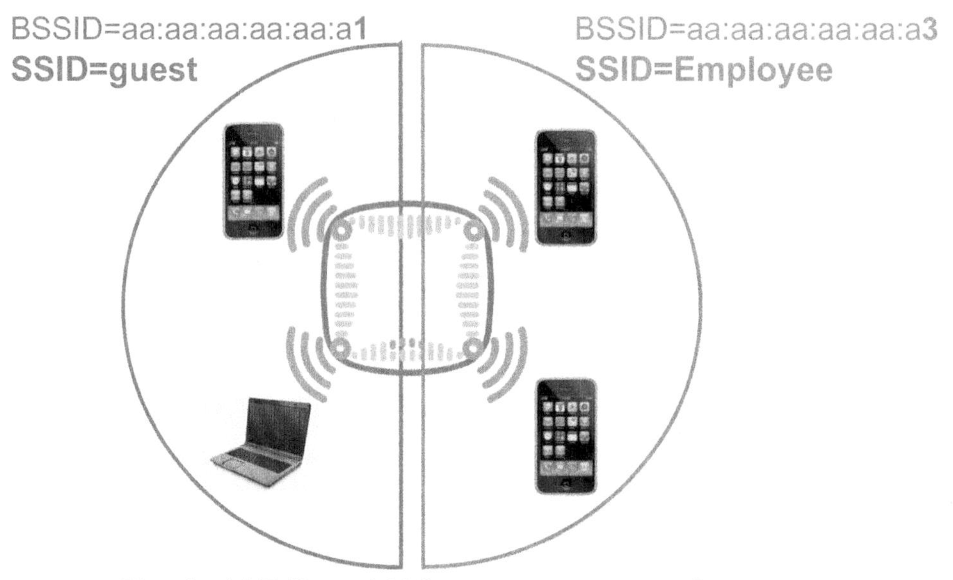

Figure 1-34 Basic Service Set (BSS)

A Basic Service Set (BSS) is defined as an AP WLAN, and all associated clients in that AP WLAN coverage area. Each BSS is identified by its BSSID, which is based on the AP radio MAC address.

Figure 1-34 shows a single physical AP radio, which has configured to support two logical WLANs. The physical AP radio's MAC address is aa:aa:aa:aa:aa:a0, as set from the manufacturer. Suppose that you first define an SSID named "guest." This SSID will be assigned a unique BSSID, based on the radio's physical MAC address. As shown, the BSSID to be used for the guest SSID ends in ":a1." Next, you defined a second SSID, to be supported on the same AP radio. This SSID might be assigned a BSSID ending ":a3." In this way, a single physical radio can support multiple WLANs and therefore, multiple BSSs.

All WLAN clients associated to the AP radio's guest SSID are considered part of the same BSS. Therefore, when these stations transmit 802.11 frames, their own MAC address is placed in the "Transmitter Address" frame field, and the connected guest BSSID is used as the "Receiver Address."

Others may connect to the Employee WLAN on the same physical radio. This is a unique BSS, with a unique BSSID. They will transmit their frames to receiver address aa:aa:aa:aa:aa:a3.

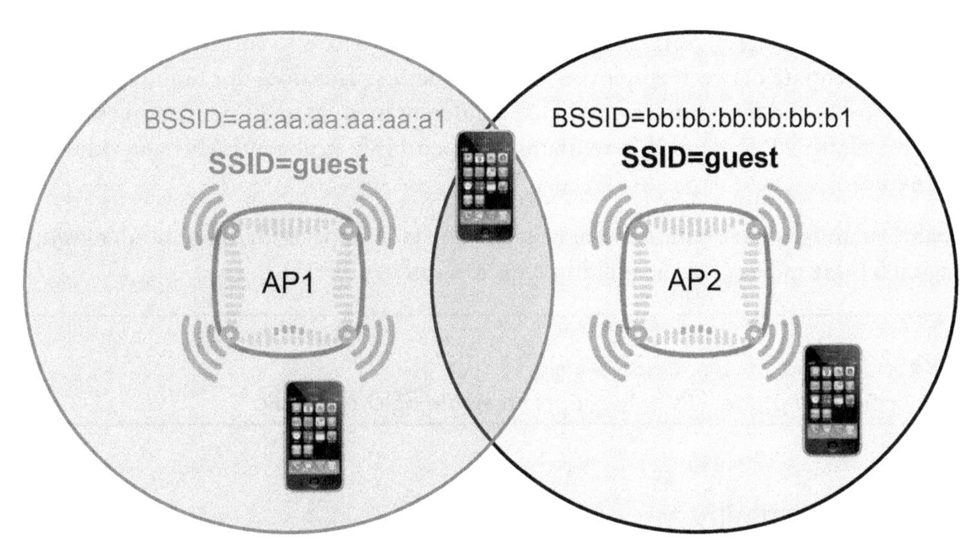

Figure 1-35 Extended Service Set (ESS)

An Extended Service Set (ESS) is defined as all clients associated with the same logical network name, often configured across multiple APs. This logical network name is technically called the ESSID, but the de facto term used is SSID. The SSID name is case sensitive and identifies the WLAN to the client. APs each transmit their own unique BSSID and perhaps a common set of SSIDs. These are sent over the air in beacon and in probe frames.

When you define an (E)SSID, each AP assigns this (E)SSID a unique a 48-bit MAC address. This MAC address is derived from the AP radio's physical MAC address and is referred to as the BSSID.

Figure 1-35 shows the BSSID for the previously discussed scenario, and another physical AP has now been added. Of course, this AP radio has a unique MAC address assigned from the manufacturer—in this case, bb:bb:bb:bb:bb:b0. This radio has been configured to support (E)SSID guest as well, and so has a BSSID ending in ":b1." Both APs support the "guest" SSID. They do so by making their unique BSSID known. Thus, the Basic Service Sets of each AP is Extended, across two APs, in the form of the common ESSID named "guest."

Of course, both AP radios may also support the previously discussed Employee SSID. This is not shown in the figure, to make the figure easier to interpret.

At this point, you may wonder, "Why would I define multiple SSIDs?" Perhaps the most common reason revolves around differing security needs—specifically, authentication mechanisms. You probably want users that connect to the guest SSID to use a captive portal web login, and only have Internet access.

You will often want to create another SSID for employees, which uses a more secure 802.1X-based security, and allows limited access to corporate resources. In an HPE Aruba WLAN, the 802.11 authentication type must be the same for all users connecting to the same SSID. However, their access to resources may be different, through the use of Role Based Access Control (RBAC) and Firewall Policies. Roles and Firewall Policies will be covered in more detail in a later module.

Perhaps you need a "managers" SSID, which allows unlimited access to corporate resources, or maybe you need to differentiate between employees and contractors. This does not require a different SSID unless the employees and the contractors will use a different method of authenticating such as 802.1X or Captive Portal. This is a major differentiator between HPE Aruba WLANs and other enterprise WLAN vendors.

Occasionally, you might even want to separate clients by radio type, with different user groups on the 2.4 GHz and 5 GHz radios. This would also require a different SSID.

Note
Many consumer-grade APs only support a single SSID per radio.

Wireless device mobility

Wireless devices may be either fixed or in motion while they access the WLAN.

An example of a fixed (stationary) WLAN device is a wireless printer. A wireless printer does not move while accessing the network. An example of a WLAN device in motion that is considered a highly mobile device (HMD) would be an iPad running a YouTube video while the user is walking down a hallway. An example of a WLAN device in motion that is considered a somewhat mobile device (SMD) would be a laptop.

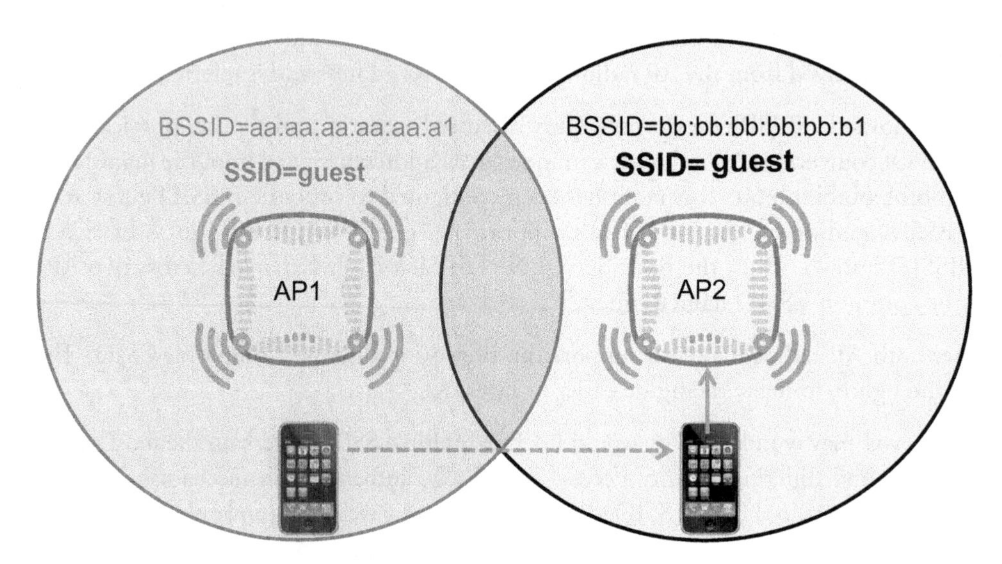

Figure 1-36 WLAN Client Mobility

WLAN clients choose when to disconnect from one access point and reconnect to another access point, a concept known as client mobility. This is shown in Figure 1-36.

The endpoint originally connected to the guest SSID on AP1. Thus, frames were transmitted to the BSSID ending in ":a1." Then they walked down the hallway. As the signal from AP1 faded, the signal to AP2 strengthened. At some point, the device decides to disconnect from AP1 and connect to AP2. The client is still serviced by the guest (E)SSID, but is now transmitting to the BSSID ending in ":b1."

Client mobility is largely based on signal quality. However, some clients also use SNR, noise, retry rate, and other values transmitted in the AP's beacon, such as AP client count.

With a properly designed WLAN, the APs provide good RF coverage, with overlapping service areas that support the same SSID. Thus, an extended WLAN is created, through which end users can seamlessly roam.

Roaming

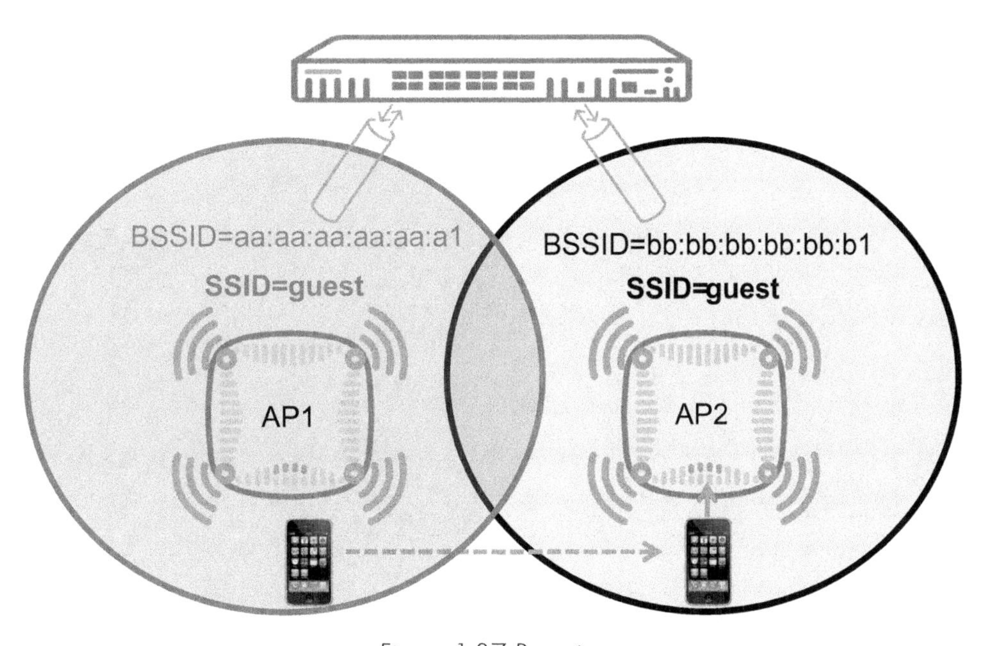

Figure 1-37 Roaming

When 802.11 clients roam from one AP to another, they change their point of attachment to the network (new AP/ BSSID) while remaining in the same logical WLAN (connected to the same (E) SSID) (Figure 1-37). To facilitate this roaming, the controller maintains client authentication, state, and firewall session information. This ensures that roaming is seamless to the users and the applications they use.

Learning check

8. Laptops are which type of mobile device?
    a. Stationary
    b. Somewhat mobile
    c. Highly mobile

Answers to learning check

8. Laptops are which type of mobile device?
    a. Stationary
    b. **Somewhat mobile**
    c. Highly mobile

2 Mobile First Architecture

LEARNING OBJECTIVES

- ✓ Now that you have learned about general WLAN concepts, you will be introduced to the HPE Aruba product line and controller portfolio. You will understand the various modes of operation for these controllers, as well as their architecture and features.
- ✓ You will be able to describe the Aruba solution and deployment models. Finally, you will learn how Aruba licensing works, and how to add licenses to a typical HPE Aruba deployment.

Product line and portfolio

Product line

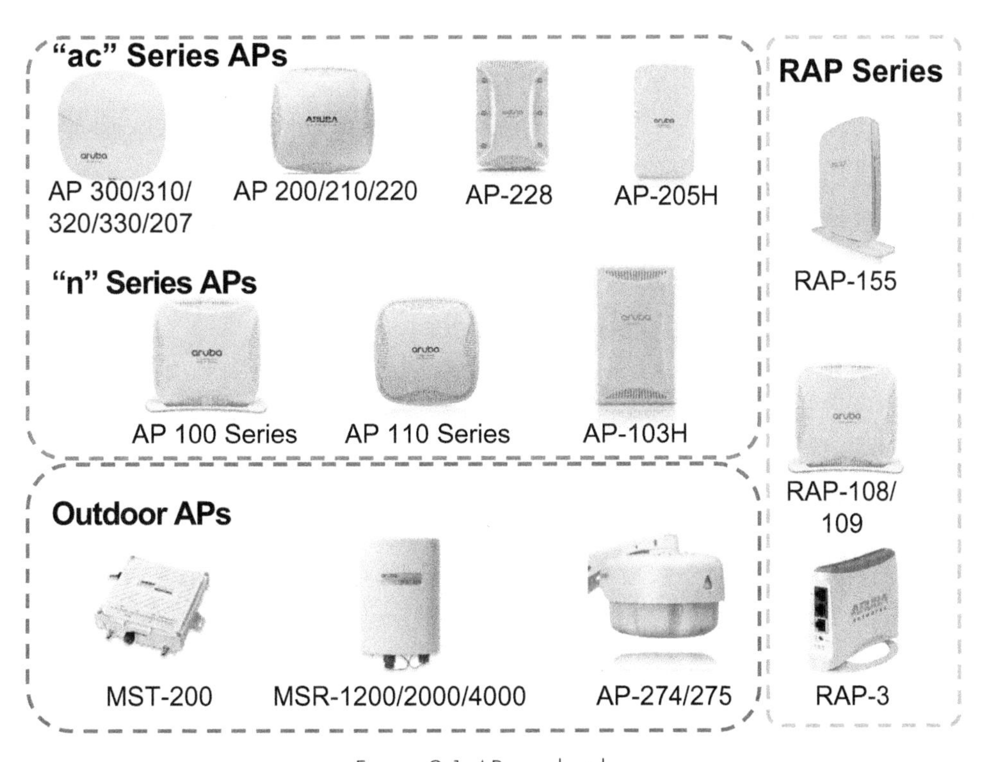

Figure 2-1 AP product line

Following is an overview of the Aruba products. Figure 2-1 highlights the Aruba AP product lines.

- **100 series** APs support 802.11n. Each comes with different specifications and capabilities. Some support 2 x 2 spatial streams, while others have 3 x 3 stream support.
- **200 series** APs support 802.11ac. Most APs can be purchased as either Campus APs (CAP), which are controller-based, or as Instant APs (IAP), which operate without a Controller appliance, and act as a virtual controller. IAPs can be converted to CAPs when desired and returned to IAPs if required. A Remote AP (RAP) always boots up initially as an IAP, but can automatically or manually be converted back to a Remote AP (RAP). AP 207 includes integrated Bluetooth Low-Energy (BLE).
- **300 Series** APs support 802.11ac wave 2, and so support MU-MIMO. BLE is also integrated.
- **AP 175** supports 802.11n, while the **APs 274/275** support 802.11ac. The 175/274/275 are outdoor APs and are built to survive the extremes of an outdoor environment.
- **MSR APs** are outdoor mesh APs. Mesh WLANs are deployed in scenarios where you can get power to each AP, but it is not feasible to run Ethernet cabling to every AP. This is often the case for outdoor areas like parking lots, parks, mining operations, and the like.

Aruba Controller Portfolio

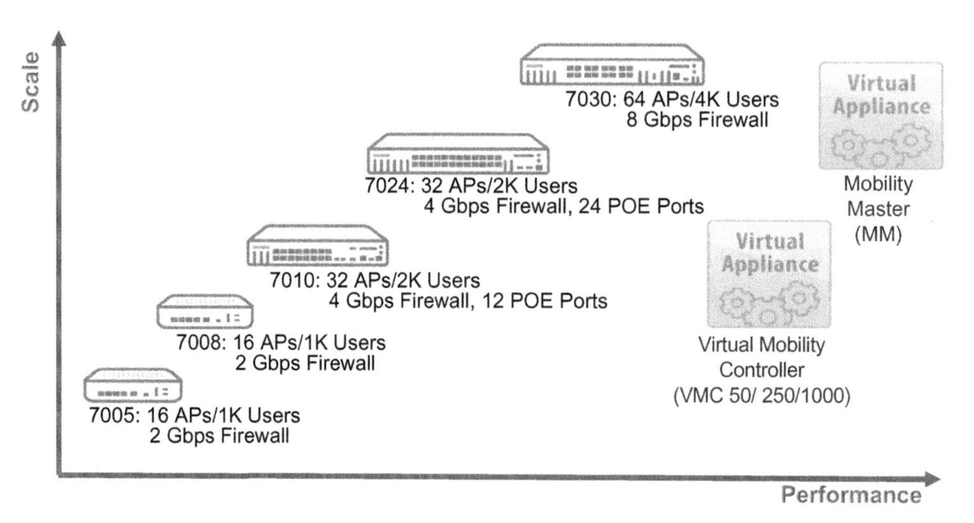

Figure 2-2 Aruba 70xx Series Controller, Mobility Master, and VMC Portfolio

- **Mobility Master**—Aruba Mobility Master is the next generation of master controller that runs ArubaOS 8.x and is deployed as a virtual machine (VM) or hardened server appliance for more memory and compute power. With the Mobility Master, major ArubaOS 8 features can be updated via software without requiring any planned network outages (Figure 2-2).
- **Virtual Mobility Controller**—The VMC runs on ArubaOS 8.x, providing a flexible deployment alternative to the mobility controller hardware appliance.
- **70xx series controllers**—Aruba 7000 series Mobility Controllers optimize cloud services and secure enterprise applications for hybrid WAN at branch offices, while reducing the cost and complexity of deploying and managing the network. This series of controllers combine wireless, wired and hybrid WAN services, support up to 64 APs and 24 Ethernet ports, and feature integrated WAN compression, health checks, zero-touch configuration, and policy-based routing.

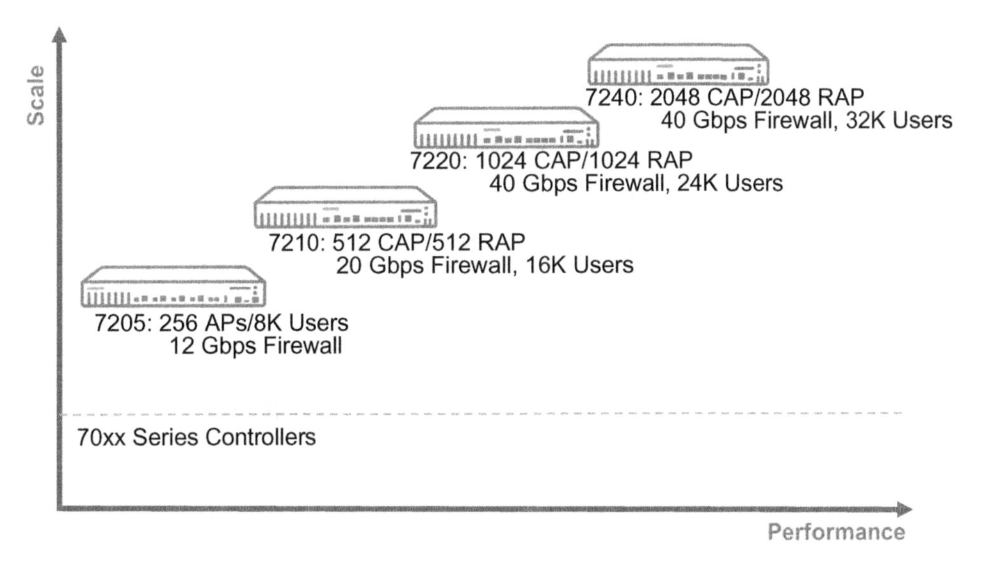

Figure 2-3 Aruba 72xx Series Controller Portfolio

- **72xx series controllers**—Aruba 72xx series controllers are powered with a new central processor that employs up to eight cores with four threads each, it is like having a total of 32 virtual CPUs. As a result, the 7200 series supports up to 32,000 mobile devices and performs stateful firewall policy enforcement at 40 Gbps—plenty of capacity and speed for Bring Your Own Device (BYOD) deployments and 802.11ac devices. New levels of visibility offered by AppRF™ technology allow IT personnel to see applications by user, prioritize them, and control access based on policies (Figure 2-3).

All the controllers can execute the same functionality and can be configured, managed, and monitored in exactly the same way. The difference between the controller models is in network capacity and scalability. The smallest capacity controller is the 7005, which is capable of supporting 16 APs and 1000 users. The 7240 controller is the largest capacity controller and can handle 2048 APs and 32,000 users.

Figure 2-4 Management, access control, and location services

- **Airwave** is a management platform for monitoring and managing the networks. Airwave also has other capabilities such as Reporting, Visual RF, and rogue detection. Airwave can monitor Aruba networks and other vendors as well. There is also a cloud-based management system called **Aruba Central** (Figure 2-4).
- **Clearpass** is used for network control, access security and advanced features such as Captive Portal, guest login, self-registration, and onboarding employee-owned devices.
- The **Meridian System** is used for location awareness and advertising features.

Learning check

1. An IAP can also be deployed as a Campus AP or a Remote AP?
   a. True
   b. False

2. How many APs will an Aruba 7024 Mobility Controller support?
    a. 16
    b. 32
    c. 64
    d. 128

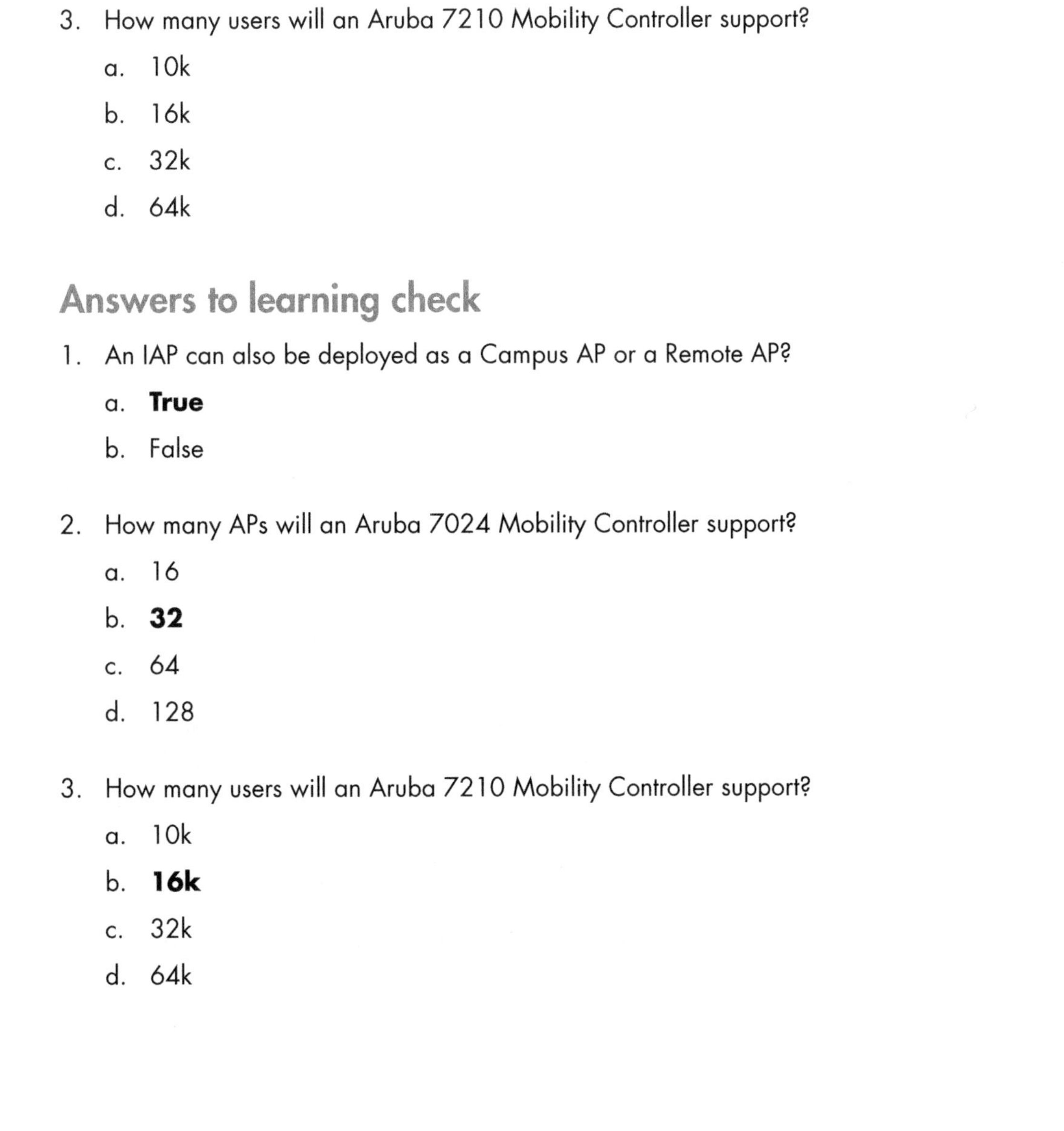

3. How many users will an Aruba 7210 Mobility Controller support?
    a. 10k
    b. 16k
    c. 32k
    d. 64k

Answers to learning check

1. An IAP can also be deployed as a Campus AP or a Remote AP?
    a. **True**
    b. False

2. How many APs will an Aruba 7024 Mobility Controller support?
    a. 16
    b. **32**
    c. 64
    d. 128

3. How many users will an Aruba 7210 Mobility Controller support?
    a. 10k
    b. **16k**
    c. 32k
    d. 64k

Controller modes

AP terminology

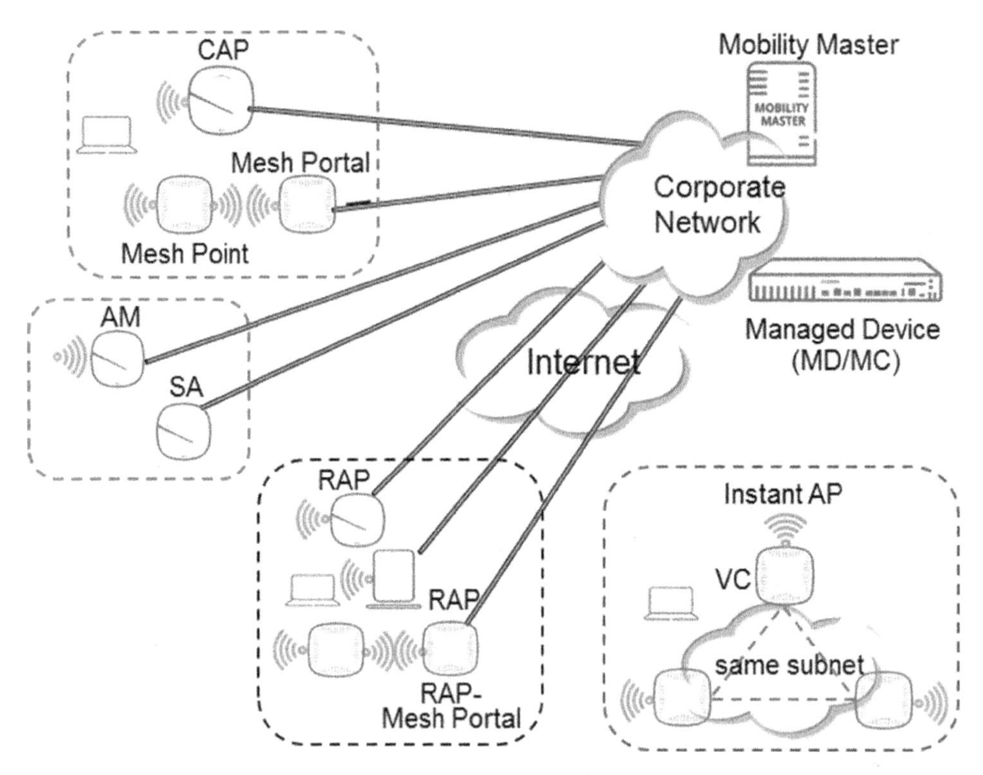

Figure 2-5 AP terminology

APs can be configured in many ways to take on specific roles in the WLAN architecture. Terminology related to these roles and modes of operation is introduced below (Figure 2-5).

- **CAP**—Campus APs are also referred to as simply an AP or regular AP. This is the typical AP mode of operation. The CAP or AP connects to a controller, gets its configuration, and begins to offer WLAN services to end users.
- **Mesh AP**—Mesh APs operate similar to CAPs. The difference is that many of them do not have an Ethernet connection. A "Mesh Portal" AP *does* have a physical connection to the corporate network because it is feasible to run Ethernet cabling to them. The Mesh Point APs needed to be installed where it is not feasible to run Ethernet cabling. They must use a radio interface to get an uplink to the corporate network.
- **AM**—Air Monitor APs do not support user connectivity. They are dedicated to constantly scan the radio environment, gathering Intrusion Detection System (IDS) and Layer 2 information.
- **SA**—Spectrum APs have been setup (temporarily or permanently) to capture radio signals for analysis. APs that gather spectrum data across all operating channels, but do not service clients, are called Spectrum Monitors (SM). APs that DO service clients and also gather spectrum data on the channel of operation are said to be operating in hybrid mode.

- **RAP**—Remote APs act similar to Campus APs but are used to extend the WLAN to a remote location. RAPs communicate with an Aruba controller across the Internet. To do so, the RAP must setup a VPN tunnel to the controller. A RAP can also be setup as a Remote Mesh portal. This is basically a RAP with Mesh portal capabilities.
- **IAP**—Instant APs operate in an autonomous or standalone mode, and so do not need a controller appliance. All IAPs in the same subnet will form a swarm, elect a Virtual Controller (VC), and provide WLAN services to clients.

Controller Modes

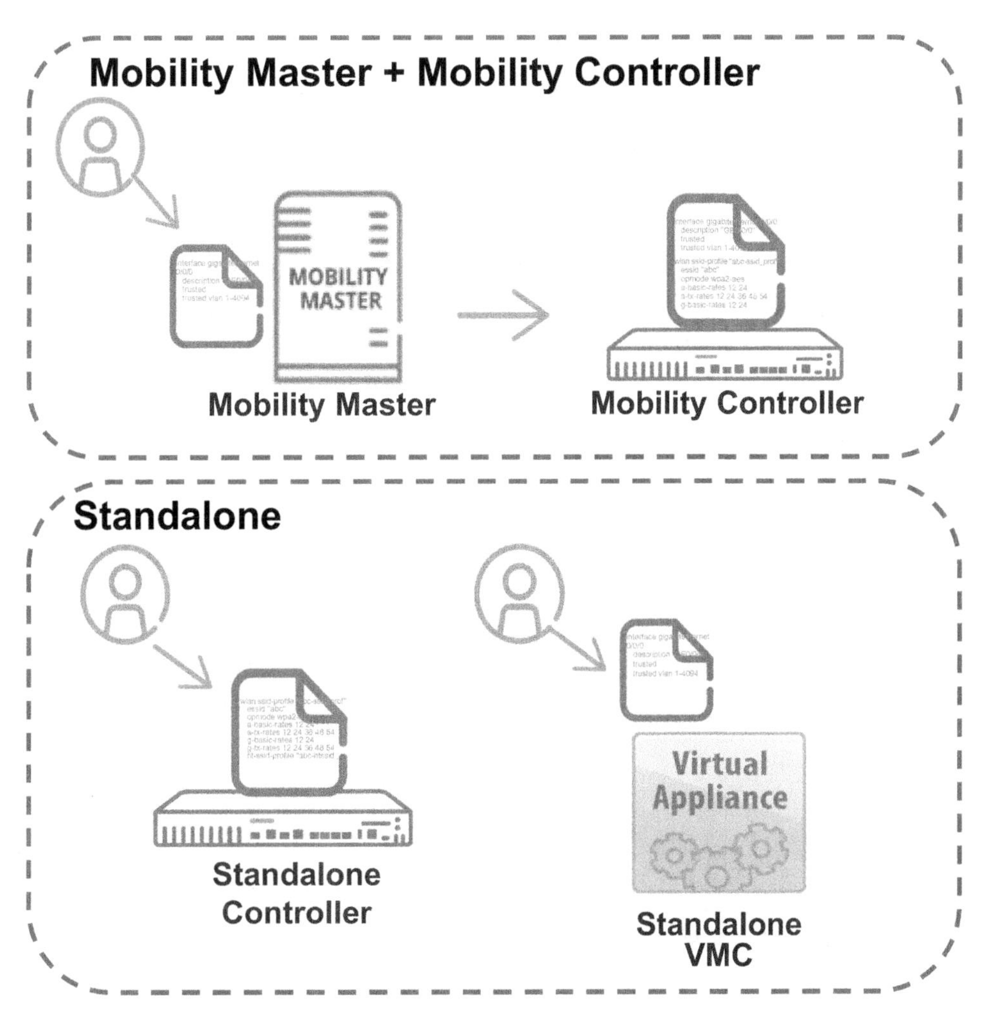

Figure 2-6 Controller modes

Mobility Master (MM) Mode—This mode uses a centralized, multitier architecture, supported as of the 8.x release UI. This UI provides a clear separation between management, control, and forwarding functions (Figure 2-6).

- The entire configuration for both the Mobility Master and managed devices is set up from a centralized point, thereby simplifying and streamlining the configuration process.
- Mobility Master consolidates all-master, single master-multiple local, and multiple master-local deployments into a single deployment model.
- Mobility Master takes the place of a master controller in the network hierarchy.

Mobility Controller Mode—The Mobility Controller is managed by the Mobility Master and is referenced as a Managed Device (MD) or Managed Node (MN). AP and Client traffic is handled by Managed Devices. In fact, *multiple* MDs working together can provide high availability for all clients, ensuring service continuity during failover events.

Standalone Mode—A Standalone controller cannot be managed by a Mobility Master and is deployed by itself. When controllers operate in Standalone mode, some 8.x features are lost.

Master-Local modes

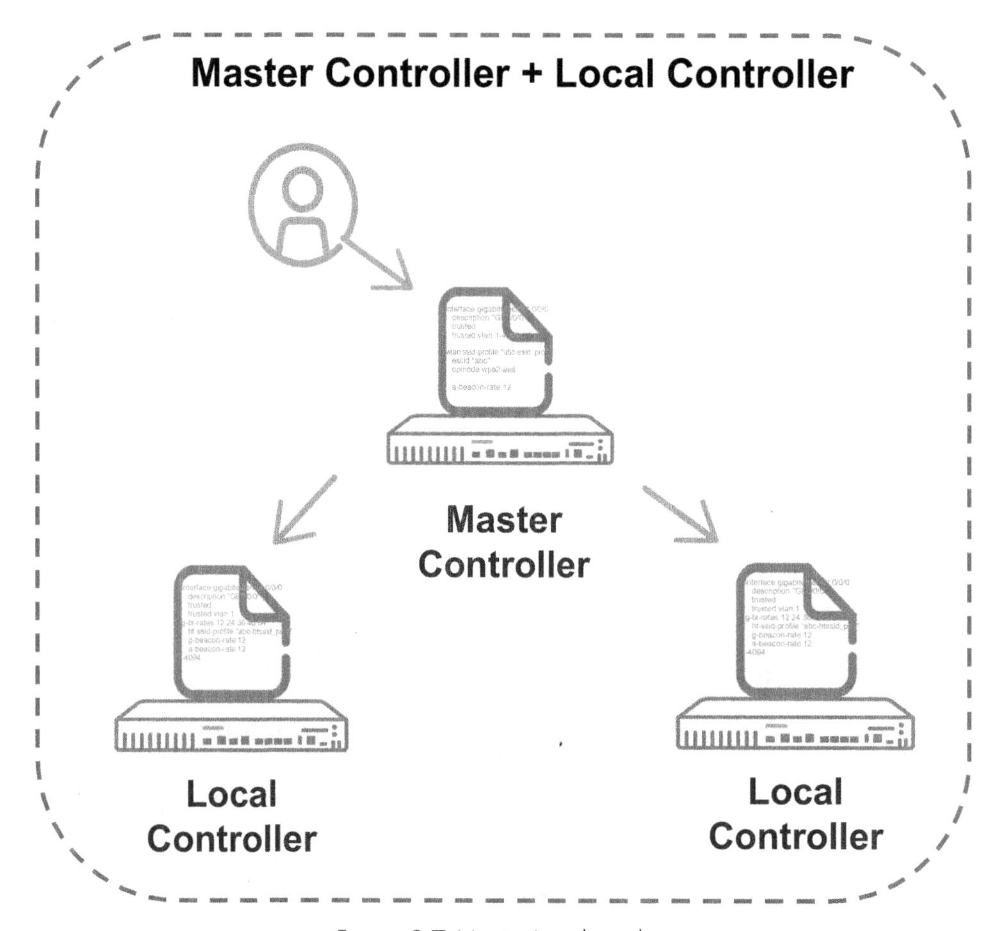

Figure 2-7 Master-Local modes

A "Master-Local" setup is used to migrate from AOS 6.x to AOS 8.x, without changes to existing infrastructure and topology. You can use HPE Aruba 72xx and 70xx for this Master-Local scenario (Figure 2-7).

The **Master Controller** is responsible for all global configuration. This includes the AP groups, WLAN settings, firewall roles and policies, RF radio settings, and much more. The Master pushes configurations to all local controllers periodically, or when you save changes on the Master.

Note

The Master controller will not push any L2 and L3 configurations such as VLANs, VLAN IP, IP routes, DHCP, and so forth, to local the controllers.

Use the Master's GUI to monitor the entire WLAN environment. The Master can also act as a centralized license server and hold the whitelist for RAPs. It can also terminate APs, but this does reduce scalability.

When APs boot, they seek a controller and are initially directed to the Master Controller. This is not a requirement but is the most typical method. The Master Controller acts as the Control-Plane-Security (CPSec) trust anchor. It issues certificates to local controllers, which in turn issue certificates to APs.

Local Controllers terminate APs, based on scalability limits. All user traffic is encrypted/decrypted, firewalled, and switched or routed in and out of the local controller.

8.x Architecture

Aruba OS 8.x Architecture

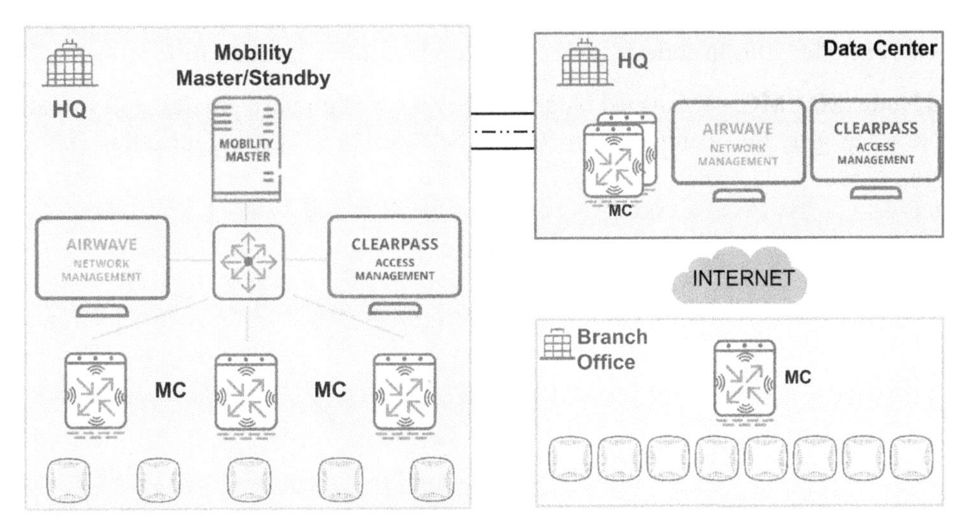

Figure 2-8 Aruba OS 8.x Architecture

The Mobility Master is central to the HPE Aruba solution. This is where you setup the entire configuration —for both the Mobility Master itself, and all managed devices (Figure 2-8).

- Mobility Master consolidates all-master; single-master, multiple-local, and multiple master-local deployments into a single deployment model.
- Common configurations (default configuration) across devices are extracted to a shared template. This then merges with device-specific configurations to generate the configuration for an individual device.
- Traffic from APs and clients are handled by Managed Nodes (MC/MD).
- You can deploy Branch Office Controllers (BOC) using Zero Touch Provisioning (ZTP)

Master Controller versus Mobility Master

Master Controller (AOS 6.x)—An MC can be deployed only on a physical controller. MCs are used to terminate APs. When you configure as Master Local, the master only pushes *partial* configuration (AP-Groups, Local DB, Whitelist DB, and so forth) to all the locals, and also will not validate the configuration (Syntax and range of parameter values).

Mobility Master (AOS 8.x)—You can deploy MMs on both physical and virtual (VM) appliance. MMs cannot terminate any APs. MMs can push *full* configuration to all the managed devices (MD/MC) and will validate the configuration before distribution to MD.

Local Controller versus Managed Node (Mobility Controller)

Local Controller (AOS 6.x)—Local Controllers can be deployed only on a physical appliance and gets partial configuration from the Master Controller. L2 and L3 configuration should be configured on the local controller. This includes VLANs, interface IP address, and so forth.

Managed Node (MD/MC)—Managed Nodes can be deployed on both physical and virtual (VMC) appliances. They get full configuration from the Mobility Master, including L2 and L3 configurations.

Partial versus hierarchical configuration

Partial configuration model

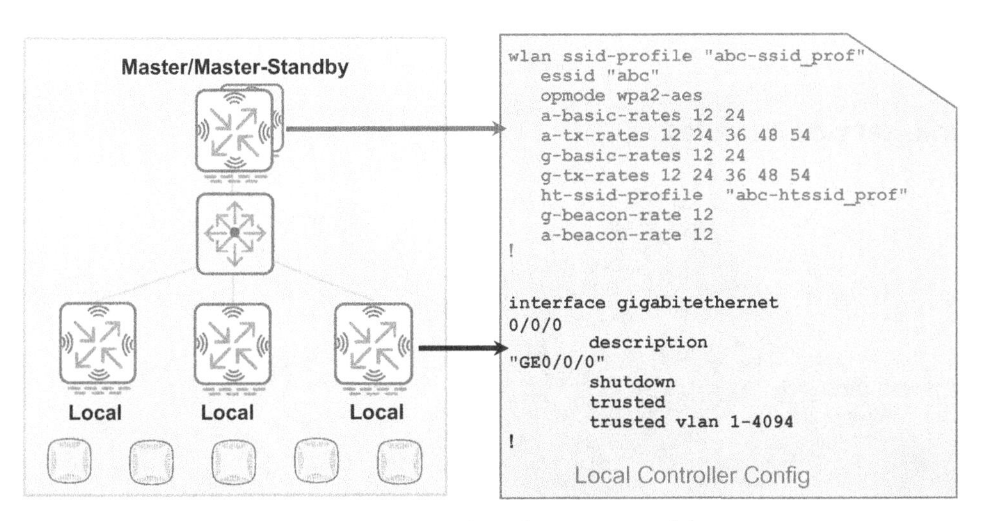

Figure 2-9 Partial configuration model

Partial configuration (6.x): In this model, the master pushes only WLAN configuration and database information, as shown in Figure 2-9's lighter gray text, at the top of the configuration.

It cannot push L2 and L3 configuration. This should be configured separately on each local controller, as shown in dark text, near the bottom of the configuration.

Hierarchical configuration model

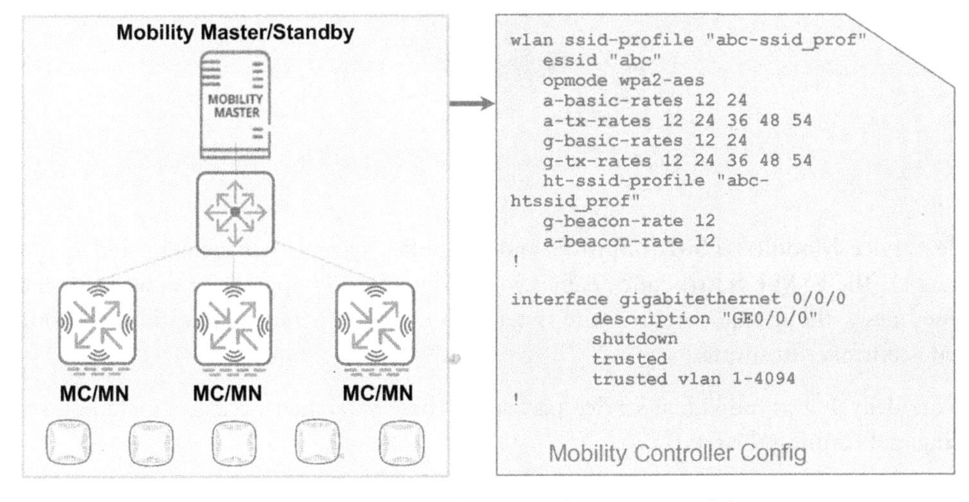

Figure 2-10 Hierarchical configuration model

In this configuration model, the Mobility Master can push full configuration to managed devices. The configuration shown in Figure 2-10 is light gray, indicating that the entire configuration is distributed.

8.x features

Loadable service modules

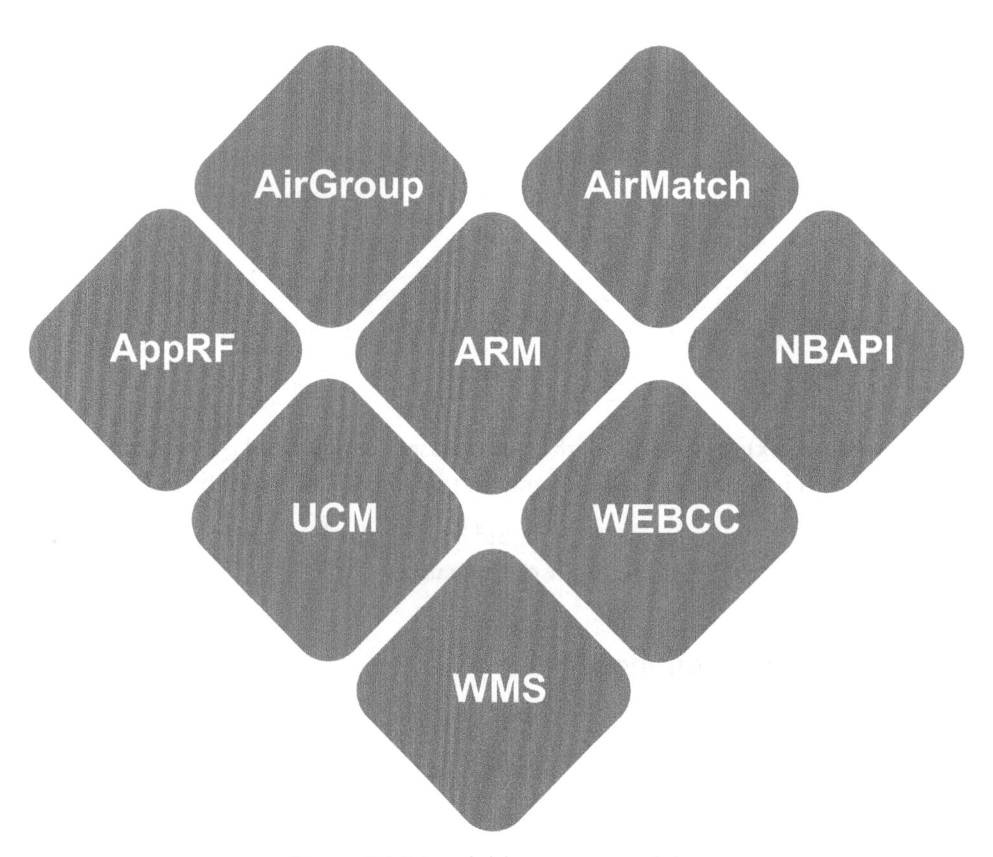

Figure 2-11 Loadable service modules

Loadable Service Modules (LSM) improves and simplifies system management and maintenance (Figure 2-11). The LSM infrastructure enables you to dynamically upgrade or downgrade individual service modules without requiring an entire system reboot, and upgrade applications without having to upgrade controller firmware.

Services are delivered as individual service packages. These versioned packages include instructions for loading and running the service.

Every service module has a corresponding service package, which can be downloaded from the Aruba support site and installed on the Mobility Master.

For example, the file for the "AirMatch" feature is currently named "ArubaOS_MM_8.0.1.0-svcs-ctrl_airmatch_56862." The file for the AppRF service is currently called "ArubaOS_MM_8.0.1.0-svcs-ctrl_appRF_56862."

In summary, key advantages of LSMs include:

- Upgrade applications without upgrading controller firmware
- Every application has its own compressed image
- Upgrades are done in runtime, and do NOT require controller reboot

Clustering

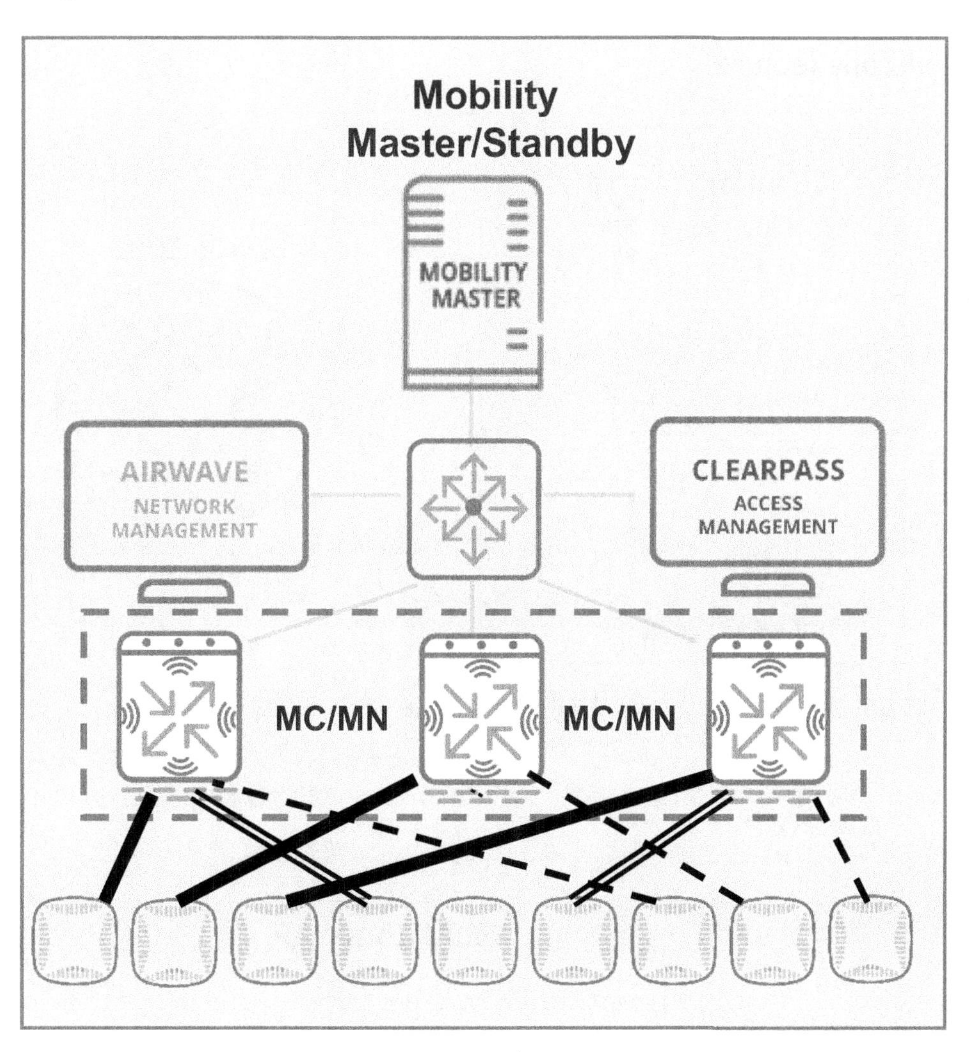

Figure 2-12 Clustering

A cluster is a combination of multiple managed devices, working together to provide high availability to all the clients (Figure 2-12). This ensures service continuity during failover events.

ArubaOS 8.0 supports a 12-node cluster. The managed devices need not be identical and can be either L2-connected or L3-connected, with a mixed configuration. However, clients are deauthenticated for managed, L3-connected devices in a cluster.

It supports 7200 Series, 7000 Series, and VM platforms. Cluster setup supports a single cluster with a mix of 7200 Series and 7000 Series controllers. A mix of VM and hardware platforms is not supported, and all controllers must run the same firmware version.

To reduce excessive workload among managed devices, the cluster load-balancing feature helps balance the stations among multiple User Anchor Controllers (UAC).

The MultiZone feature

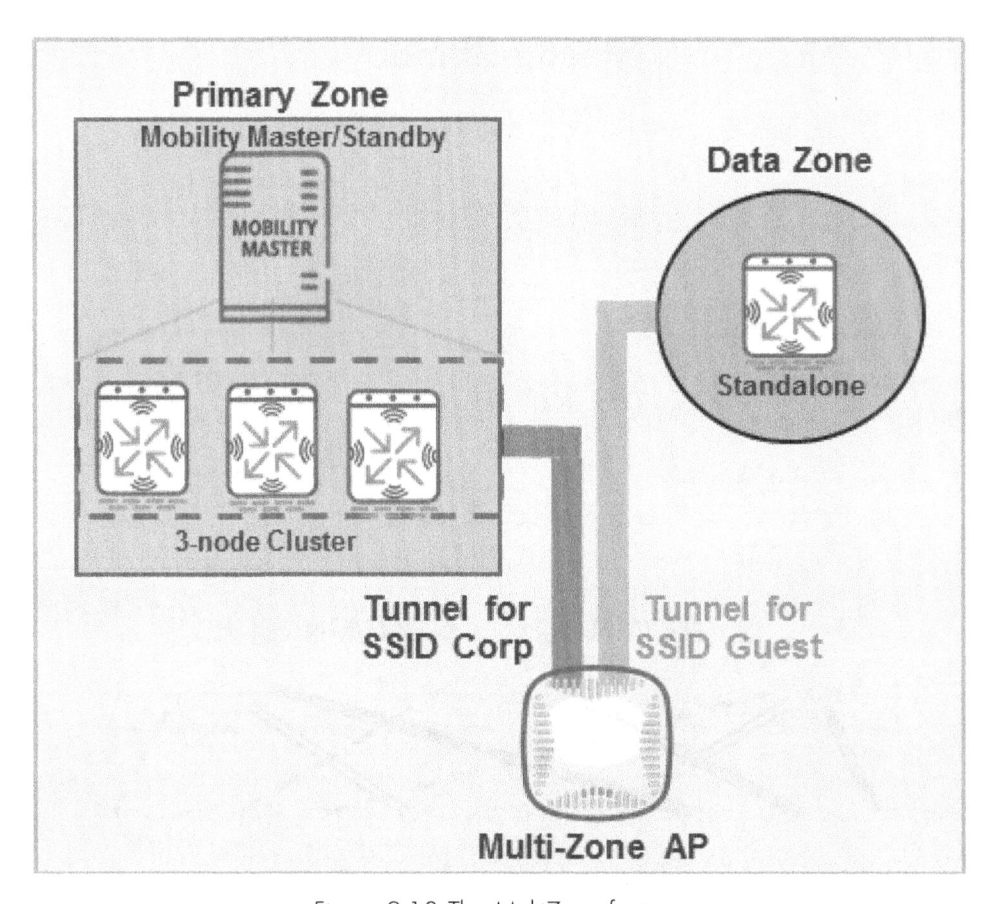

Figure 2-13 The MultiZone feature

The MultiZone feature enables you to manage guest and corporate WLANs on the same APs, but under separate management and traffic zones, managed by separate controllers. Introduced in AOS 8 on Mobility Master, MultiZone allows you to improve separation between say guest and corporate traffic. They have different SSIDs, with tunnels terminated to different controllers, perhaps managed by different entities, with an "Air Wall" between them.

Key to this feature is the concepts of Zones and Multi-Zone APs.

A **zone** is a collection of controllers under a single administrative domain. This can be a single controller or a cluster of controllers.

A **Multi-Zone AP** is an AP that can terminate its tunnels on controllers in different zones. For example, AP tunnels that support corporate employee WLANs can terminate in one zone, while tunnels for guest traffic can terminate to controller(s) in a different zone. The controllers in each zone can be managed by different people and have different configurations. The managed devices in different zones do not communicate with one another.

When an AP boots up, it contacts the Primary Zone. In the example shown, the Primary Zone includes the Mobility Master with a three-node cluster. Once connected, the AP receives MultiZone configuration. It thus learns which zone should terminate specific WLAN tunnels. In Figure 2-13, tunnels for internal, corporate SSIDs terminate in the primary zone, while tunnels for guest SSIDs terminate in the Data zone.

The Aruba solution

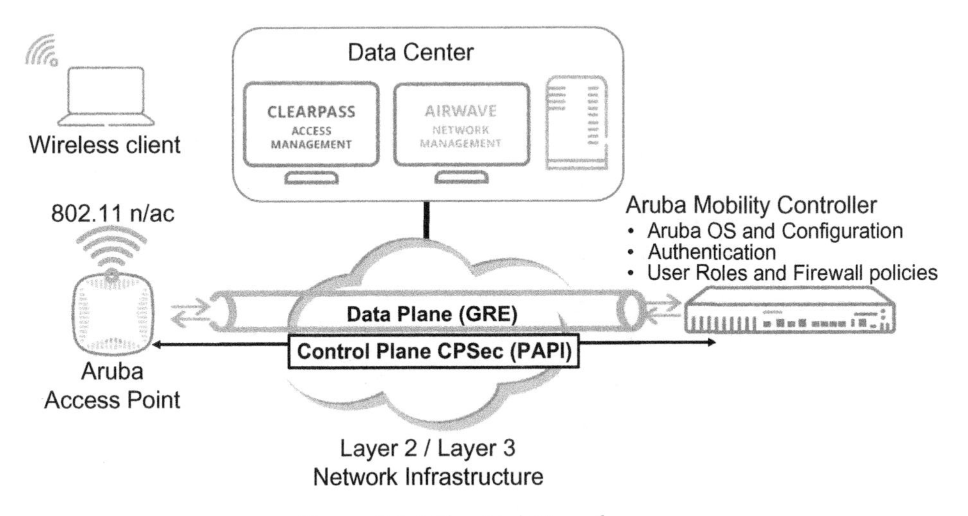

Figure 2-14 The MultiZone feature

APs connect to controllers to be configured and managed. Tunnels are created between them—some for control/management purposes and some for sending end user WLAN data traffic (Figure 2-14).

When an AP receives WLAN user data frames, it simply forwards them to the controller via the appropriate GRE tunnel. Data received from corporate users is typically encrypted.

The controller decrypts and firewalls the frames. It must then switch or route them onto the network. Of course, users must authenticate before their data traffic is forwarded.

Typically, control traffic is based on the Programming Application Program Interface (PAPI) protocol, which uses UDP port 8211. Client traffic is encapsulated into a GRE tunnel, which uses protocol 47. PAPI and GRE do not provide inherent security or encryption by themselves. As you will see later, additional security mechanisms are used to provide user data security and management traffic encryption.

When you configure a cluster, the AP forwards user traffic to the User Anchor Controller (UAC) via a GRE tunnel. The AP forwards control traffic to its AP Anchor Controller (AAC) over PAPI.

Note

APs do not use GRE tunnels to forward user traffic in bridge mode, as discussed in a later Chapter.

GRE tunnels

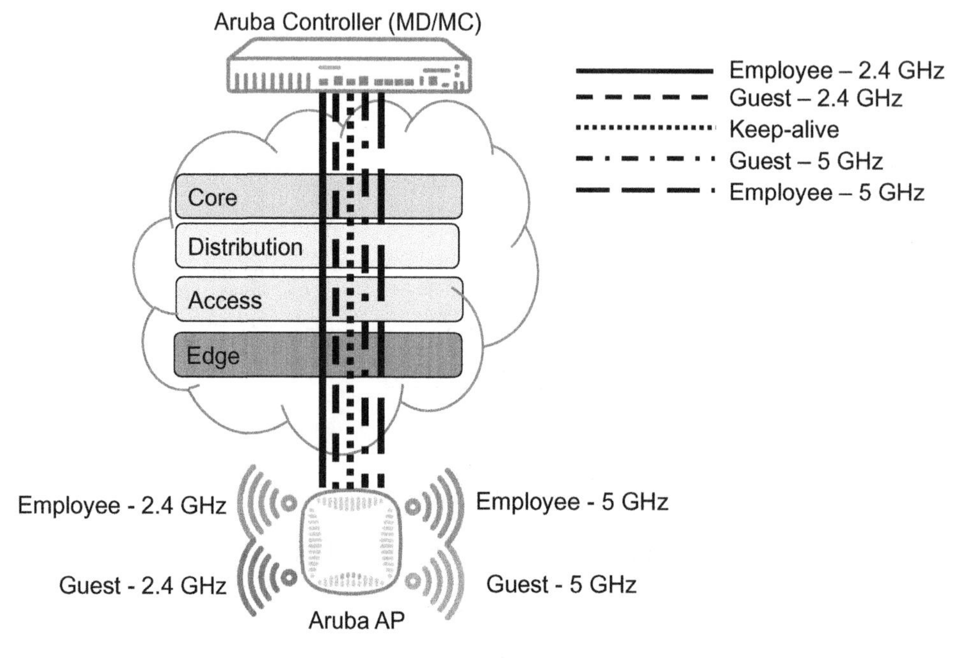

Figure 2-15 GRE tunnels

APs receive their configuration from the controller and then create GRE tunnels for user traffic and keepalives. Legacy APs and 802.11n-capable APs create one tunnel per SSID, per Radio. They also create one additional GRE tunnel for keepalives. This minimizes the amount of GRE keepalives generated.

For example, perhaps a model 7240 controls 2048 APs, each with three SSIDs on each of their two radios...2048 x 6 = 12,288 GRE keepalives per second. Using a single tunnel for keepalives reduces this number to 2048.

Campus AP deployment model

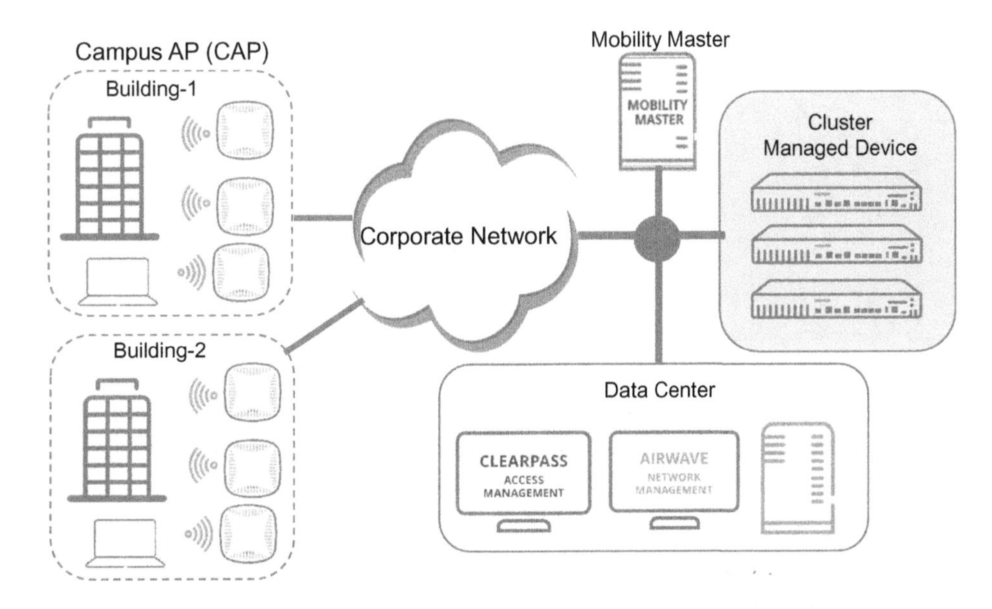

Figure 2-16 Campus AP deployment model

You will now learn about various Aruba deployment models, starting with a Campus AP model, as shown in Figure 2-16. In this model, APs are deployed in campus mode (CAP) on all the floors of all buildings in an entire campus. APs should be powered via PoE as a best practice. Another best practice is to create a cluster for AP failover and load balancing. You should plan how AP anchor Controllers (AAC) and User Anchor Controllers (UAC) should be deployed for optimal load balancing.

Many networks today also use Airwave and ClearPass servers for added functionality, visibility, security, and network monitoring. It is always recommended to deploy at least one Air Monitor (AM) per floor.

Remote AP and VIA deployment model

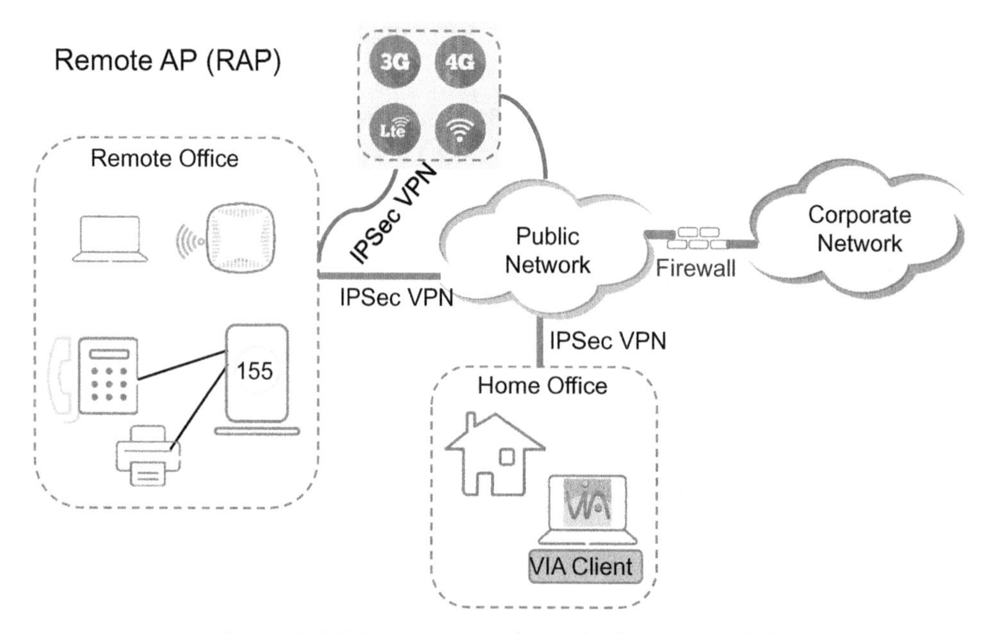

Figure 2-17 Remote AP and VIA deployment model

This model relies on RAPs or Virtual Intranet Access (VIA) client to provide secure and flexible corporate network access at small remote or home offices. This provides wired or wireless corporate network access for wireless users from any location with an Internet connection (Figure 2-17).

RAPs are quite popular. You can place them in small branch offices or home offices—anywhere with Internet access. RAPs can also have a backup uplink using an Evolution Data Optimized (EVDO) or Global System for Mobile communication (GMS) USB stick.

The RAP relies on the remote site's local provider to send packets via the Internet and terminate on the DMZ controller. Some customers use a RAP with a USB-EDVO/GSM USB stick as they travel.

VIA is an Aruba VPN client that is totally integrated in the Aruba solution. VIA is dormant when the client is on the corporate network but is automatically activated when the client is on a foreign network.

Most RAP/VIA deployments use a "Split-tunnel" forwarding mode. Using this mode, corporate traffic is forwarded back to the controller via an IPsec VPN tunnel. Non-corporate traffic is locally bridged directly to local resources or the Internet.

This traffic segregation is done according to the split-tunnel firewall policy, which you will learn about in a later module.

Figure 2-17 does not show the items in the corporate network. The infrastructure inside the corporate network is shown in Figure 2-16 on the previous page.

Branch deployment model

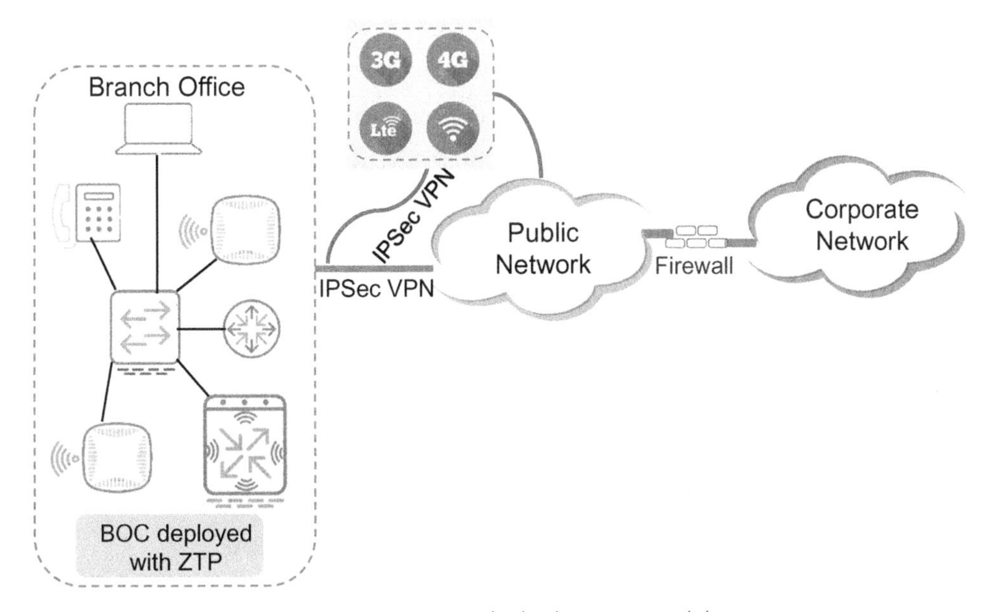

Figure 2-18 Branch deployment model

This deployment model is suitable for distributed enterprises with branch and remote offices. Often, these remote locations use cost-effective, hybrid WAN connectivity solutions such as DSL, 4G, and LTE technologies.

The 7000 Series Cloud Services Controllers are optimized for such locations. The sites are more likely to use cloud security architectures instead of dedicated security appliances. Also, clients are likely to access applications in the cloud, rather than on local application servers.

In this model, a Branch Office Controller (BOC) can be deployed in Zero Touch Provisioning (ZTP) mode. The managed BOC connects to the network, establishes a secure tunnel to the master, and downloads its global configuration. ZTP automates this deployment, giving you a plug-n-play solution.

Another feature of AOS 8.x is Authentication Survivability. This allows managed BOCs to provide client authentication and authorization survivability when remote authentication servers are not accessible.

The Health-Check feature uses ping-probes to measure WAN availability and latency on selected uplinks. Based on these health-check results, managed BOC devices will either continue to use its primary uplink, or failover to a backup link.

Figure 2-18 does not show the items in the corporate network. The infrastructure inside the corporate network has already been shown.

Learning check

4. Which of the following statements are true about the AP and Controller communication?
   a. Management traffic (Control Plane) is always PAPI encapsulated
   b. User traffic (Data Plane) is always GRE encapsulated (in Tunnel mode)
   c. All traffic will be encrypted (IPSec) when AP is deployed as a RAP
   d. Management traffic will be encrypted when CPSec is enabled
   e. All the above (Correct)

Answers to learning check

4. Which of the following statements are true about the AP and Controller communication?
   a. **Management traffic (Control Plane) is always PAPI encapsulated**
   b. **User traffic (Data Plane) is always GRE encapsulated (in Tunnel mode)**
   c. All traffic will be encrypted (IPsec) when AP is deployed as a RAP
   d. **Management traffic will be encrypted when CPSec is enabled**
   e. All the above (Correct)

Licensing

License in AOS 8.x

This topic focuses on the Aruba OS 8.x licensing concepts, features, and configuration. It will discuss centralized licensing, global and dedicated pools, and calculating minimum licensing requirements.

Types of licenses

Each license can be either an evaluation or a permanent license. A **permanent license** permanently enables the desired software module on a specific Aruba controller. You can only obtain permanent licenses through the sales order process. Permanent software licenses keys are sent to you via email.

With an **evaluation license**, you gain unrestricted functionality of a software module, on a specific controller, for 90 days (in three 30-day increments). You add evaluation licenses to the Mobility Master, and make them sharable within a licensing pool. Expired evaluation licenses remain in the license database until the controller is reset, using the command **write erase all.** After issuing this command, all license keys are removed.

The WebUI displays information about evaluation license status. To determine remaining time on an evaluation license, select the Alert flag ( ) in the WebUI titlebar. When an evaluation period expires, the controller automatically backs up the startup configuration and reboots itself at midnight (according to the system clock). All permanent licenses are unaffected. The expired evaluation licensed feature is no longer available and is displayed as Expired in the WebUI.

Subscription Licenses are feature-specific and should be installed for each additional feature you require. For example, the Web Content and Classification (WebCC) license enables WebCC features for the duration of the subscription—one, three, five, seven, or ten years. The subscription time period starts when a license key is generated from the HPE Aruba licensing website.

Thirty days before the license period expires, an alert appears in the Mobility Master's WebUI banner. This banner warns you that the license will soon expire. After the expiration date, the license continues to operate as an active license for an extended grace period of 120 days. After this final grace period elapses, the license permanently expires.

To summarize, the three types of license are Evaluation, Permanent, and Subscription.

Centralized licensing

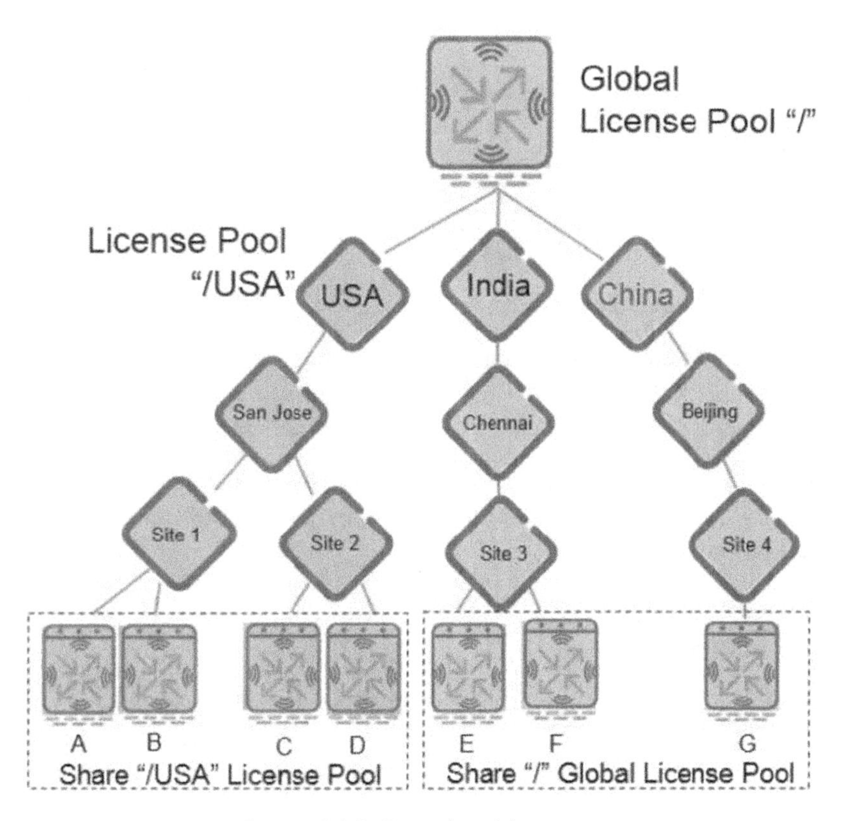

Figure 2-19 Centralized licensing

AOS 8.x supports a centralized licensing architecture in which a group of managed devices share a pool of licenses. A primary and backup Mobility Master can share a single set of licenses. This eliminates the need for a redundant license set on the backup server (Figure 2-19).

Managed Devices maintain information sent from Mobility Master, even if the managed device and Mobility Master can no longer communicate.

License pools

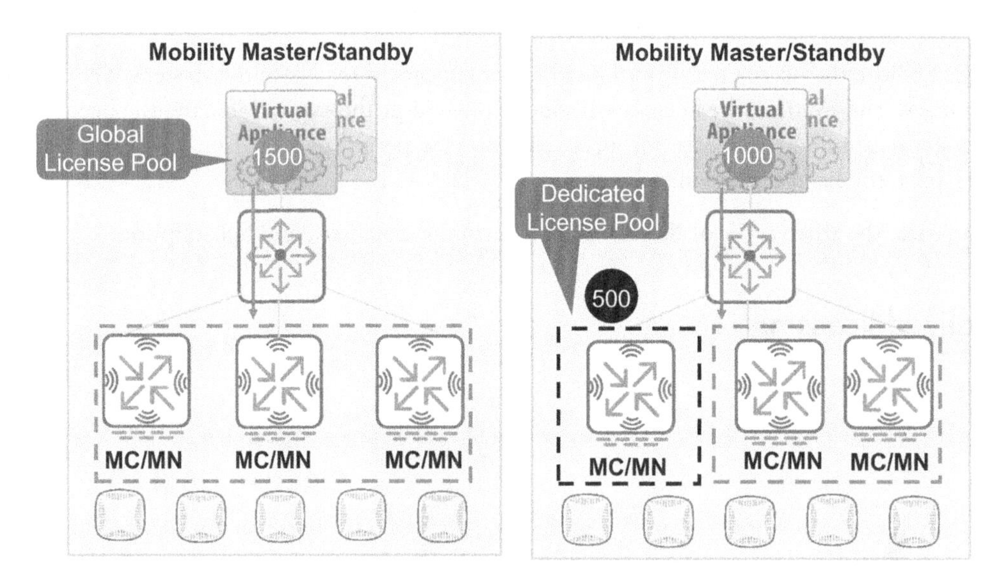

Figure 2-20 License pools

Two distinct types of license pools are shown in Figure 2-20, and described below.

Global pool—Mobility Master uses licensing pools to distribute licenses to a large number of managed devices across geographic locations. By default, all managed devices associated with Mobility Master share a single global pool. This pool contains all sharable licenses added to that Mobility Master.

Dedicated pool—AOS allows you to create additional licensing pools at a configuration node, allowing groups of managed devices at or below that configuration level to share licenses among themselves, but not with other groups.

Generating a license key

As of ArubaOS 8.0, the only way to install a license in a Mobility Master deployment is to install that device on the Mobility Master and then associate that license with either a specific managed device, or a shared pool of licenses. Licenses cannot be added directly to a managed device via the managed device UI.

Before you can use the Aruba Software License Management website to generate a license for Mobility Master, or a managed device installed on a VM, you must obtain the following:

1. A License Certificate ID, which you can request from your sales account manager or authorized reseller.
2. A MM serial number and ArubaOS passphrase. To determine the Mobility Master serial number, use the command **show inventory**. To identify the Mobility Master passphrase, use the command **show license passphrase**.
3. To generate a license, access to the Aruba licensing website:
   - My Networking Portal (MNP): http://hpe.com/networking/mynetworking/

ArubaOS 6.x licenses cannot be directly ported to ArubaOS 8.x. You can migrate existing licenses to the 8.x Mobility Master through the MyNetworkingPortal online.

Enabling sharable license features

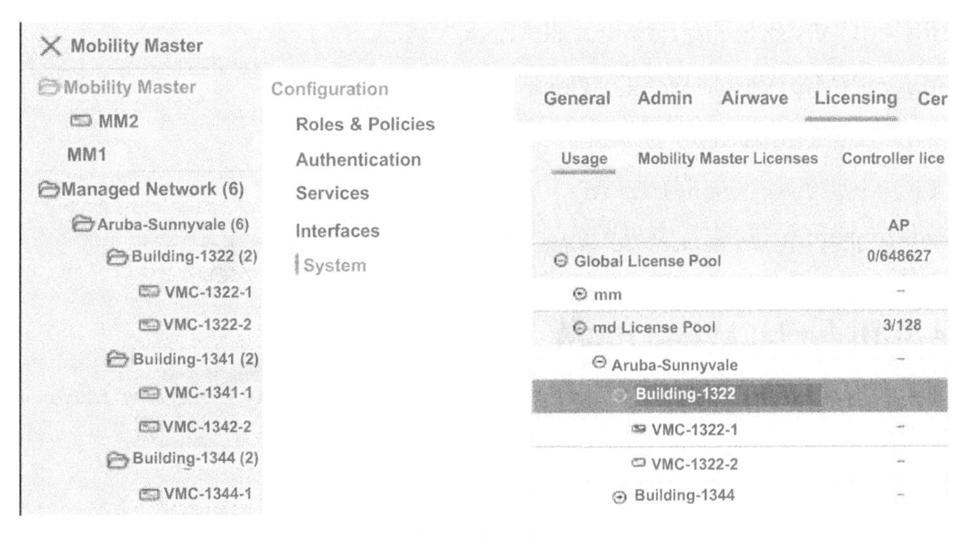

Figure 2-21 Enabling Sharable License Features

To enable a sharable license via the Mobility Master WebUI (Figure 2-21):

1. Access the Mobility Master WebUI and click the configuration menu in the upper left corner to select **Mobility Master (mm).**
2. Navigate to **Configuration** > **System** > **Licensing**.
3. Select the **Usage** subtab. The **Global License Pool** and **Usage for Global License Pool** tables appear.
4. In the **Usage for Global License Pool** table, click the check box by each license type to enable features supported by those licenses.

License SKUs for VMM and VMC

The following list describes license SKUs appropriate for both Virtual Mobility Masters (VMMs) and Standalone Virtual Mobility Controllers (VMCs).

- Types of Licenses
  - LIC-MM-VA-xx should be installed on VMM
  - LIC-MC-VA-xx should be installed on VMC or VMM if VMC is an MD
  - Install AP/PEFNG/RFP licenses on MM or Master or Standalone VMC/Master
- VMM (Virtual Mobility Master)
  - All licenses installed on MM
  - LIC-MM-VA-xx license (count = All APs all MDs)
  - AP/PEF/RFP licenses (count = All APs in Network—includes AP on 7000's series and VMC)
  - LIC-MC-VA-xx license (count = all APs on VMC)
- Standalone VMC (Virtual Mobility Controller)
  - All Licenses Installed on VMC
  - LIC-MC-VA-xx (count = All APs)
  - AP/PEFNG/RFP (count = All APs)

License SKUs for MCM and HMM

The following list describes license SKUs appropriate for both Mobility Controller Master (MCM) and Hardware Mobility Master (HMM).

- MCM License SKUs
  - All licenses installed on MC
  - On MCM, that is, legacy master, MM/MC licenses are not required (VMC is not supported on MCM)
  - Install AP/PEFNG/RFP licenses on MCM
  - MDs do not consume any license since they will be hardware
  - APs on MDs will consume AP/PEF/RFP licenses
  - Based on AP count

- HMM License SKUs
  - All licenses installed on MM
  - LIC-MM-VA-xx is in-built, you do not need to install it.
  - Install LIC-MC-VA-xx if VMCs are MDs. (count = all APs on VMC)
  - Install AP/PEFNG/RFP licenses on HMM. (count = all APs in Network—includes AP on 7000s series and VMC))
  - MDs will consume LIC-MM-VA-xx license
  - APs on 7xxx MDs will consume LIC-MM-VA-xx license and AP/PEF/RFP licenses.
  - APs on VMC MDs will consume LIC-MM-VA-xx license, LIC-MC-VA-xx license, and AP/PEF/RFP licenses

Calculating license requirements

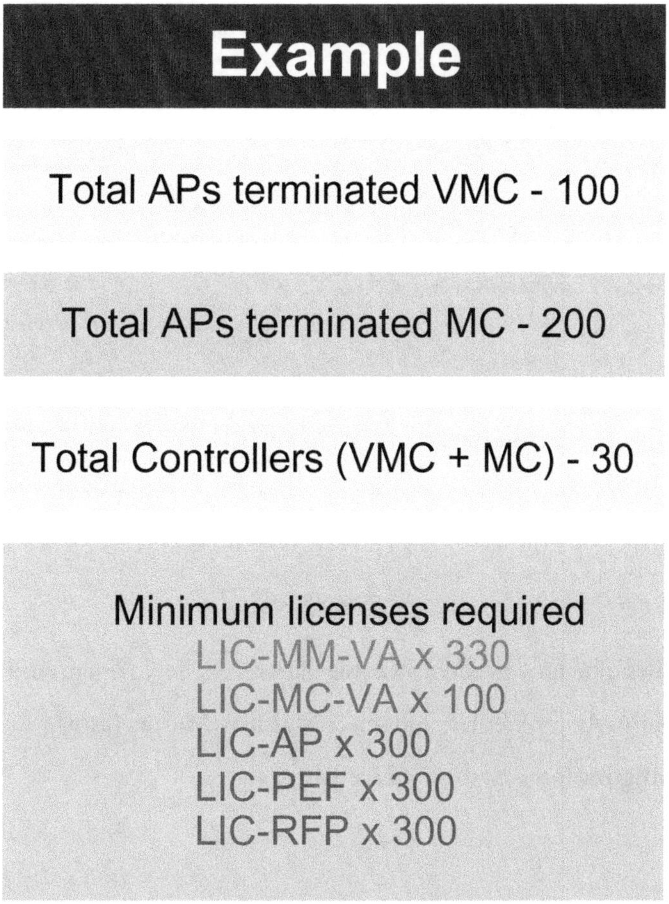

Figure 2-22 Calculating license requirements

To deploy APs, you must install MM and AP licenses (Figure 2-22).

For example, if you plan to deploy 30 MCs, with 300 APs, the minimum licenses required are:

1. MM licenses: 300 + 30 (MC + AP)
2. AP licenses: 300
3. MC licenses: 100
4. PEFNG: 300

Other potential needs include licensing for features such as RFProtect, WebCC, and PEFV, but these are optional. An in-depth discussion of these features is beyond the scope of this book. However, a brief description is provided here:

- **RFProtect** software prevents denial-of-service and man-in-the-middle attacks and mitigates over-the-air security threats.
- The **WebCC** feature extends application visibility, classification, and control features. Specifically, you get reputation-based URL filtering, IP reputation, and Geolocation filtering capabilities.
- Policy Enforcement Firewall for VIA clients (**PEFV**) enables role-based policy enforcement for remote, VIA-connected clients.

Adding Licenses

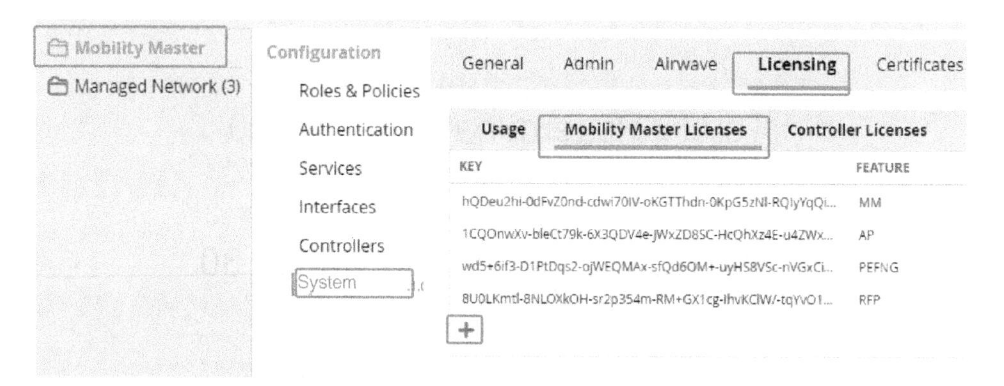

Figure 2-23 Adding Licenses

The following steps describe how to add a sharable license to a deployment, using the WebUI.

1. Access the Mobility Master WebUI, and select **Mobility Master (mm).**
2. Navigate to **Configuration > System > Licensing.**

3. Select the **MM Licenses** subtab. The **Key** table appears.
4. Click + below the **Key** table. The **Install Licenses** window appears.
5. In the **Install Licenses** window, enter the serial number for one or more licenses. Each license key must be on a separate line.
6. Click **OK**.

Learning check

5. Which are applicable to a Mobility Master (MM)?
   a. It can be a centralized licensing server
   b. It can terminate APs
   c. It will do Configuration validation
   d. It will push full configuration to managed node (MD/MC)
   e. It must be deployed only on a Physical server

6. Choose the three types of HPE Aruba licenses.
   a. Permanent
   b. Semipermanent
   c. Subscription
   d. Mobility Controller Limited
   e. Evaluation
   f. Mandatory
   g. AP-only

Answers to learning check

5. Which are applicable to a Mobility Master (MM)?
   a. **It can be a centralized licensing server**
   b. It can terminate APs
   c. **It will do Configuration validation**
   d. **It will push full configuration to managed node (MD/MC)**
   e. It must be deployed only on a Physical server

6. Choose the three types of HPE Aruba licenses.
   a. **Permanent**
   b. Semipermanent
   c. **Subscription**
   d. Mobility Controller Limited
   e. **Evaluation**
   f. Mandatory
   g. AP-only

3 Mobility Master Mobility Controller Configuration

LEARNING OBJECTIVES

- ✓ This book closely follows how you would implement a WLAN solution in the real world. In Chapter 1, you prepared yourself by understanding fundamental concepts. In Chapter 2, you learned how to select products and gained an understanding of the architecture of those products.
- ✓ Now you will learn how to deploy and configure these products. This starts with a discussion of Virtual Machine (VM)-based installations, before moving to a discussion of hierarchical groups and subgroups.
- ✓ Next, you will discover concepts related to Mobility Controller setup and how the MC joins an MM.
- ✓ With a system now in place, you can begin to apply your knowledge of hierarchical groups and subgroups toward configuration, and the hierarchical nature of these configurations.
- ✓ Finally, you will learn about local MC configuration, including IP addressing, Trunking, and more.

VM installation

VM requirements

To support an HPE Aruba Mobility Master deployment, the VM must meet certain criteria and be installed with specific minimum settings:

- Aruba MM deployments require vSphere Hypervisor 5.1 or 5.5.
- For minimum hardware requirements, you must have Quad-core Core i5 1.9 GHz processors with hyper threading enabled, and at least 8 GB RAM. 16 GB of RAM is preferred. Of course, the hypervisor host should not be oversubscribed.
- You need a minimum of two physical NICs on the ESXi host, but four is preferred. One is shared with the ESXI Management, and the others are used for the data path.

The Aruba OS has the following minimum VM requirements:

- Minimum three vCPU's
- Minimum 4 GB RAM
- Minimum 10 GB with two separate drives. Flash is to be stored on a separate disk.
- Three virtual NICs
- Two virtual disks
- No oversubscription

Again, these are *minimum* requirements. The exact numbers depend on the number of devices managed by the MM.

In addition to VMware, there is also Kernel-based Virtual Machine (KVM) support for Linux. KVM is a virtualization infrastructure for the Linux kernel. This turns the Linux kernel into a hypervisor, creating an environment similar to that described above, for VMware's vSphere hypervisor.

VM Setup

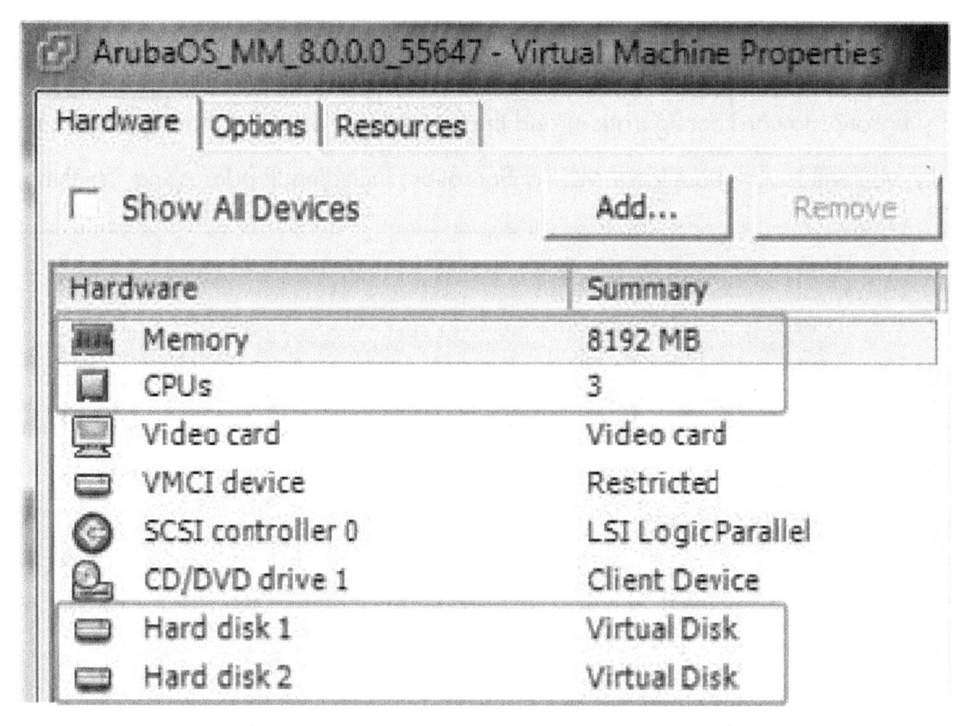

Figure 3-1 VM Setup

This example shows 8 MB of RAM, three CPUs, and the required two disks. The number of logical processors on the ESXi host should be greater than or equal to the sum of vCPUs allocated to each of the VMs setup in that host (Figure 3-1).

The memory allocated to each VM should not exceed the overall host memory capacity.

The total CPU utilization, memory usage, and network throughput should not exceed 80% host capacity.

Network Adapters

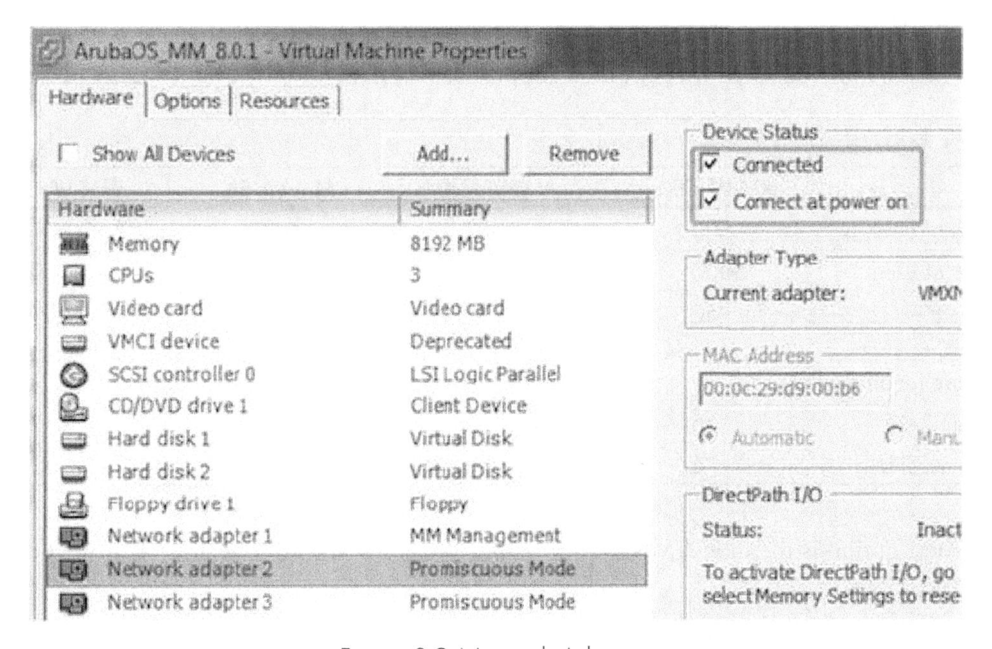

Figure 3-2 Network Adapters

Since the Mobility Master essentially replaces a Mobility Controller in Master mode, there are similarities in setup (Figure 3-2). These similarities extend to the network interfaces. An Aruba Mobility Controller has multiple interfaces. On some models of Mobility Controller, there is an out-of-band management port. The new Mobility Master VM replicates this feature.

The Mobility Master VM comes configured with three virtual network adapters, separated into two general types of interfaces (Management and Traffic). This is represented by Network adaptors 1, 2, and 3 in the VM.

By default, only the first two Network adapters in the VM are enabled. Network adapter 1 is a management interface used for out-of-band management. Network adapter 2 is the Gig0/0/0 port.

Network adapter 3 is disabled by default. Thus, a mgmt-interface and Gig0/0/0 are on by default in the initial configuration.

It is important to realize that the Management and the Traffic Interfaces should not be connected together—they should not be on the same interface or network/port group. This is to prevent loops.

If you are not using out-of-band redundancy, then disable the Device Status connected and power on. These settings are highlighted with a box, near the top-right corner of the figure.

Promiscuity mode and Forged Transmits must be enabled for MM redundancy.

MM Sizing

Table 3-1 shows the four MM models, their support capabilities, and their hardware requirements:

Table 3-1 MM Sizing

	MM-VA-500	MM-VA-1K	MM-VA-5K	MM-VA-10K
Number of devices	500	1000	5000	10,000
Number of users	5000	10,000	50,000	100,000
Number of Controllers	50	100	500	1000
Hardware requirements				
Minimum RAM	8	12	16	28
Minimum vCPU	6	8	10	16

The smallest option is the model MM-VA-500, which can manage 500 devices. This includes controllers and APs.

The largest MM is the MM-VA-10K. This has a minimum requirement of 28 G Ram and 16 vCPUs to manage 10,000 devices and 100,000 users.

If your network has more than 10,000 devices, at 8.0.1, you must have two MM networks. Each MM-VA has a specific license.

Remember, MMs cannot manage APs directly. All APs terminate to physical mobility controllers or VMCs.

VMC Sizing

Table 3-2 shows the VMC models, their support capabilities, and their hardware requirements:

Table 3-2 VMC Sizing

	VMC-VA-50	VMC-VA-250	VMC-VA-1K
Number of APs	50	250	1000
Number of users	4000	8000	24,000
Hardware requirements			
Minimum RAM	6	6	8
Minimum CPU	4	5	6

The three VMC versions are sized according to the number of APs and users that can be serviced. APs terminate on the VMC. Each model of VMC has a specific license.

MM configuration—VM console

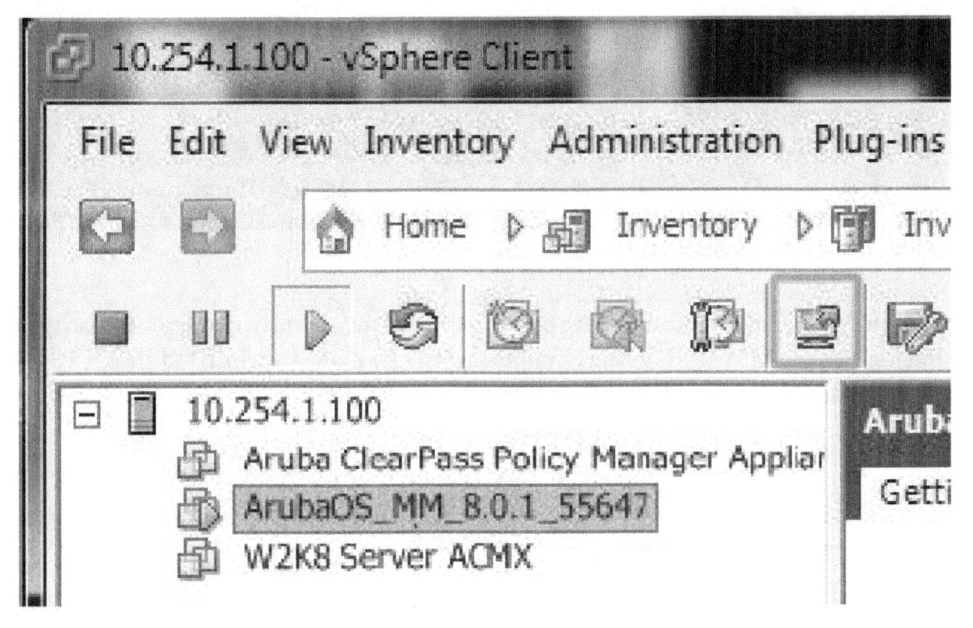

Figure 3-3 MM configuration—VM console

You must configure your VM environment as previously described. Then you will obtain the Open Virtual Appliance (OVA) file for the MM. An OVA file is merely a group of files, packaged as a single file, in a format suitable for VM installation (Figure 3-3). In this case, the packaged files are those required to deploy an MM. You must load this OVA file to your VM. This process is beyond the scope of this book.

Once the VM has been loaded with the MM .ova file and configured according to specification, the VM may be started. You can then access and configure the MM.

To do this in vSphere, open a console connection to the VM MM, and execute a script. This script prompts you for various information including the management VLAN, port, IP address, and default gateway. You will also specify the regulatory country, time zone, and date.

The script will then prompt you to configure an admin password. Figure 3-4 shows part of the script.

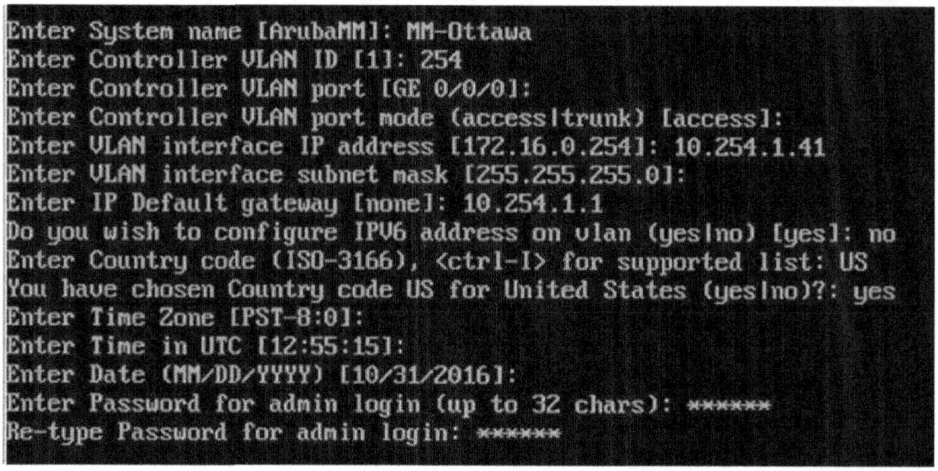

```
Enter System name [ArubaMM]: MM-Ottawa
Enter Controller VLAN ID [1]: 254
Enter Controller VLAN port [GE 0/0/0]:
Enter Controller VLAN port mode (access|trunk) [access]:
Enter VLAN interface IP address [172.16.0.254]: 10.254.1.41
Enter VLAN interface subnet mask [255.255.255.0]:
Enter IP Default gateway [none]: 10.254.1.1
Do you wish to configure IPV6 address on vlan (yes|no) [yes]: no
Enter Country code (ISO-3166), <ctrl-I> for supported list: US
You have chosen Country code US for United States (yes|no)?: yes
Enter Time Zone [PST-8:0]:
Enter Time in UTC [12:55:15]:
Enter Date (MM/DD/YYYY) [10/31/2016]:
Enter Password for admin login (up to 32 chars): ******
Re-type Password for admin login: ******
```

Figure 3-4 MM Configuration—script

After the script completes, you can point your browser to the recently configured management IP address and login to the MM GUI. In the example shown, the management IP address is 10.254.1.41. For credentials, use a username of "admin" and the password you just defined in the script.

Learning check

1. An MM can be installed on a 7240 Controller
   a. True
   b. False

Answers to learning check

1. An MM can be installed on a 7240 Controller
   a. True
   b. **False**

Hierarchy

GUI hierarchy

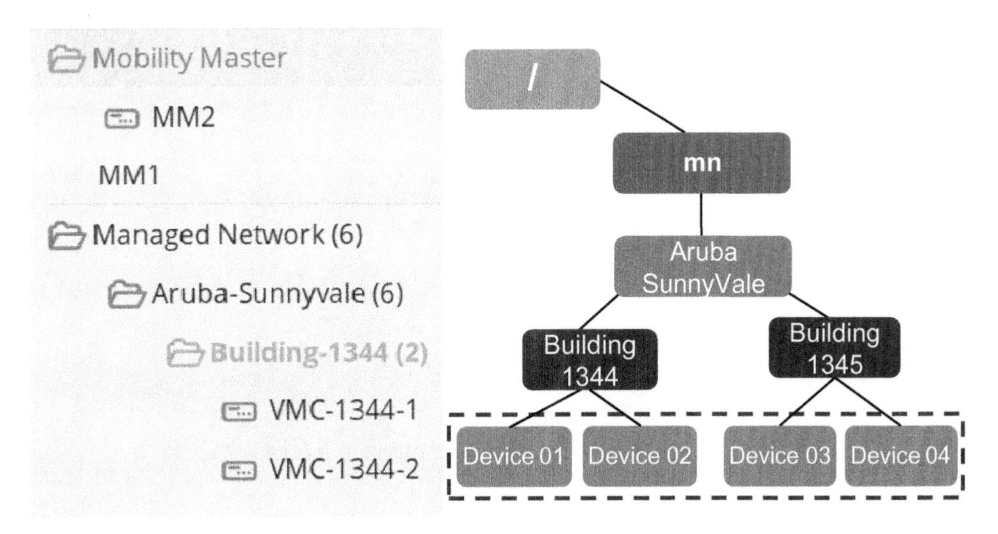

Figure 3-5 GUI hierarchy

Aruba's GUI layout is hierarchical in nature for both configuration and monitoring functions. For configuration, groups and subgroups are created. The managed devices are installed in one of the groups (Figure 3-5).

Any configuration applied to a group affects all managed devices within the hierarchy of that group.

When you select a group for monitoring, you will only see devices in the hierarchy of that group. Devices can be placed in a subgroup but cannot be moved to another group or subgroup. For this and many other reasons, it is best to plan your hierarchy properly in the beginning.

The left side of Figure 3-5 shows a screenshot from an actual deployment, while the right side diagrams the conceptual tree-like construct of this deployment.

The first group below root is the system group Managed Networks (MN). Figure 3-5 shows a scenario in which groups are created based on physical location. This is a very typical approach, and likely the one you will use.

Hierarchical configuration model

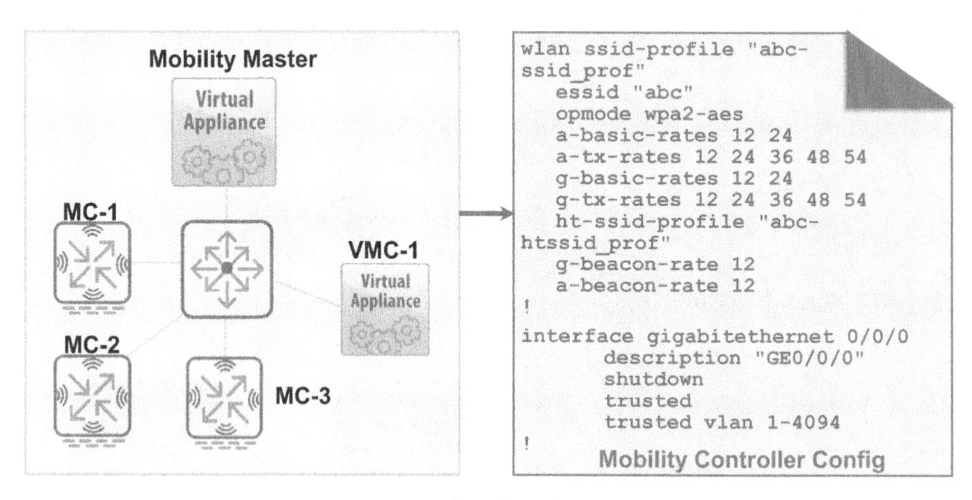

Figure 3-6 Hierarchical configuration model

All configuration that is done on the Mobility Master is pushed down to its managed devices (Figure 3-6). This centralized configuration applies both global and local configurations. The configuration can be organized in a hierarchy. Common configuration is placed near higher points in the hierarchy. More specific configuration is placed on a subgroup, or even on the managed controller.

The mm System group

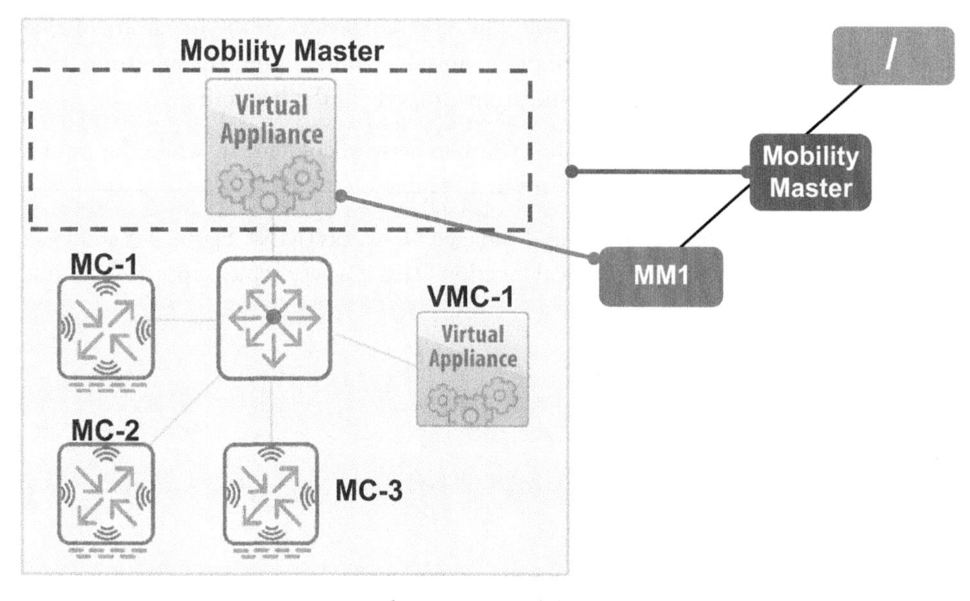

Figure 3-7 Configuration model core architecture

Figure 3-7 shows some group hierarchy, as it relates to the MM. The "mm" group is created by the system to act as a container for MMs. This is where you define various settings. This includes roles, policies, authentication servers, clustering, redundancy, and AirGroups. You also define VPN and Firewall global settings. All configuration is pushed down to the MM1 device within that group.

The mm group is also where you install licenses. Since the MM1 device is in the MM group, it inherits the MM system licensing and group configuration. You can also configure directly on the MM1 node. If you do so, this configuration takes precedence over the system Group MM configuration.

The mn System group

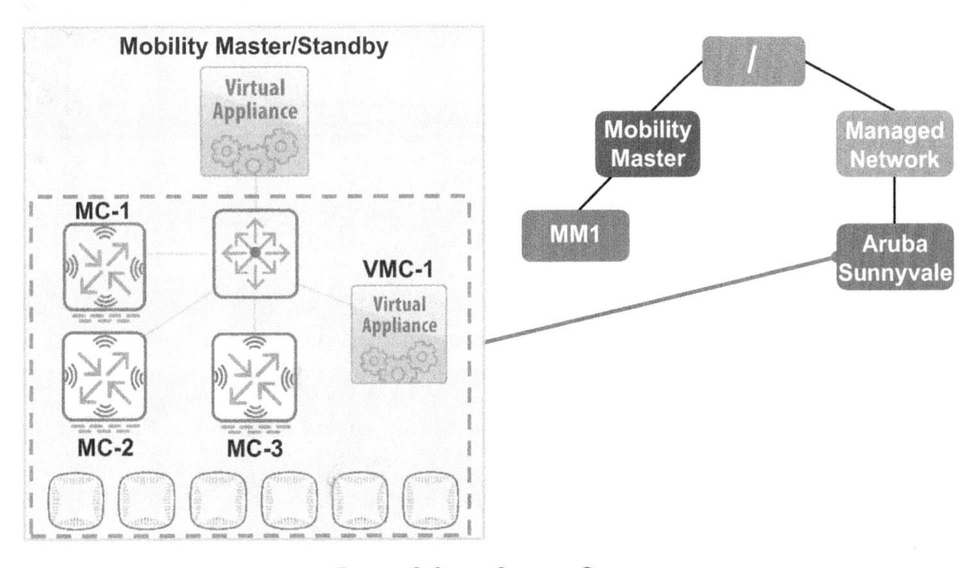

Figure 3-8 mn System Group

Managed Network (mn) is a system group for devices (Figure 3-8). This is where you can start a hierarchy of groups and subgroups. The figure shows the mn system group, and an administratively defined group named Aruba-Sunnyvale.

Within the mn you can configure Roles and Policies, WLANs, AP-groups, Authentication and much more. The entire configuration is pushed down to Groups and sub-groups, and finally to the actual MCs.

If desired, you can configure services directly on a group or subgroup. This configuration is sent down the hierarchy. It supersedes any configuration flowing down from the hierarchy above.

Subgroups

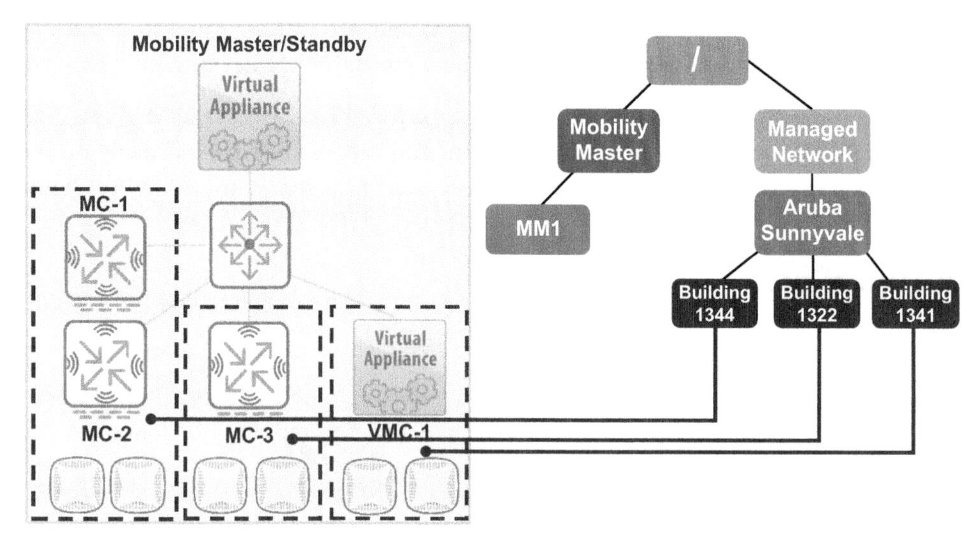

Figure 3-9 Subgroups

Figure 3-9 expands on this scenario, showing subgroups named Building 1344, 1322, and 1341. As you can see these are subgroups of the Aruba-Sunnyvale group, underneath the mn system group.

The mn configuration is pushed to the group Aruba Sunnyvale and on down to the subgroups.

Managed devices

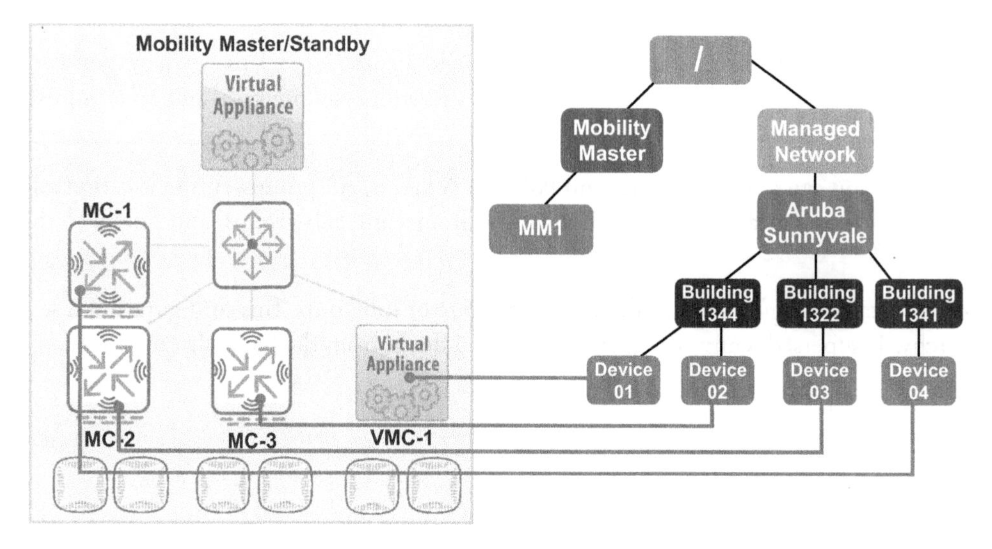

Figure 3-10 Managed devices

Devices are added to groups, subgroups, or even the system group. As you add groups and subgroups, you gain more granular control over configurations. You can create a hierarchy that fits your specific needs (Figure 3-10).

In this example, the global configuration is pushed down to the Sunnyvale group. If needed, you can configure unique settings on the Sunnyvale group. As you know, these settings are pushed down to the subgroups below it—Building 1344, 1322, and 1341.

You can modify any subgroup configuration as needed, and this will flow down into the devices contained therein. Finally, you can modify individual device configurations, and all of this configuration is pushed down to the MCs. This includes both physical Mobility Controllers and VM-based, Virtual Mobility Controllers (VMC).

Adding groups

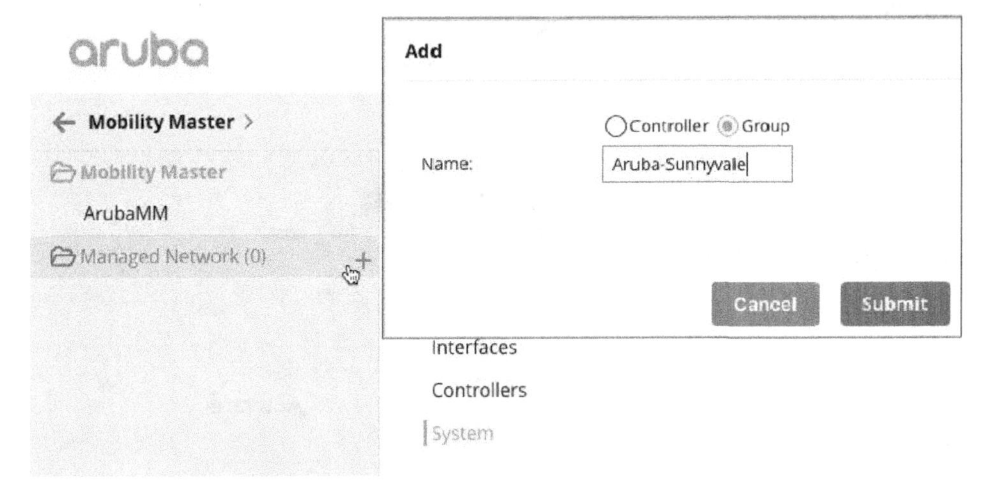

Figure 3-11 Adding groups

To add a Group, click the "+" symbol next to the "Managed Networks" system group. Then choose the Group radio button, and enter the group name. Figure 3-11 shows creation of a group named Aruba-Sunnyvale.

Once this group is created, you can click the "+" next to it, and create a subgroup. Continue using this technique to create subgroups of subgroups, as desired.

Learning check

2. The MM and the MN are what type of groups?
    a. These are not groups
    b. System groups
    c. Top group
    d. Subgroup

Answers to learning check

2. The MM and the MN are what type of groups?
    a. These are not groups
    b. **System groups**
    c. Top group
    d. Subgroup

Mobility Controller setup

MC Zero Touch Provisioning

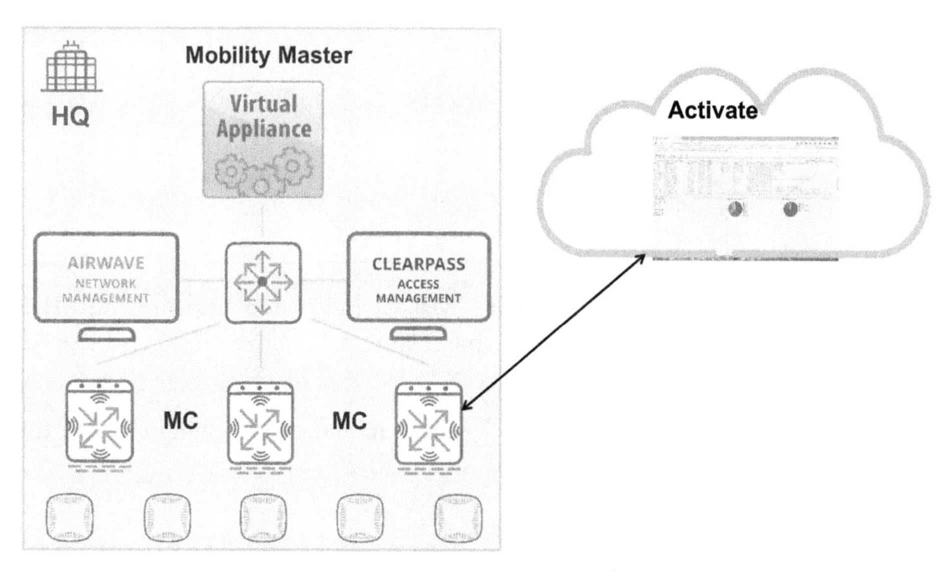

Figure 3-12 MC Zero Touch Provisioning

All controllers can be configured locally. However, if the MC has Internet connectivity, the HPE Aruba "Activate" cloud-based service can direct the MC to the MM. All controllers support Zero Touch Provisioning (ZTP) with the aid of Activate (Figure 3-12).

The MM manages all of the local and global configuration. Once the MC is in communication with the MM, you can no longer configure the MC directly. This eases and scales your management and configuration efforts because you have a single point of configuration for the entire deployment. In a large deployment, the MM acts as a single touch point.

Adding MCs to the hierarchy

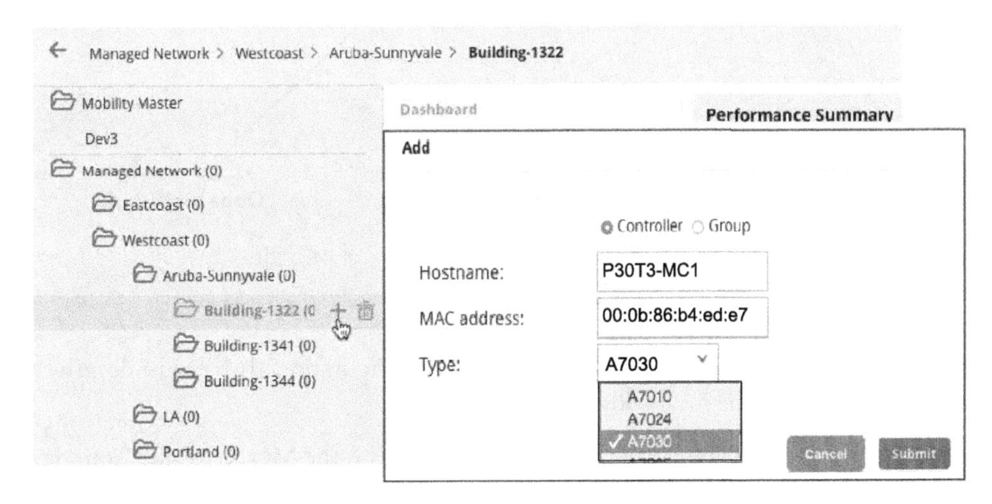

Figure 3-13 Adding MCs to the hierarchy

Before configuring MCs, you should create your group hierarchy. It is very difficult to move an MC from one group or subgroup to another. Remember, this hierarchy is used for both configuration and management (Figure 3-13).

To add an MC to a (sub)group simply click on the group's "+" sign, add the MC host name and MAC address. Specify the device type and then click Submit.

When the MC is directed to the MM, it joins in the appropriate group and receives its configuration.

Direct MC to MM via ZTP or Script

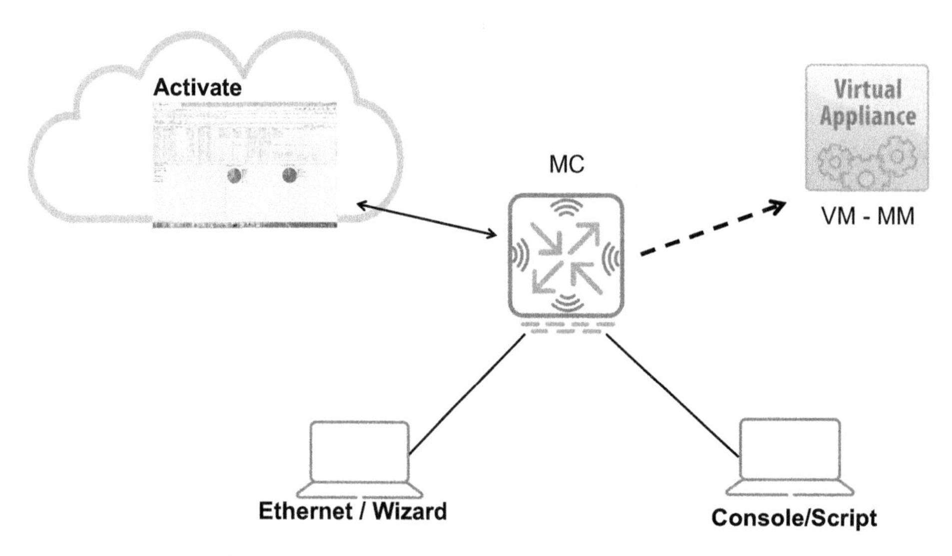

Figure 3-14 Direct MC to MM via ZTP or Script

All controllers can be directed to their MM via their local configuration. This can be done by running the full startup script from the local console (Figure 3-14).

Another option is to connect your laptop to an Ethernet port on the MC controller. Your device gets a DHCP-assigned IP address, so you can connect to the MC and run the initial configuration wizard. The wizard lets you configure the MM's IP address, thus directing it to the MM.

If the MC has access to the Internet, then the Activate cloud-based service can direct the MC to the MM. The MM is fully responsible for the configuration of the MCs and VMCs. This gives a single touch point for all the configuration of the MC. This includes VLAN and IP addresses as well as WLAN and firewall configurations.

Wizard installation

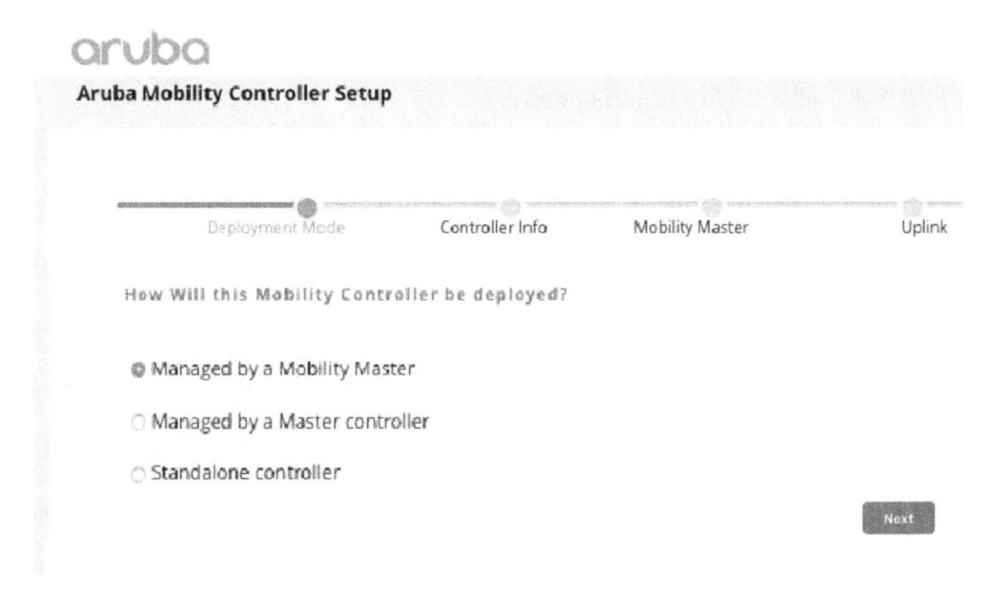

Figure 3-15 Wizard installation

To direct your MC to the MM using the wizard, connect an Ethernet cable from a laptop to the MC. You will get an IP address. Simply open a browser page and you will be automatically placed in the configuration wizard (Figure 3-15).

There will be five windows to complete the initial configuration. The first window is Deployment Mode. From this window, select the deployment method from the following:

- Standalone will setup one MC independently.
- Managed by a Master controller will setup to talk to a Master MC in a Master Local mode.
- Managed by a Mobility master will setup this MC to communicate with the VM mobility master.

The next window is named Controller Info. Here you will configure:

- The Name of the MC
- Country code. If this controller is shipped to USA, Israel, or Japan, then there is no choice. For all other countries, select your country code.
- admin password
- Time, NTP Server, and Time zone

Next up is the Mobility Master window, where you will configure:

- The IP address or FQDN of the MM
- The IPv6 IP address of the MM. This field is optional.
- The MM MAC address. This field is optional.
- Redundant MM MAC address. This field is optional.
- Is the master a Virtual machine. So are you going to an MM or a Master MC.
- The Pre-shared Key to be used with the MM

The fourth window is the Uplink window and will request:

- VLAN and IP addressing information
- Port mode or access or Trunk

Note

If TRUNK is selected, then the NATIVE VLAN is set to 1, and allowed VLAN will be the previously configured VLAN. This may cause issues depending on how the Uplink router/switch has been configured.

- VLAN IP address
- Default Gateway
- DNS IP address. You may need this if FQDN was previously selected.

The last window will provide you a summary of the initial setup and ask you to confirm. The Summary window would appear to the right of "Uplink." This is not shown in Figure 3-15.

MC Directed to MM Script

```
Enter Option (partial string is acceptable): full-setup

Are you sure that you want to stop auto-provisioning and start full setup dialog? (yes/no): yes
  .......

Enter System name [Aruba7030]: P30T3-MC1
Enter Switch Role (master|standalone|md) [md]:
Enter IP type to terminate IPSec tunnel (ipv4|ipv6) [ipv4]:
Enter Master switch IP address or FQDN: 10.254.1.10
Is this a VPN concentrator for managed device to reach Master switch (yes|no) [no]:
This device connects to Master switch via VPN concentrator (yes|no) [no]:
Is this a VPN concentrator for managed device to reach Master switch (yes|no) [no]:
This device connects to Master switch via VPN concentrator (yes|no) [no]:
Is Master switch Virtual Mobility Master? (yes|no) [yes]:
Master switch Authentication method (PSKwithIP|PSKwithMAC) [PSKwithIP]:
Enter IPSec Pre-shared Key: ********
Re-enter IPSec Pre-shared Key: ********
```

Figure 3-16 MC Directed to MM Script Part 1

To direct your MC to the MM using the script, connect a console cable to the MC, choose "full-setup," as shown in Figure 3-16. Follow the script to do basic configuration.

Figure 3-16 shows the first part of this process, during which you will enter the following information:

- The Name and role of the MC.
- The IP address of the MM.
- Whether this MC will be a VPN concentrator or use a VPN concentrator.
- The Pre-Shared Key (PSK) to be used with IPSec tunnels to connect to the MM.

Figure 3-17 below shows the rest of the process.

```
Enter Uplink Vlan ID [1]: 70
Enter Uplink port [GE 0/0/0]:
Enter Uplink port mode (access|trunk) [access]: access
Enter Uplink Vlan IP assignment method (dhcp|static) [static]:
Enter Uplink Vlan Static IP address [172.16.0.254]: 10.2.70.100
Enter Uplink Vlan Static IP netmask [255.255.255.0]:
Enter IP default gateway [none]: 10.2.70.1
Enter DNS IP address [none]:
Do you wish to configure IPV6 address on vlan(yes|no) [yes]: no
This controller is restricted to Country code US for United States, please confirm (yes|no)?: yes
Enter Time Zone [PST-8:0]:
Enter Time in UTC [07:16:49]:
Enter Date (MM/DD/YYYY) [10/20/2016]:
Do you want to create admin account (yes|no) [yes]:
Enter Password for admin login (up to 32 chars): ******
Re-type Password for admin login: ******
```

Figure 3-17 MC Directed to MM Script Part 2

This final portion of the script includes configuration of the following parameters:

- Basic port, VLAN, and IP addressing information. Note that if trunk mode is selected the native VLAN is VLAN1.
- Country code. If this controller is shipped to USA, Israel, or Japan then there is no choice. For all other countries, select your country code.
- Time zone
- admin password

IPSec Keys

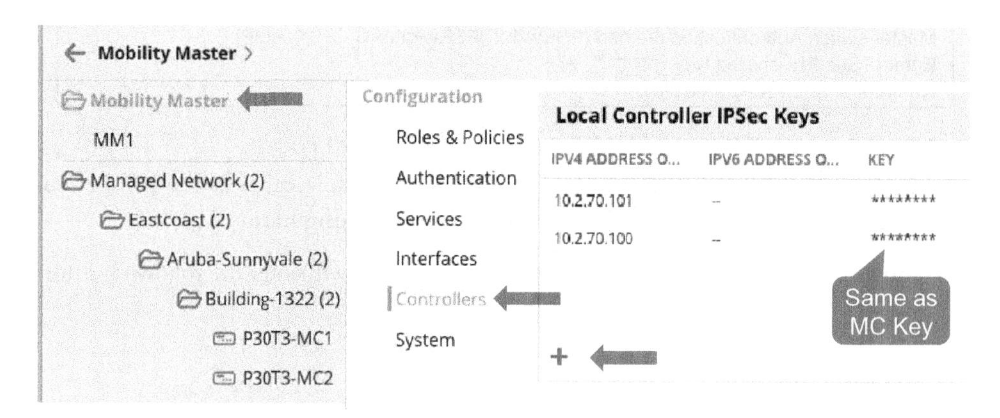

Figure 3-18 IPSec Keys

Recall that MCs and MMs connect by establishing a secure IPSec tunnel. If your IPSec configuration uses PSKs, the MC and MM must have identical PSKs. Figure 3-18 shows how to configure the MM with the same PSK you just configured on the MC. Point your browser to the MM GUI. Under configuration, click Controllers. Then click the "+" sign and configure the MC's IP address and Pre-Shared Key.

ZTP Activate

Create New Rule

Input for Rule

Rule Type:	Provisioning Rule
Parent Folder:	default
Provision Type:	Managed Device to Ma
Redundancy Level:	No Redundancy
Config Node Path:	
Primary Master Controller:	Select primary...
Master Controller IP:	10.254.1.10
Primary VPN Concentrator MAC:	Select VPN MAC...
VPN Concentrator IP:	Optional
Secondary VPN Concentrator MAC:	Select VPN MAC...
Country Code:	United States
Rule Name:	default.provision.mana

Done Cancel Re-Order

Figure 3-19 ZTP Activate

If the MC has access to the Internet, it will attempt to communicate with the cloud-based Activate service. Activate will direct the MC to the MM to receive its configuration. Figure 3-19 shows an example configuration for the ZTP Activate service.

Note

Activate is a kind of "kick-start" service for Aruba deployments. Activate enables a network installer to simply plug a new MC into an Internet connection anywhere in the world and connect power. The MC will automatically be directed to the appropriate MM and acquire its configuration. For more information on Activate, do an Internet search on "Aruba Activate."

VPN concentrator

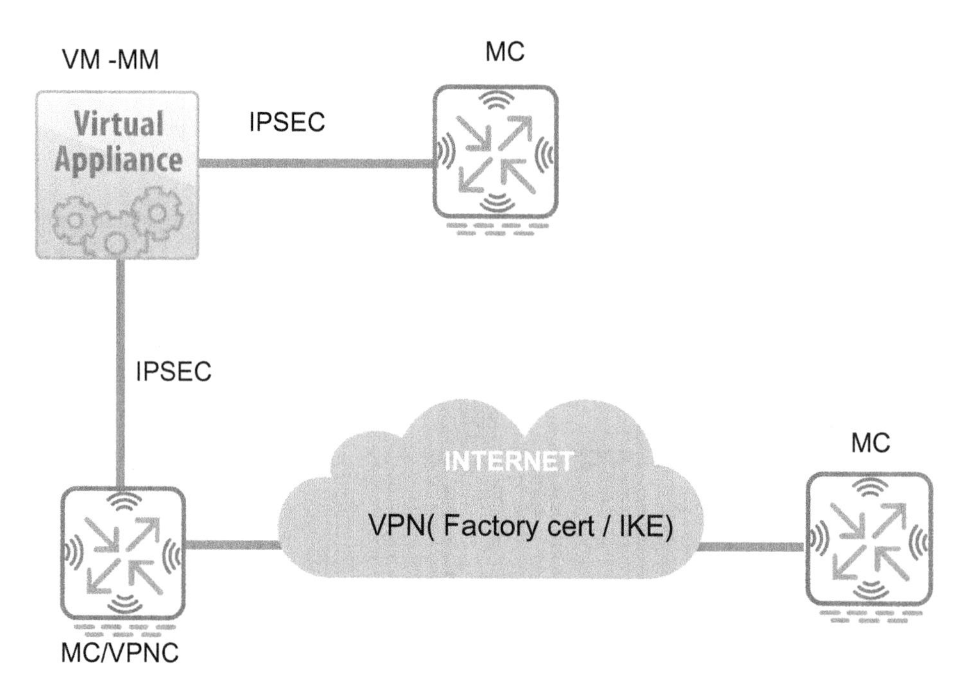

Figure 3-20 VPN concentrator

Figure 3-20 shows two MCs connected to an MM. The top-most MC connects to the MM via the corporate infrastructure, and an IPSEC tunnel forms between them.

The bottom MC is at some remote site and must connect to the MM via the Internet, using a site-to-site VPN tunnel. To accommodate this scenario, you deploy an MC with routable connectivity to the MM. You configure this MC as a VPN Concentrator (VPNC). Remote-site MCs traverse the Internet, create a VPN connection to this MC/VPNC, and gain access to the MM.

All MCs using a site-to-site VPN must terminate on a MC/VPN. The MC/VPN can also terminate APs.

Note

IPSec is a standardized suite of protocols used to create secure Virtual Private Networks (VPN). IPSec VPN services ensure that a VPN endpoint only allows connectivity to authorized devices. Also, once the tunnel is formed, all packets are encrypted and hashed. This requires compatible keys on the VPN endpoints. As indicated in Figure 3-20, the devices ship with a factory certificate which may be used for authentication, among other methods. Also the devices use the standard Internet Key Exchange (IKE) to help form the connection and exchange key information.

MC VPN concentrator

```
Enter Option (partial string is acceptable): full-setup


Are you sure that you want to stop auto-provisioning and start full setup dialog? (yes/no):
yes

Enter System name [Aruba7030]: A7030-VPN-MC
Enter Switch Role (master|standalone|md) [md]:
Enter IP type to terminate IPSec tunnel (ipv4|ipv6) [ipv4]:
Enter Master switch IP address or FQDN: 10.254.1.10
Is this a VPN concentrator for managed device to reach Master switch (yes|no) [no]: yes
Enter IPSec Pre-shared Key: ********
Re-enter IPSec Pre-shared Key: ********
Enter Master switch MAC address: 00:0C:29:59:F8:32
Enter Redundant Master switch MAC address [none]:
Enter Uplink Vlan ID [1]: 80
Enter Uplink port [GE 0/0/0]:
```

Figure 3-21 VPN concentrator

```
Enter Option (partial string is acceptable): full-setup


Are you sure that you want to stop auto-provisioning and start full setup dialog? (yes/no):
yes

Enter System name [Aruba7030]: Aruba7030-U
Enter Switch Role (master|standalone|md) [md]:
Enter IP type to terminate IPSec tunnel (ipv4|ipv6) [ipv4]:
Enter Master switch IP address or FQDN: 10.254.1.10
Is this a VPN concentrator for managed device to reach Master switch (yes|no) [no]:
This device connects to Master switch via VPN concentrator (yes|no) [no]: yes
Enter VPN concentrator IP address or FQDN: 67.90.90.56
VPN concentrator Authentication method (FactoryCert|PSKwithMAC) [FactoryCert]: PSk
VPN concentrator Authentication method (FactoryCert|PSKwithMAC) [FactoryCert]:
PSKwithMAC
Enter IPSec Pre-shared Key: ********
Re-enter IPSec Pre-shared Key: ********
Enter VPN concentrator MAC address: 00:0C:29:59:F8:32
Enter Redundant VPN concentrator MAC address [none]:
```

Figure 3-22 MC Using VPN concentrator

Site-to-site VPNs can be setup via the cloud-based Activate server, or by using the MC full-setup script.

Figure 3-21 shows using the full setup script to set up an MC as a VPNC. Figure 3-22 shows using the full setup script to set up an MC to use that VPNC for MM access.

Spend a few moments to compare and contrast the two figures. Do you see the main difference between them? The device named A7030-VPN-MC is configured as a VPN concentrator. The device named Aruba7030-U is NOT configured as a concentrator. However, it is configured to connect to the A7030-VPN-MC concentrator, using a Pre-Shared Key (PSK) for authentication.

Note

The MC/VPNC must have routable connectivity to the MM.

Learning check

3. Name three types of ways to direct an MC to an MM
    a. Local console
    b. Using AirWave
    c. Using Activate
    d. Wizard configuration
    e. Using ClearPass

Answers to learning check

3. Name three types of ways to direct an MC to an MM
    a. **Local console**
    b. Using AirWave
    c. **Using Activate**
    d. **Wizard configuration**
    e. Using ClearPass

Hierarchical configuration

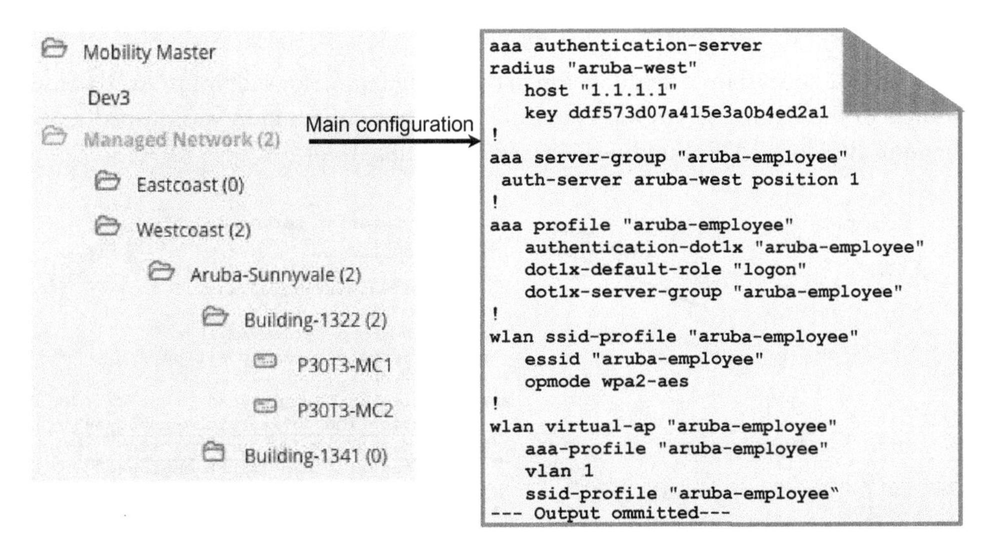

Figure 3-23 mn Configuration

Figures 3-23 through Figure 3-26 show what you can do with hierarchical configuration. You can see the command-line syntax configuration, which was completed in the mn group.

For consistent configuration among all MCs, begin your configuration efforts in the mn group. You can start at a lower group if desired. Configuration created in the Eastcoast group is only propagated to subgroups and devices below that hierarchy. The Westcoast group would not be affected.

In this example, the bulk of the configuration was created in the Managed Network system group.

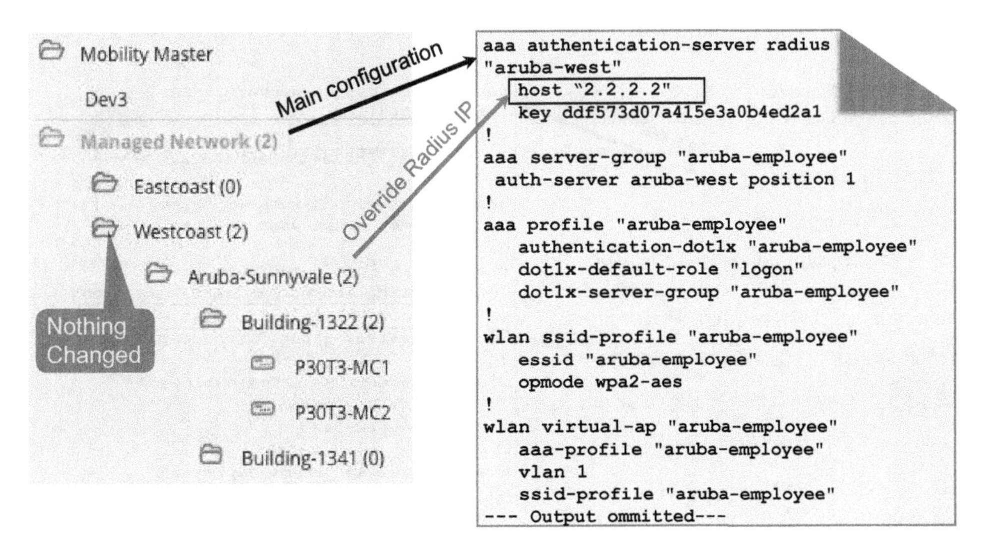

Figure 3-24 Subgroup configuration override

Figure 3-24 shows the syntax created in mn, which was pushed down to all groups and devices below it. Since no configuration was done in Eastcoast and Westcoast, no changes were made to these groups, or the subgroups below them.

However, Figure 3-24 does show a modification to the configuration for subgroup "Aruba-Sunnyvale." The IP address of the Radius Server was changed. This new configuration was pushed down to the three subgroups Building-1322, Building-1341, and Building-1344.

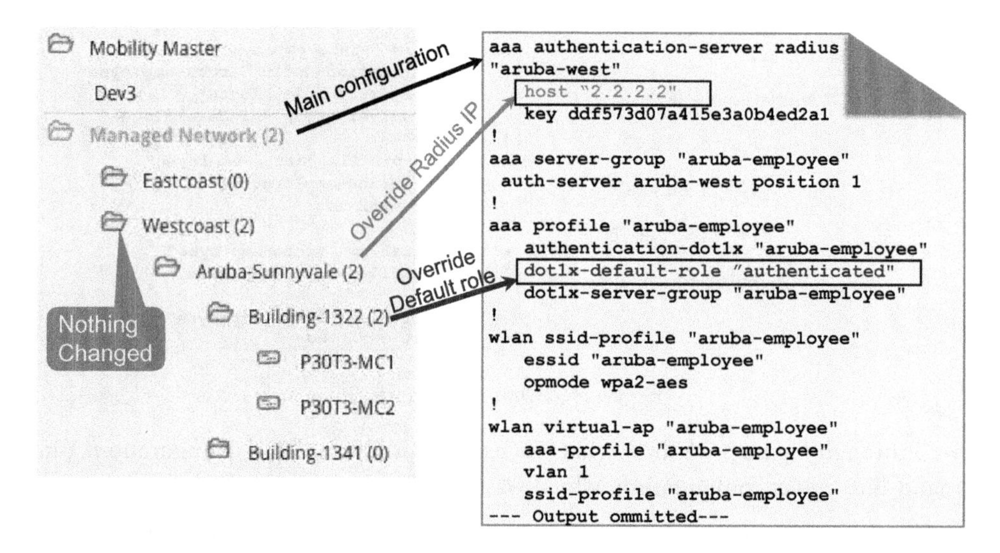

Figure 3-25 MC VPN concentrator

In this scenario, another change has been made, this time in the subgroup Building-1322. Specifically, the dot1x default role was changed from "logon" to "authenticated."

This new configuration was then pushed down to the MCs P30T3-MC1 and P30T3-MC2.

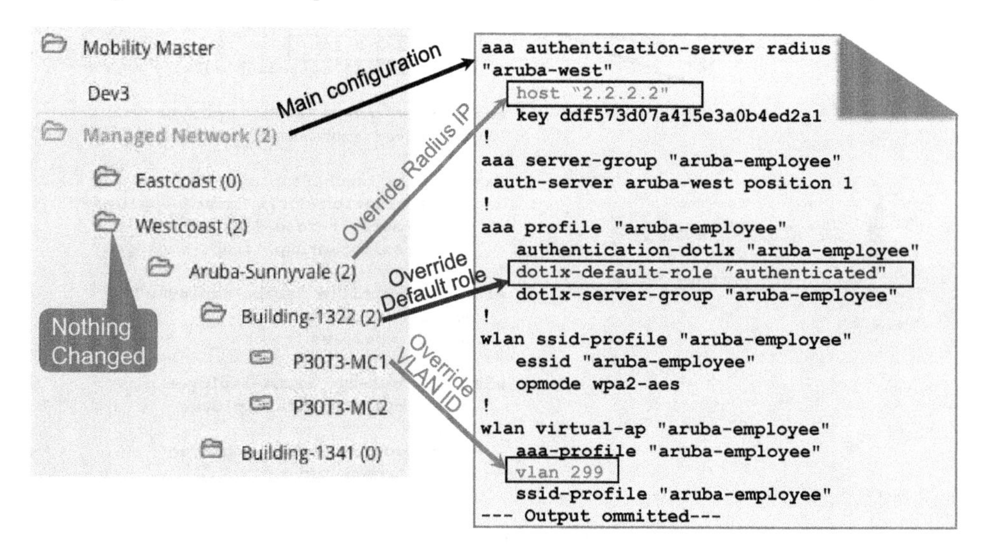

Figure 3-26 MC VPN Concentrator

Finally, a configuration change has been made for a specific MC: the VLAN ID was modified.

This change is only pushed down to a specific MC device named P30T3-MC1. This change will not affect any other devices, including MC P30T3-MC2.

Figure 3-26 shows the configuration that will be applied to P30T3-MC1. This example shows how you can use hierarchy to quickly apply changes to all devices, by configuring the mn group. However, you still have fine-grained control over device-level configurations.

Configuration validation and distribution

Before the MM synchronizes configurations to other devices, three validation checks are performed.

1. **Syntax validation**—Ensures correct command syntax, data types, and value ranges.
2. **Semantic validation**—Performs dependency checks across commands or instance count limits. If a profile refers to another profile, it ensures the referenced profiles exist.
3. **Platform validation**—Determines which features are supported on the platform. Also verifies the number and type of ports on the platform.

Validation failure—Invalid commands are rejected and are not included in the pending configuration. If a new device is added that cannot support an existing configuration, the device is rejected.

Validated configurations are then distributed, as described in the following list.

- **Partial Configuration**—Sent to pre-existing devices in the deployment. They already have the settings from previous configuration efforts, and so only a partial configuration is sent to the device. These changes are distributed down the hierarchy, as previously described.

Full Configuration—Sent to new devices and to devices that have lost their configurations.

MC local configuration

MC configuration

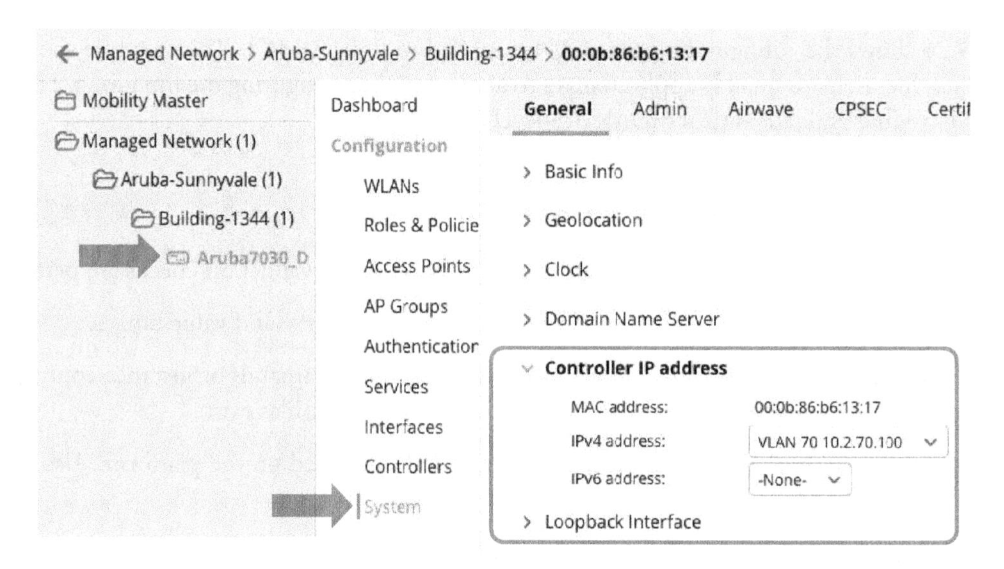

Figure 3-27 MC Configuration

To configure a controller's local management VLAN (Figure 3-27):

1. Click on the controller
    a. Click on "System"
    b. Select the "General" Tab
    c. Expand the section "Controller IP Address"

Note

Changing the Controllers management IP will cause the controller to reboot.

MC Geolocation

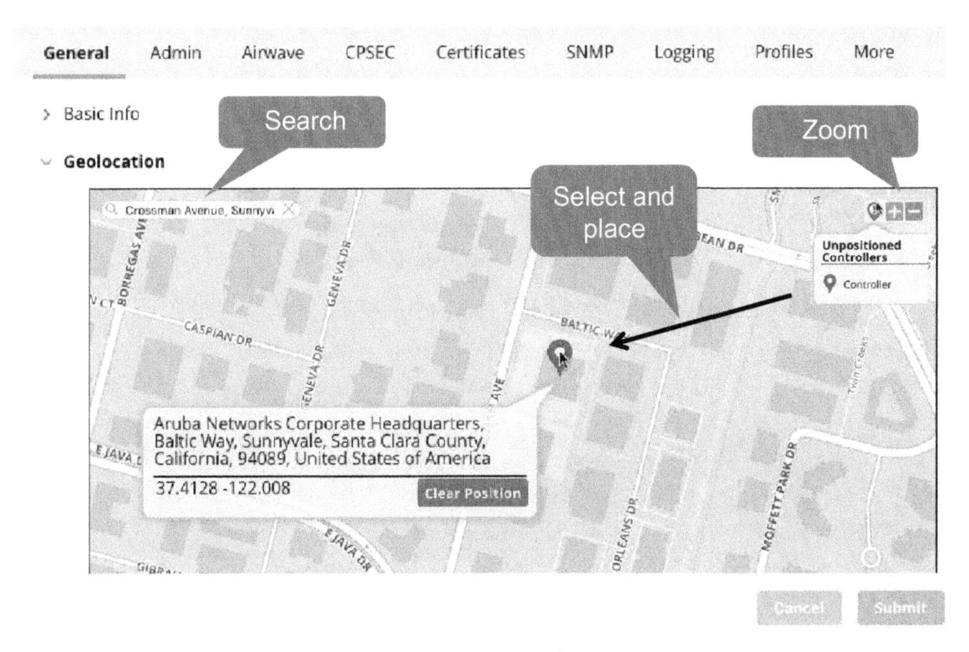

Figure 3-28 MC Geolocation

You can use the Geolocation map to place your devices (Figure 3-28). This provides a convenient visual representation of your network. You can search by address and zoom in and out. Once you have the desired location simply select the device from the list and drag onto the map.

To access the Geolocation map:

1. Click on the controller
   a. Click on "System"
   b. Select the "General" Tab
   c. Expand the section "Geolocation"

MC interface VLANs and ports

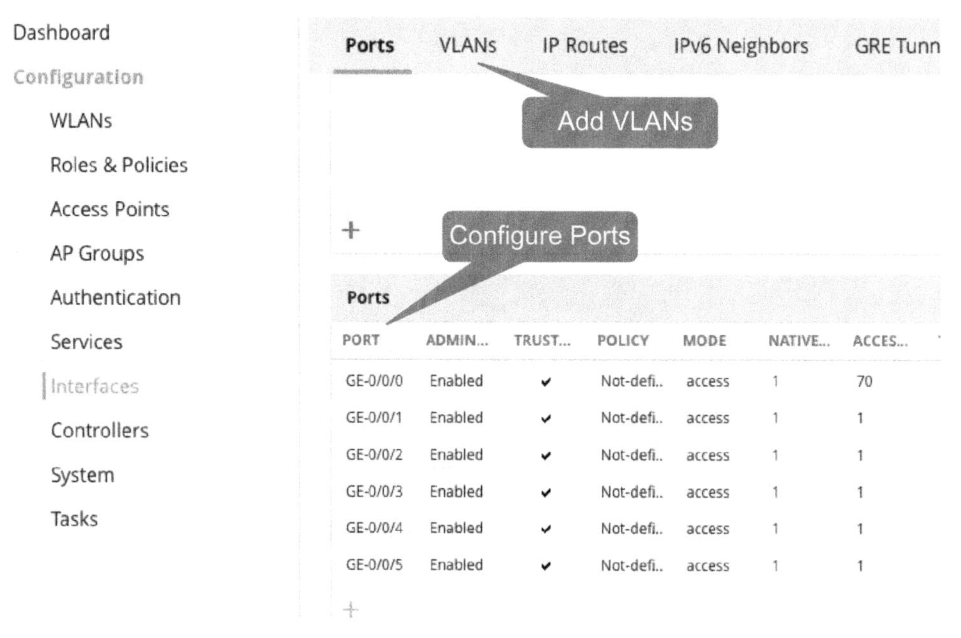

Figure 3-29 MC interface VLANs and ports

You can perform local configuration on a device, such as adding VLANs, IP addresses, or port configuration. To do so, select the device and then select the Interfaces submenu (Figure 3-29).

MC Interface port parameters

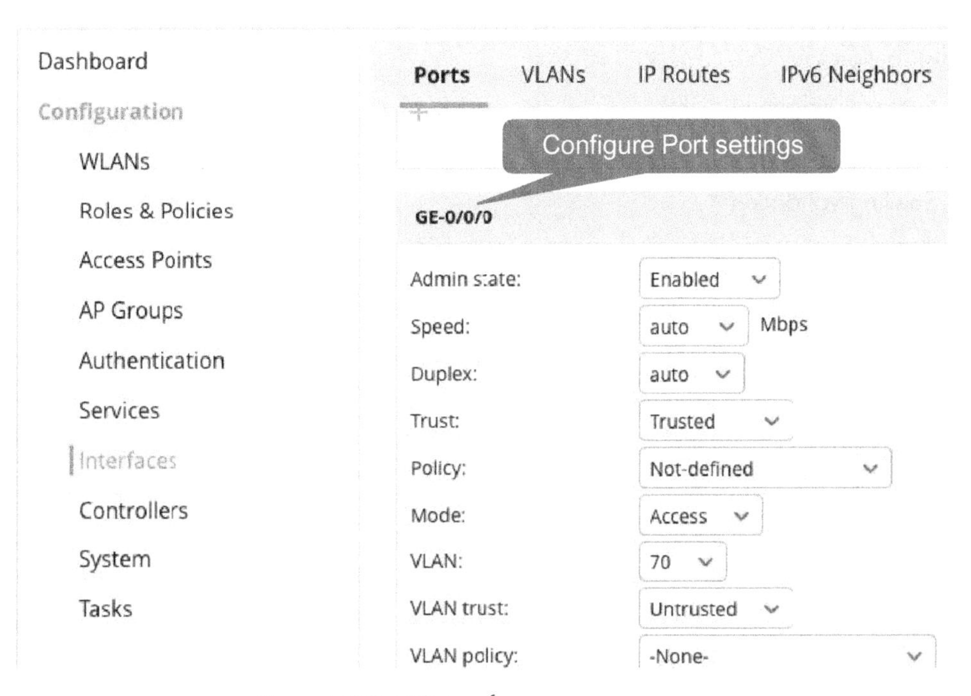

Figure 3-30 MC Interface port parameters

You can change port parameters such as VLAN, trunk or access mode, trusted/untrusted configuration, and so forth (Figure 3-30). To do so, click on the device, click on the Interfaces submenu, and select the Ports tab.

Learning check

4. Which four choices are relevant to configuration validation, as performed by the MM, prior to distribution to MCs?
   a. Syntax validation
   b. Semantic validation
   c. Spelling validation
   d. Platform validation
   e. Validation failure

5. In what group is the MC controller IP address configured?
   a. System group
   b. The first Subgroup
   c. Last subgroup
   d. At the device level

Answers to learning check

4. Which four choices are relevant to configuration validation, as performed by the MM, prior to distribution to MCs?
   a. **Syntax validation**
   b. **Semantic validation**
   c. Spelling validation
   d. **Platform validation**
   e. **Validation failure**

5. In what group is the MC controller IP address configured?
   a. System group
   b. The first Subgroup
   c. Last subgroup
   d. **At the device level**

4 Secure WLAN Configuration

LEARNING OBJECTIVES

✓ In the last chapter, you learned about the hierarchical nature of ArubaOS 8 configurations. In this chapter, you will leverage that knowledge by actually configuring a controller to support end users. While learning about configuration techniques and options, you will gain a deeper understanding of WLAN components, such as SSIDs, Radios, and VLANs.

✓ You will learn about the advantages of creating AP group structures and how to create them. You will explore typical WLAN configurations to support end user connectivity.

WLAN components

WLAN deployment prerequisites

Before you can configure a WLAN, you must gather some information and make some decisions. You must decide the SSID name and which methods of authentication and encryption are appropriate. You need to select which radios will support this SSID—2.4 GHz, 5 GHz, or both.

You typically want APs to broadcast a guest SSID by including it in their beacon and probe response frames. You must decide whether users can easily find and associate with a particular SSID or whether users must have this preconfigured on their client device.

Once users associate with an AP's SSID, they must authenticate. You decide whether this authentication is merely an unsecured "handshake" or based on a Pre-Shared Key (PSK) or requires some 802.1X/EAP method. 802.1X/EAP is a more secure form of username/password or certificate-based authentication.

After successful authentication, user endpoints are assigned an IP address on a particular VLAN. Yes, you need to decide which VLANs and IP addresses are to be associated with particular SSIDs and their connected users.

All of these WLAN requirements are configured and stored in a particular AP Group. AP groups control which APs advertise a particular SSID, Radio band, RF settings, and so forth. This is an effective method for organizing APs and SSIDs. It can streamline your workflow and improve WLAN scalability.

In summary, prior to deployment, you need to plan and to choose the following WLAN options:

- SSID name
- Encryption type (open, WPAv1, or WPAv2)
- Authentication (none, PSK, or 802.1x)
- Supported radio bands (will the WLAN be supported on the 2.4 GHz band, 5 GHz band, or both)
- Assigned VLAN

This section will review these options, starting with the concept of AP groups.

AP group structure

AP-group

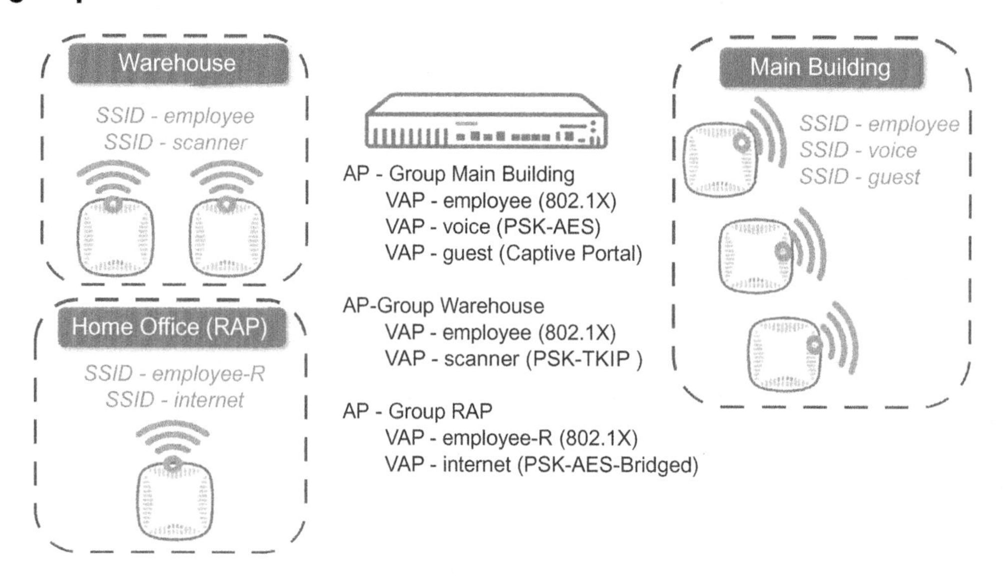

Figure 4-1 AP-Group

APs advertise SSIDs based on their AP-group membership and how that group is configured. An AP can only belong to one AP-Group at a time. If you want all APs to advertise the same SSIDs, then you only need one AP-group. On medium and large deployments, it is more typical to need multiple AP-groups.

Figure 4-1 shows an AP-group named Main Building. Of course, all APs physically installed in the Main Building are members. In this scenario, you decided that these APs must support three SSIDs—employee, voice, and guest.

There is no need for voice services in the warehouse, and guests are not allowed to be there. You are wise and create a separate AP-group named Warehouse. Its AP members only support the SSIDs named employee and scanner.

Aruba RAPs have been setup in certain employee homes, and so you created a group named RAP. This group supports an SSID named employee-R. This SSID is configured to securely tunnel corporate traffic back to the office. The Internet SSID is configured to directly bridge traffic the user's local traffic on to the Internet.

AP group and profile structure

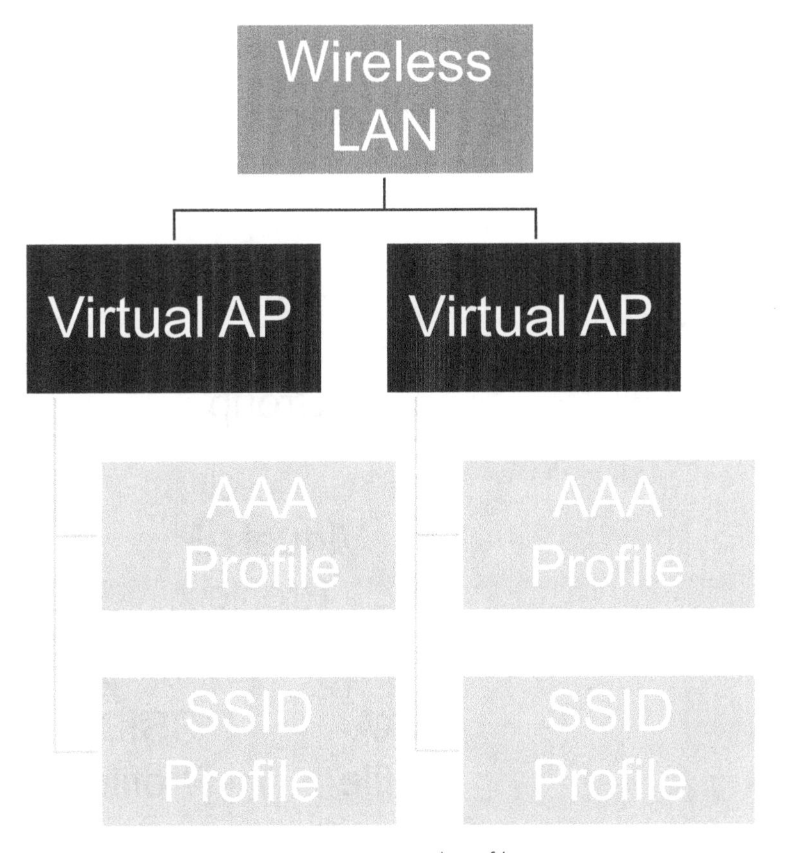

Figure 4-2 AP group and profile structure

AP groups contain all the required AP configuration. This configuration is organized into profiles. This discussion is focused on the WLAN-specific configuration, which is shown in Figure 4-2. The WLAN-specific configuration has all of the profiles for WLAN, the name of the network (SSID), authentication, encryption, supported data rates, and more.

In addition to WLAN-specific profiles, AP groups also contain profiles for other types of configurations, including the following:

- The RF management section has radio configuration.
- The AP section is for port configuration, regulatory domain, and SNMP. The system profile determines which controller will execute this AP Group.

- The QOS section contains parameters to configure QOS, often for voice and video applications.
- The IDS profile controls Intrusion Detection System options for the group.
- The Mesh section defines the mesh domains and any encryption needed between mesh nodes.

Note

Most profiles have a default profile. *It is highly recommended to never change a default profile.* It is better to create a new profile. By changing a default profile, you may unintentionally change other AP-groups, which happen to be using the very same default profile.

AP group profile hierarchy

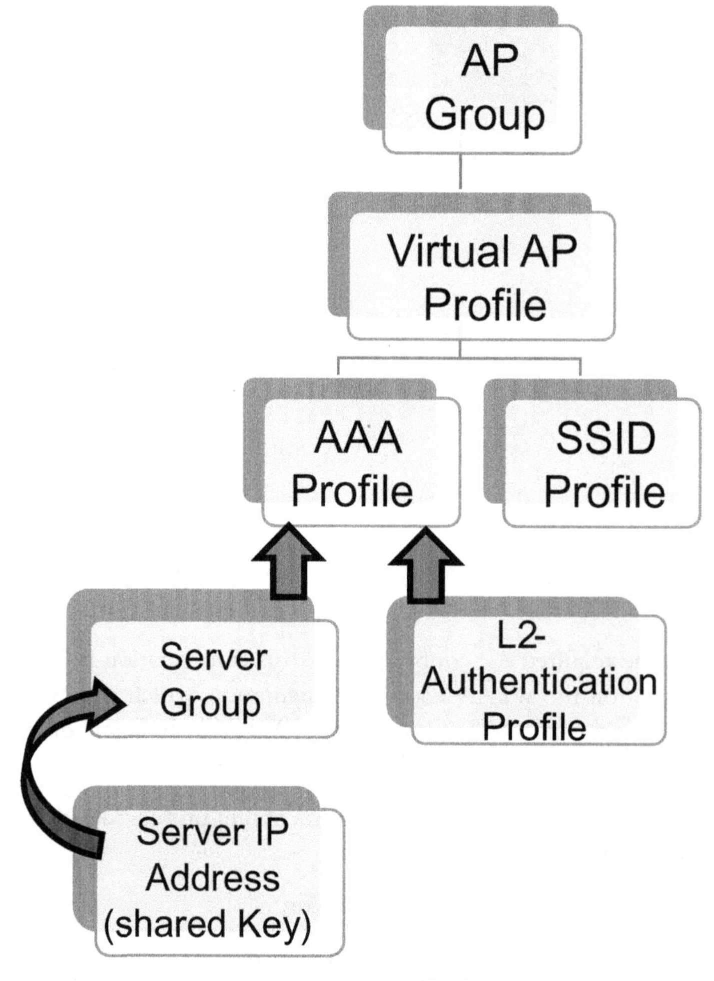

Figure 4-3 AP group profile hierarchy

Figure 4-3 shows AP group profile hierarchy focused on the Wireless LAN section as previously described. Near the top is a Virtual AP (VAP) profile, which contains the client VLAN and traffic-forwarding mode.

The SSID profile specifies the SSID name, authentication and encryption methods, and supported RF Channels and data rates.

The AAA profile contains the roles and other authentication-specific settings. External AAA servers might be used for this SSID, such as a RADIUS server. In this case, the AAA profile references a server group profile to get information about the server, such as IP address and shared-key information. The AAA profile can also reference a Layer 2 authentication profile, for things like 802.1X-related configuration.

AP group and profiles

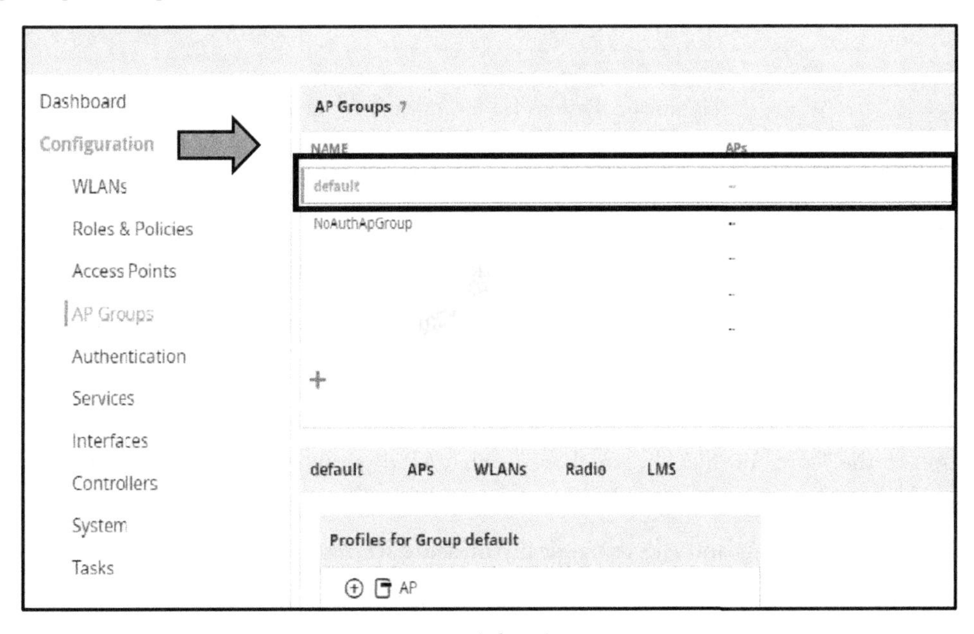

Figure 4-4 Predefined AP groups

Aruba controllers have two predefined AP groups, named default and NoAuthAPGroup, as shown in Figure 4-4. To see the AP groups, navigate to **Managed Network> Configuration > AP Groups**.

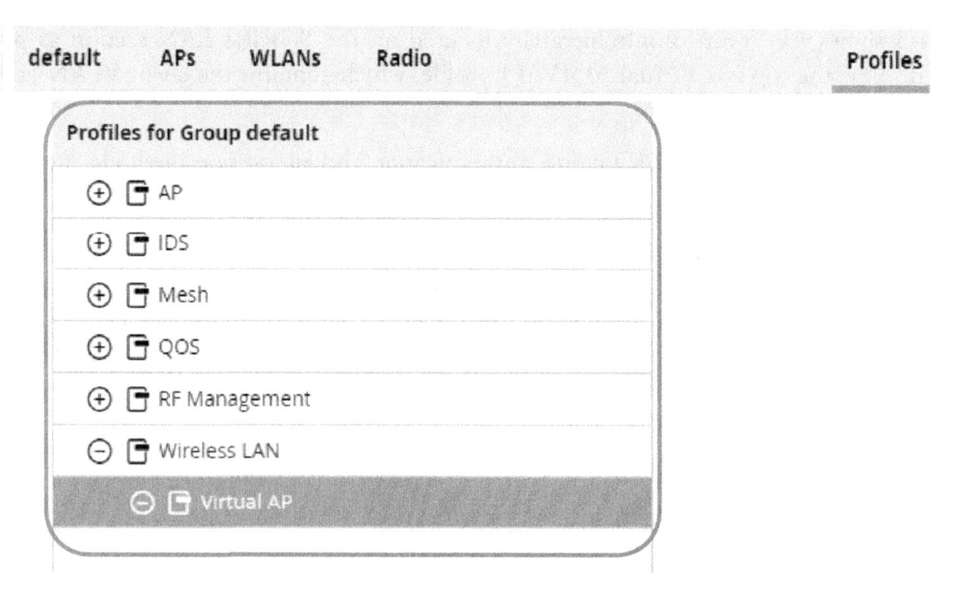

Figure 4-5 Profiles for the default AP group

Aruba controllers also have default profiles.

You just learned about one of the key profiles for a functional WLAN—the Virtual AP (VAP) profile. Each Virtual AP is effectively one WLAN, largely defined by its SSID and AAA profiles:

- **SSID profile**—Determines SSID name, required authentication and encryption type, and data rates.
- **AAA profile**—Determines authentication method, roles, and which servers to use.

In addition to the "core" profiles shown above, each VAP directly references one of each of the following profiles types:

- **802.11k**—How clients and APs dynamically measure RF resources to facilitate roaming.
- **AnySpot**—Suppress client probe requests to free up network resources and improve network performance.
- **HotSpot 2.0**—Help clients identify which APs in your 802.11u hotspot are suitable for their needs.
- **WWM traffic management**—Wi-Fi Multi-Media (WMM) prioritizes voice and video traffic

Each AP group can have as many virtual APs (SSIDs) as necessary, but four or less is best. Too many SSIDs wastes radio bandwidth by causing larger and/or more numerous management frames. An AP group can also consist of other profiles with possible parameters the administrator may wish to change.

For an analogy, some readers may be familiar with the Ethernet switching concept of Virtual LANs (VLANs) along with "trunking" or "802.1q tagging" mechanisms. These technologies enable a single, physical Ethernet port to support multiple Virtual LANs. Similarly, each physical AP radio can support multiple Virtual WLANs. This occurs by applying multiple VAP profiles to an AP group.

 **Note**

Aruba APs can support up to 16 SSIDs on dual radio APs.

Note

Most profiles have a default. *It is highly recommended to never change a default profile.* It is better to create a new profile. Default profile modifications may cause unintended changes to other AP-groups.

Creating AP-groups

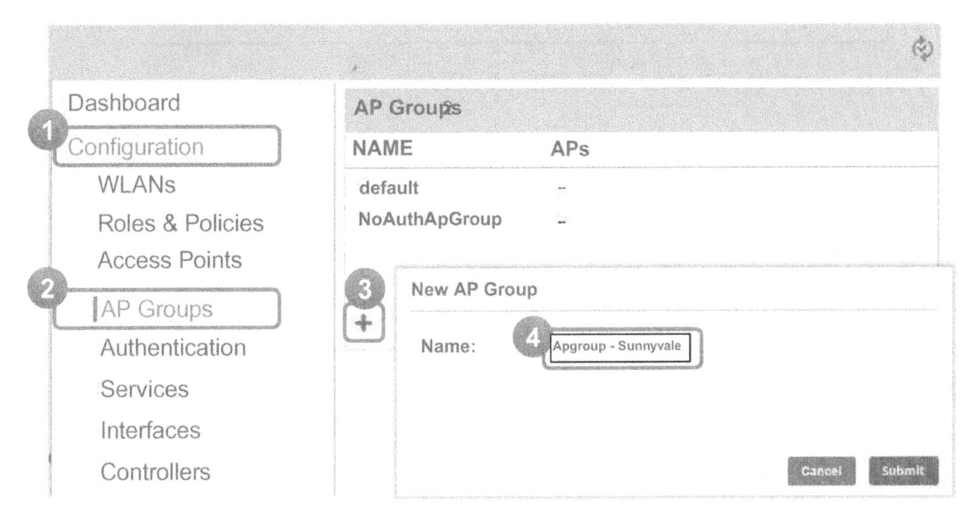

Figure 4-6 Creating AP groups

To create an AP group, navigate to **Managed Network> Configuration > AP Groups** and click the + symbol to create a new AP group. Enter the New AP group name in the AP group name field and then click Submit (Figure 4-6).

Committing your configuration changes

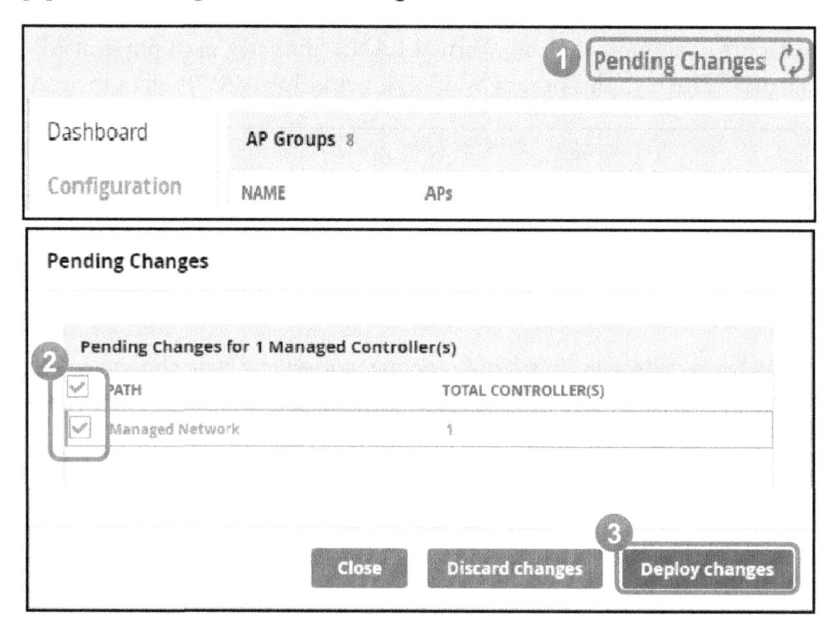

Figure 4-7 Committing your configuration changes

It is **very important** that you commit your configuration changes! Do not forget to click on Pending Changes near the top-right, as shown. At the end of the configuration process, you must commit changes for the configuration to be validated and pushed down the hierarchy group/subgroups (Figure 4-7).

Learning check

1. Which of the following is part of the AP-group structure?
   a. IDS
   b. AP
   c. Mesh
   d. RF Management
   e. QoS
   f. Wireless LAN
   g. All of the above

2. An AP-group is a container that holds all of the AP configuration profiles within it?
   a. True
   b. False

Answers to learning check

1. Which of the following is part of the AP-group structure?
   a. IDS
   b. AP
   c. Mesh
   d. RF Management
   e. QoS
   f. Wireless LAN
   g. **All of the above**

2. An AP-group is a container that holds all of the AP configuration profiles within it?
   a. **True**
   b. False

Use AOS UI to create a WLAN

Create a new WLAN

The recommended method for creating WLANs is through the WLAN wizard, available as of AOS 8.0. To use this WLAN wizard, browse to the Mobility master or stand-alone controller WebUI, and navigate to **Configuration > Tasks.** Then choose **Create a new WLAN**.

The wizard automatically creates a new Virtual AP profile, a AAA profile, along with 802.1X, Server group and SSID profiles, with the same name as the WLAN. The configuration settings and values are defined via the wizard. These profiles support additional, advanced features that are not configurable via the WLAN wizard. This is also available from the **Configuration > Tasks** page.

Another option is to configure WLANs manually, using WebUI or CLI. While this method allows for more granular control and customization, it is only recommended for advanced users.

New WLAN general information

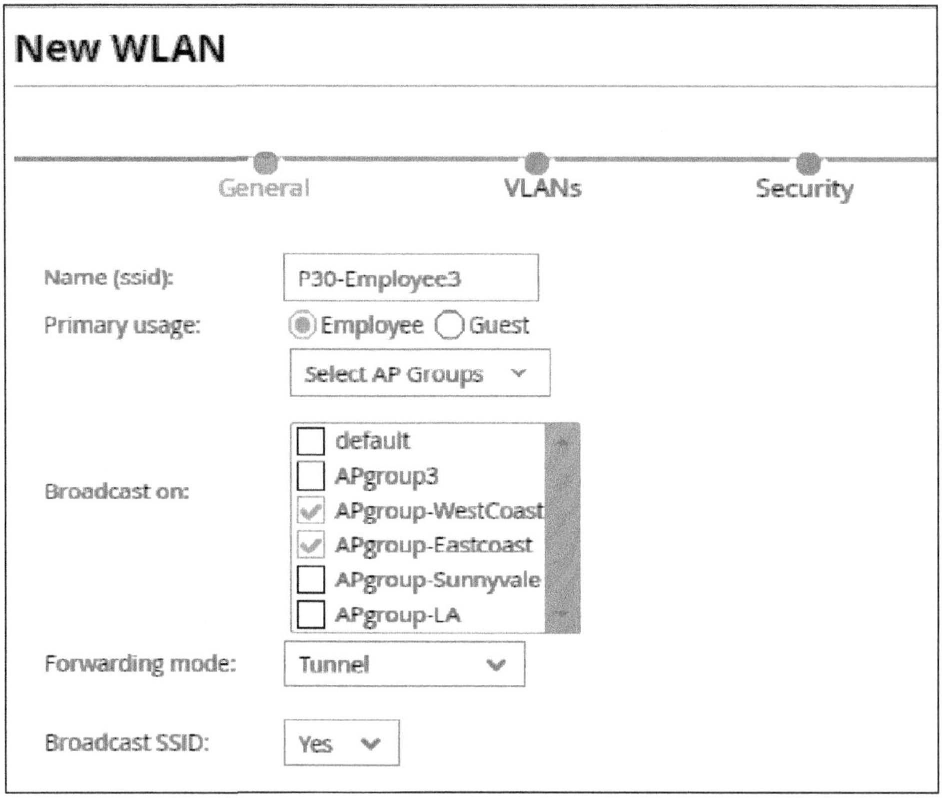

Figure 4-8 New WLAN general information

The WLAN wizard prompts you for many parameters, arranged in an easy, logical workflow. The first set of parameters is grouped within the General information area. You configure the SSID name and then choose other setting options or leave them at their defaults as appropriate. Figure 4-8 shows these options, which are described below:

- **Name (SSID)**—Assign a WLAN name. This SSID name can be broadcast to WLAN clients.
- **Primary Usage**—Select whether the WLAN will support Employees or Guests.
  - Employee—You are presented with enterprise options for authentication and encryption, as covered in this module. (Default setting)
  - Guest—You are presented with captive portal options to be covered in Chapter 9.

- **Select AP groups**—Use drop-down to select from a list of configured groups. (default = All APs)
- **Broadcast on**—Decide whether to broadcast the SSID on all APs associated with the MC or the MM, or only for select AP groups. If you choose the **Select AP Groups** option, you are prompted to select one or more AP groups.
- **Forwarding mode**—Select how the APs are to forward this traffic onto the LAN.
  - Tunnel mode—tunnels *data traffic* to the Mobility Controller using GRE. PAPI *control traffic* is not tunneled—it is transported via UDP port 8211. (Default)
  - Decrypt-Tunnel mode—the AP decrypts and decapsulates all client 802.11 frames. It then translates them to 802.3 frames and sends them via a GRE tunnel to the MC, where firewall policies are applied. For data traveling from a wired network to a client, the MC sends 802.3 frames via GRE tunnel to the AP. The AP translates these to 802.11 frames, performs encryption and hashing functions and sends them to the client. PAPI control frames arrive at the controller via a secure IPSec tunnel, such as CPSec.
  - Other forwarding modes (Split-tunnel and Bridge)—to be discussed in later chapters.
- **Broadcast SSID**—By default, APs broadcast the new WLAN SSID upon creation. To avoid this, use the drop-down list and select No**.** Our focus here is on Employee SSIDs, which are typically not broadcast. (Default = yes)

Features not supported (Forwarding Mode—Bridge Mode on Campus APs or Remote APs)

- Bridge Mode on Campus APs or Remote APs
- Firewall—SIP/SCCP/RTP/RTSP Voice Support
- Firewall—Alcatel NOE Support
- Voice over Mesh
- Video over Mesh
- Named VLAN
- Captive portal
- Rate Limiting for broadcast/multicast
- Power save—Wireless battery boost
- Power save—Drop wireless multicast traffic
- Power save—Proxy ARP (global)
- Power save—Proxy ARP (per-SSID)
- Automatic Voice Flow Classification

- SIP ALG
- SIP—SIP authentication tracking
- SIP—CAC enforcement enhancements
- SIP—Phone number awareness
- SIP—R-Value computation
- SIP—Delay measurement
- Management—Voice-specific views
- Management—Voice client statistics
- Management—Voice client troubleshooting
- Voice protocol monitoring/reporting
- SVP ALG
- H.323 ALG
- Vocera ALG
- SCCP ALG
- NOE ALG
- Layer 3 Mobility
- IGMP Proxy Mobility
- Mobile IP
- TKIP countermeasure management
- Bandwidth-based CAC
- Dynamic Multicast Optimization

New WLAN VLANs

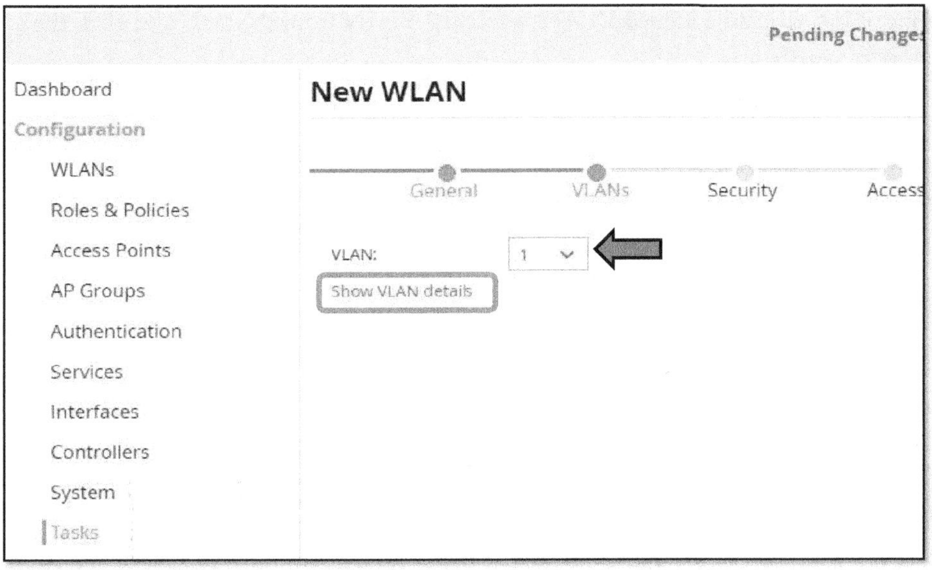

Figure 4-9 New WLAN VLANs

After you complete the General options, you are directed to VLAN options (Figure 4-9). This is where you decide the VLAN(s) into which users are placed. This, of course, affects their IP address assignment. This is one of the main ways of isolating guest and corporate traffic. Users associated with a guest SSID are placed in a separate, more tightly controlled VLAN. From the VLAN drop-down list, select your VLAN ID/VLAN Name. Then you can click on **Show VLAN details**.

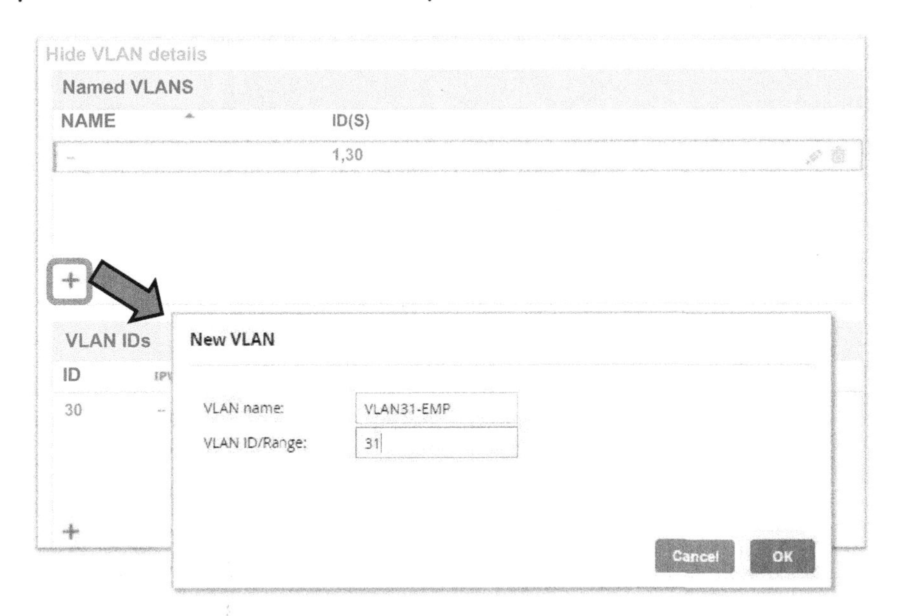

Figure 4-10 Show VLAN details

You click **Show VLAN Details** to display currently available VLANs, as configured on the MC or the MM. To add a new VLAN, click + below the **Named VLANs** table, then enter a VLAN name and ID or range of IDs. To edit a named VLAN, select the VLAN from the table and click the edit (pencil) icon (Figure 4-10).

To create a range of multiple VLAN IDs, specify the first and last number in the range, separated by a hyphen, for example, 55–58.

From the **Show VLAN Details** window, you can configure or edit the following settings:

- **VLAN IDs**—Identification number for the VLAN.
- **Admin State**—Enable or disable the VLAN interface.
- **IPv4 Address**—The IPv4 address and netmask assigned to the VLAN interface.
- **Enable NAT**—Enables source Network Address Translation (NAT) for all traffic routed from this VLAN. All ports on the Mobility Controller are assigned to VLAN 1 by default. Do not enable the NAT option for VLAN 1, as this will prevent IPsec connectivity between the Mobility Controller and its IPsec peers.
- **Local Link Address**—Configures the specified IPv6 address as the link local address for this interface.

Global Unicast Address—Specify the IPv6 address prefix to configure the global unicast address for this interface.

New WLAN security

Figure 4-11 New WLAN security

In the General settings, the Employee use-case was chosen (Figure 4-11). Therefore, you are only presented with enterprise authentication and encryption options in the Security area. These options are described below:

- **Enterprise** (802.1X/EAP-based security)—This option supports the following configuration parameters:
  - **Auth servers**—Click + to open the **Add Existing Server** window and select a preconfigured server from the list. To define a new LDAP or RADIUS server, click + on the Add Existing Server window. Figure 4-12 shows the window you would use to add a new LDAP or RADIUS server.

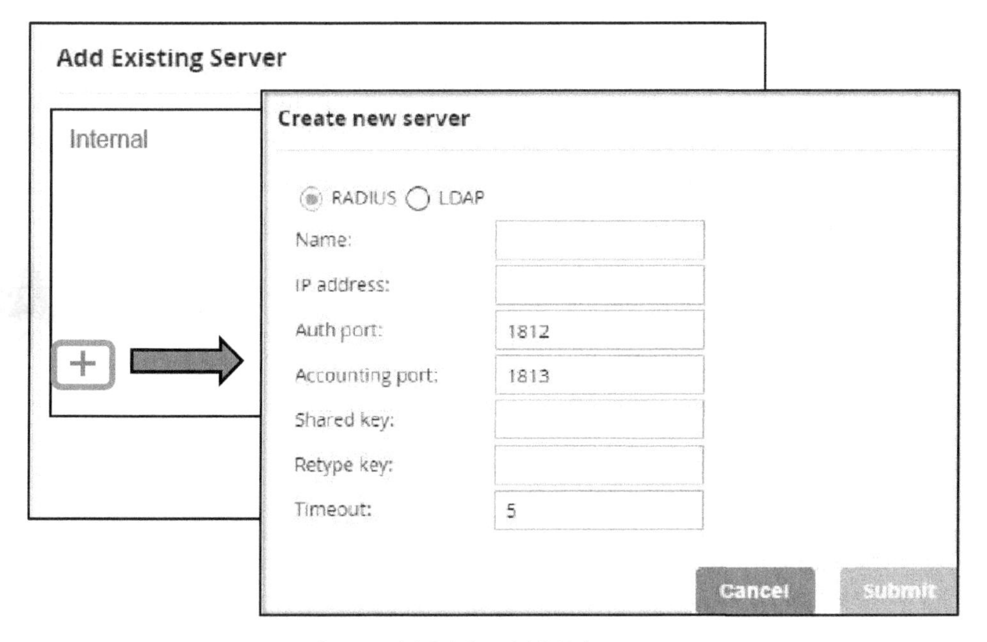

Figure 4-12 New WLAN security

  - **Reauth interval**—The interval, in seconds, between re-authentication attempts. Select this option to force clients to do 802.1X re-authentication at these intervals. The default value is 24 hours. If the user fails to re-authenticate with valid credentials, the state of the user is cleared. This setting can be overridden by defining "derivation rules," which are used to classify 802.1X-authenticated users.
  - **Machine authentication**—This option enforces machine authentication before user authentication. Either the machine-default-role or the user-default-role is assigned to the user, depending on which authentication is successful.

- **Blacklisting**—Blacklists the client if authentication fails a specified number of times.
- **Max authentication failures**—If Blacklisting is enabled, this parameter defines the number of times a user can try to login with wrong credentials before being blacklisted as a security threat.

- **Personal** (PSK-based security)—This option supports the following configuration parameters:
  - **Key management**—Use this setting to select the Layer-2 encryption type to be used on this WLAN SSID. Select either **WPA-2 personal** (AES-based encryption—recommended) or **WPA personal** (TKIP-based encryption).
  - **Passphrase**—Enter the WPA-2 or WPA Pre-Shared Key (PSK).
  - **Retype**—Retype the PSK.
  - **MAC authentication**—Select this option to enforce machine authentication before user authentication. If selected, either the machine-default-role or the user-default-role is assigned to the user, depending on which authentication is successful.
  - **Blacklisting**—Blacklists the client if authentication fails a specified number of times.
  - **Max authentication failures**—If Blacklisting is enabled, this parameter defines the number of times a user can try to login with wrong credentials before being blacklisted as a security threat.
  - **Open**—This option supports the following configuration parameters:
    - **MAC authentication**—Select this option to enforce machine authentication.
    - **Blacklisting**—Blacklists the client if authentication fails a specified number of times.
    - **Max authentication failures**—If Blacklisting is enabled, this parameter defines the number of times a user can try to login with wrong credentials before being blacklisted as a security threat.

Note

You can add other server types by manually configuring Profiles, either via the GUI or the CLI.

Note

You can choose other authentication and encryption types by manually configuring Profiles, either via the GUI or CLI.

Note

For best practices related to timer manipulations, re-authentication interval, and other advanced settings do a web search for Aruba Validated Reference Design (VRD) guides.

New WLAN access

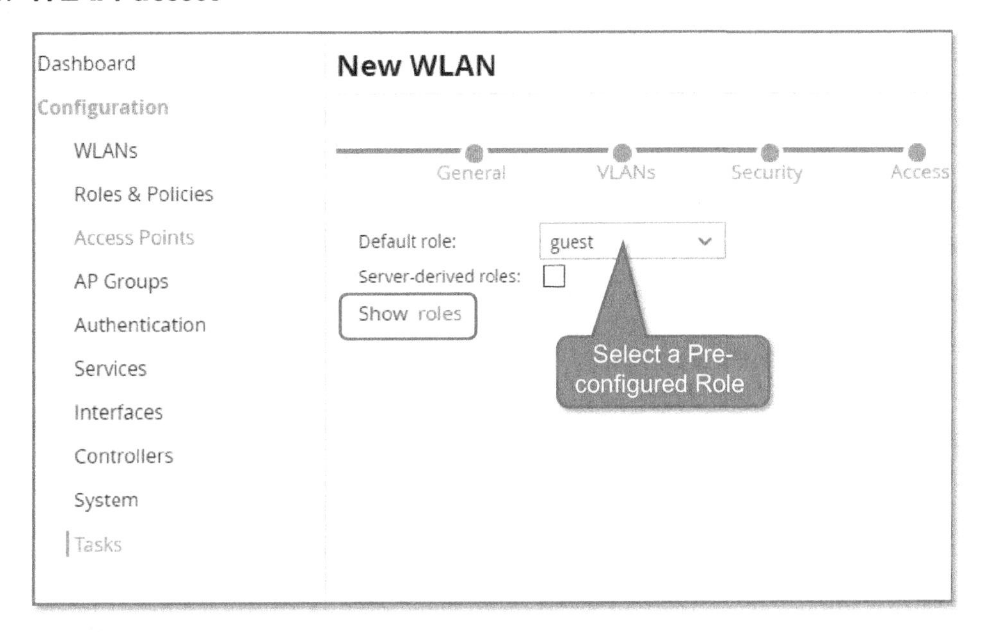

Figure 4-13 New WLAN access

Use the **Access** section to choose which Default role is assigned to successfully authenticated users. This role may have one or more firewall polices to control network accessibility. The default role is currently set to "guest," which is not appropriate for our Employee use case. Use the drop-down list to select a more appropriate existing user role, or create a new role. To create a new role, click **Show Roles** and then click the + symbol.

Note

You will learn more about creating user roles and assigning rules and policies in the modules on Roles and Firewall Polices.

Server derived roles (For employee WLANs using enterprise security)—This option provides server-derivation rules. These rules assign client user roles based on attributes returned by the server, during authentication. Derivation rules can also assign user roles based on client attributes such as SSID, even if that attribute is not returned by the server. Server-derivation rules are executed *after* client authentication.

Derivation method (For employee WLANs using enterprise security)—Select a derivation method.

Select **use value returned from ClearPass or other auth server** if users will authenticate to the WLAN via the HPE Aruba ClearPass Policy Manager or another type of authentication server.

Select **User rules** to define a custom role, based on RADIUS Server Vendor Specific Attributes (VSA). You will be prompted to define the following values:

- **Attribute**—RADIUS VSA type
- **Condition**—contains, equals, not-equals, start-with, or value-of
- **String**—Text string compared against VSA condition
- **Role**—Role assigned if the VSA condition and string match

Configuration pending

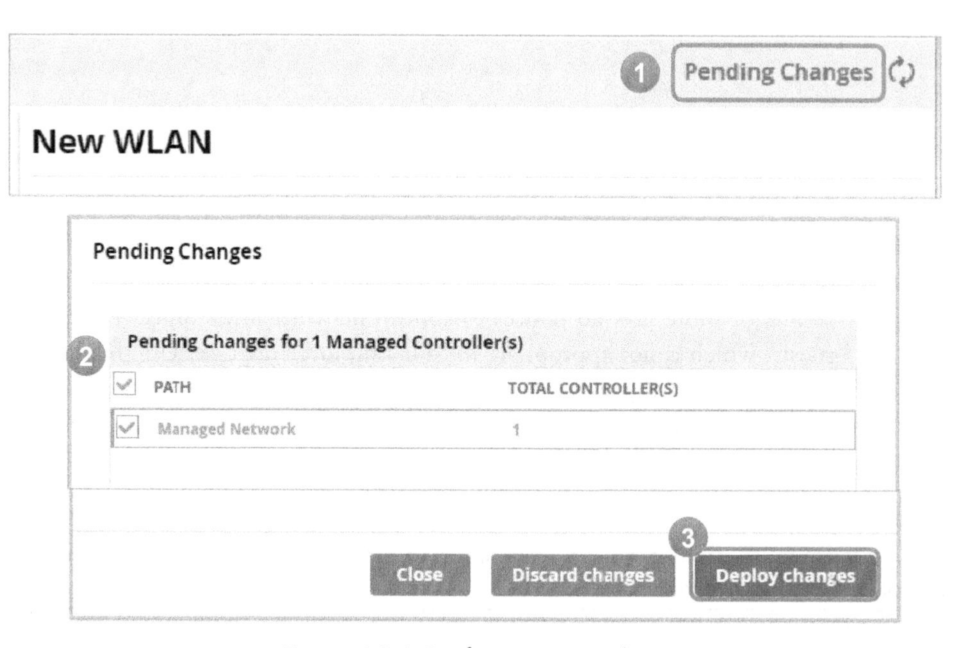

Figure 4-14 Configuration pending

It is **very important** that you commit your configuration changes! Do not forget to click on Pending Changes near the top-right, as shown. At the end of the configuration process, you must commit changes for the configuration to be validated and pushed down the hierarchy to group and subgroups.

Figure 4-14 shows this process, from clicking on Pending Changes, viewing pending changes, and deploying the changes.

The new WLAN with virtual AP profile settings

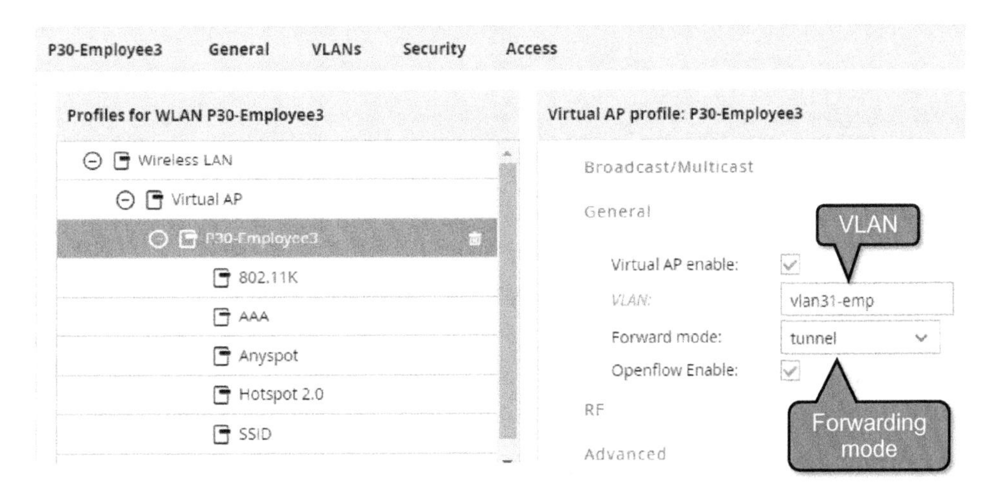

Figure 4-15 The new WLAN with virtual AP profile settings

The wizard created all the necessary profiles for the new WLAN (Figure 4-15):

- The virtual AP profile defines the VLAN and the forwarding modes.
- The AAA profile defines the type of authentication and also defines the default roles.
- The SSID profile has the SSID name and also the encryption used for this WLAN (SSID). You can also configure the regulatory channel setting.

The configured AP-group and profiles

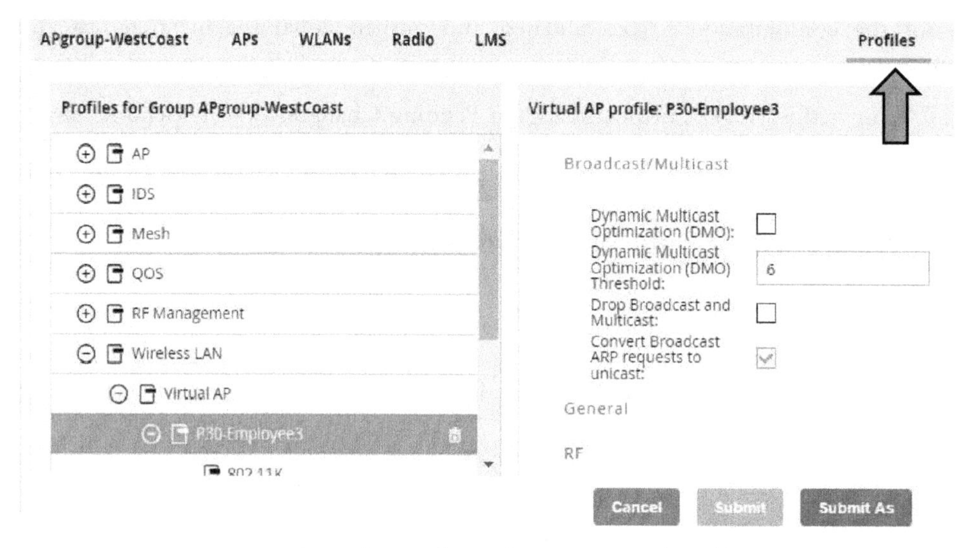

Figure 4-16 The configured AP-group and profiles

The wizard created all of the necessary profiles in one or more AP groups (Figure 4-16). This depends on which AP groups you decided to broadcast the SSID on to support the new WLAN.

Learning check

3. What are the authentication server types available when using the Tasks/WLANs configuration method?
   a. LDAP
   b. RADIUS
   c. Internal
   d. AD
   e. TACACS

4. What must you do at the end of the configuration process for your configuration to be applied, validated, and pushed down the hierarchy group/subgroups?
   a. Commit your pending changes
   b. Save configuration
   c. Testing your configuration
   d. Nothing, Configuration is saved automatically

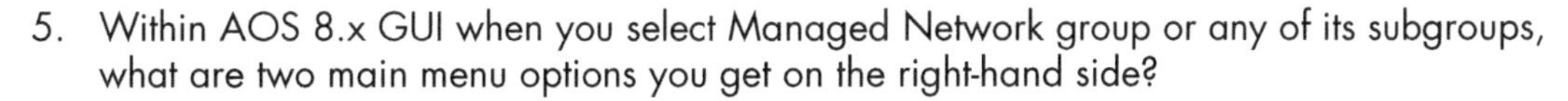

5. Within AOS 8.x GUI when you select Managed Network group or any of its subgroups, what are two main menu options you get on the right-hand side?

   a. Dashboard

   b. Configuration

   c. WLAN

   d. Profiles

6. Tunnel is the default forwarding mode of a WLAN.

   a. True

   b. False

7. VLAN ID is defined in RF management profile.

   a. True

   b. False

Answers to learning check

3. What are the authentication server types available when using the Tasks/WLANs configuration method?

   a. **LDAP**

   b. **RADIUS**

   c. **Internal**

   d. AD

   e. TACACS

4. What must you do at the end of the configuration process for your configuration to be applied, validated, and pushed down the hierarchy group/subgroups?

   a. **Commit your Pending Changes**

   b. Save configuration

   c. Testing your configuration

   d. Nothing, Configuration is saved automatically

5. Within AOS 8.x GUI when you select Managed Network group or any of its subgroups, what are two main menu options you get on the right-hand side?
   a. **Dashboard**
   b. **Configuration**
   c. WLAN
   d. Profiles

6. Tunnel is the default-forwarding mode of a WLAN.
   a. **True**
   b. False

7. VLAN ID is defined in RF management profile.
   a. True
   b. **False** (Remember, the VLAN ID is defined with WLAN Virtual AP profile settings)

5 AP Provisioning

LEARNING OBJECTIVES

✓ You have learned about organizing APs into AP groups, for ease of configuration and flexibility. Additionally, you learned about configuring WLAN profiles for Virtual APs (VAP). Before APs can be configured, they must boot up, get an IP address, find a controller, and associate with that controller.

✓ In this chapter, you will learn about the AP boot process, AP address assignment, and how APs find and associate with a controller. You will understand how AP-controller communications are secured with CPSec.

✓ You will explore options for provisioning APs, along with various packet flow scenarios. This will help you to understand how the traffic flows between the AP and controller.

AP and controller communication

Control plane and Data plane

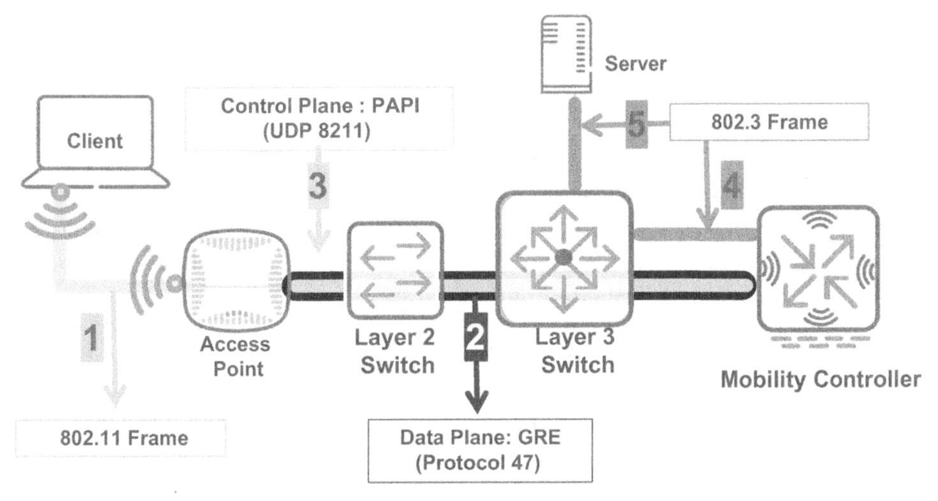

Figure 5-1 Control plane and Data plane

Figure 5-1 shows the control plane and data plane relationships and the frame flows. The AP and MC interact via the Control Plane using PAPI (UDP port 8211). If CPSec is enabled, this management traffic is encrypted using IPsec.

Note

CPSec is enabled by default in AOS 8.x but may be optionally disabled by the administrator.

To carry client data between source and destination, the AP and MC form a GRE tunnel (Protocol 47). The following steps describe this communication process, as illustrated in Figure 5-1:

1. All wireless traffic between AP and client are carried in 802.11 frames. For employee traffic, these frames are typically encrypted using AES (for WPAv2) or TKIP (WPAv1).
2. In tunnel mode, the APs do not decrypt these frames. APs simply send the traffic to the controller, via GRE tunnel.
3. AP-to-MC control traffic uses the PAPI protocol. This traffic is encrypted by IPsec when CPSec is enabled. CPSec will soon be described in detail.
4. The controller de-capsulates and decrypts the 802.11 data frames, translates them to 802.3 frames, and then switches or routes them, typically to some next-hop L3 switch.
5. The L3 switch forwards the frames on toward their ultimate destination.

Remember, users must authenticate before they gain network access. In this scenario, authentication messages pass between user and controller. If an external authentication server is in use, the controller passes appropriate authentication messages to the configured authentication server.

AP Boot process overview

To establish GRE data and PAPI control planes, the AP must acquire IP address parameters. The AP needs an IP address, netmask, default gateway, and DNS server. It also needs the controller's IP address. These parameters can be statically provisioned and saved to AP flash. This can be a laborious task, especially for large deployments with hundreds of APs.

Perhaps more commonly, APs dynamically acquire this information. APs boot and send a standard Dynamic Host Configuration Protocol (DHCP) request. You have ensured that DHCP services are setup to respond to requests from the AP's management VLAN. Thus, APs automatically get their IP address, subnet mask, default gateway, and the IP address of the DNS server.

Each AP's name defaults to its wired MAC address, and their AP-group defaults to "default."

Now the AP has met its boot requirements, it needs to discover the controller IP. Beyond static assignment, there are four methods an AP can use to learn the controller's IP address.

1. The AP can send a standard DHCP option 43 request. Assuming your corporate DHCP server has been configured to support this option, it responds to the AP with the appropriate controller IP address.
2. Aruba APs can send an Aruba Discovery Protocol (ADP) broadcast request to destination 255.255.255.255. A controller on the same subnet as the AP will respond with its own IP address. This method is great for a small, simple environment. However, in an enterprise-class deployment, you will rarely have APs on the same subnet as the controller.
3. Aruba APs can send an Aruba Discovery Protocol (ADP) multicast request to destination 239.0.82.11. A controller that receives this request will respond with its own IP address.
4. The AP learned the IP address of the DNS server during normal DHCP address assignment, assuming you have properly configured the DHCP server scope. The AP can send a reverse DNS request (resolve name to IP). If your DNS server is properly configured, it will respond with the IP address of the controller. By default, the AP uses the DNS name Aruba-Master to resolve the IP address of the Mobility Controller. You need to ensure that your DNS server is configured with an entry for the name Aruba-Master, which references the appropriate IP address. In the language of DNS, this is called an "A-record."

Once the AP has learned the controller's IP address, it can begin to communicate with it. Each AP sends its name and AP-group name to the controller. The controller responds by sending any pre-existing configuration information back to the AP that is specific to the AP's group name.

The next few pages will provide you with a deeper understanding of this topic, starting with the boot process

Booting—Factory default AP

The following steps describe AP boot and provisioning processes:

1. When power is applied, APs boot to a certain point and then send a standard DHCP request. You have ensured that corporate DHCP services are setup to respond to requests from the AP's management VLAN. Thus, APs automatically get their IP address, subnet mask, default gateway, and the IP address of the DNS server.
2. Controller IP Discovery is initiated using the static method or one of the four dynamic methods. There is no code version checkup with the AP master before the configuration is downloaded.
3. The AP sends its AP name and AP-group name to the Master controller.
4. At a minimum, the AP receives the default configuration and the IP address of controller to which the GRE tunnels should terminate (the so-called "LMS IP address") from that configuration. The AP image version must match that of the controller.

Note

The AP Master and LMS are allowed to run different versions of code. The LMS concept is introduced in a later chapter.

5. AP and controller image versions are compared. If mismatched, the AP downloads the correct OS from the LMS IP of the controller, using FTP. Now they match, and the AP reboots, starting back on Step 1. This time, the AP will make it to Step 6.

Note

APs may terminate on a controller other than the first discovered, with the use of the LMS IP parameter.

6. If CPSec is enabled, the AP's MAC address is verified against the controller's Whitelist. If rejected, the AP will not provide WLAN services. If approved, the AP moves on to Step 7.
7. The AP is then provisioned as appropriate, based upon the AP-group settings. If the AP is assigned to an AP-group, it will inherit its configuration. Otherwise, the AP joins the default group. If the AP is assigned to an undefined AP-group, it is flagged with "UG." This means "**U**nprovisioned" and "no such **G**roup."
8. When the configuration is received, the AP creates GRE tunnels to the controller and begins to provide WLAN services. At this point, the AP can provide WLAN Services to end users who connect.

Note

CPSec will be covered later in Chapter 5.

Booting after Controller provisioning

```
Starting watchdog process...
Getting an IP address...
ag7100_ring_alloc Allocated 4800 at 0x86f40000
ag7100_ring_alloc Allocated 3024 at 0x873f0000
AG7100: cfg1 0xf cfg2 0x7014
ATHRF1: Port 0, Neg Success
ATHRF1: unit 0 phy addr 0 ATHRF1: reg0 3100
AG7100: unit 0 phy is up...RGMii 1000Mbps full duplex
AG7100: pll reg 0x18050010: 0x110000  AG7100: cfg_1: 0x1ff0000
AG7100: cfg_2: 0x3ff
AG7100: cfg_3: 0x18001ff
AG7100: cfg_4: 0xffff
AG7100: cfg_5: 0xfffef
AG7100: done cfg2 0x7215 ifctl 0x0 miictrl 0x22
Writing 4
10.1.60.150 255.255.255.0 10.1.60.1
Running ADP...Done. Master is 10.1.60.101
ath_hal: 0.9.17.1 (AR5416, AR9380, REGOPS_FUNC, PRIVATE_DIAG, WRITE_EEPI
ath_rate_atheros: Copyright (c) 2001-2005 Atheros Communications, Inc, /
s Reserved
ath_rate_atheros: Aruba Networks Rate Control Algorithm
ath_dfs: Version 2.0.0
Copyright (c) 2005-2006 Atheros Communications, Inc. All Rights Reserve
ath_spectrum: Version 2.0.0
Copyright (c) 2005-2006 Atheros Communications, Inc. All Rights Reserve
ath_dev: Copyright (c) 2001-2007 Atheros Communications, Inc, All Rights
```

Figure 5-2 Preprovisioned AP boot process Part 1

```
ath_pci: 0.9.4.5 (Atheros/multi-bss)
ath_attach: scn 866d0280 sc 866e0000 ah 86700000
wifi0: Base BSSID 00:24:6c:2a:fe:f8, 8 available BSSID(s)
bond0 address=00:24:6c:ca:af:ef
br0 address=00:24:6c:ca:af:ef
wifi0: AP type AP-105, radio 0, max_bssids 8
wifi0: Atheros 9280: mem=0x10010000, irq=49 hw_base=0xb0010000
ath_attach: scn 86390280 sc 863a0000 ah 863c0000
wifi1: Base BSSID 00:24:6c:2a:fe:f0, 8 available BSSID(s)
bond0 address=00:24:6c:ca:af:ef
br0 address=00:24:6c:ca:af:ef
wifi1: AP type AP-105, radio 1, max_bssids 8
wifi1: Atheros 9280: mem=0x10000000, irq=48 hw_base=0xb0000000
ath_ahb: 0.9.4.5 (Atheros/multi-bss)

Starting FIPS KAT ... Completed FIPS KAT

Using saved lms 10.1.60.100 as the master directly
AP rebooted Sun Nov 6 08:28:47 PST 2016; reboot command execu
n given (called from ).
keep watchdog process alive for talisker (nanny will restart it)

     <<<<<       Welcome to the Access Point     >>>>>
```

Figure 5-3 Preprovisioned AP boot process Part 2

At some point, some of your provisioned, functional APs will reboot. The Figures 5-2 and 5-3 show the boot process for a preprovisioned AP, which is described below.

1. The AP broadcasts a DHCP request for an IP address, which is assigned by your DHCP server.
2. The AP uses MD address saved in local flash memory, even though the master discovery process is still running.
3. The AP sends a provisioned identity message to the Mobility Controller IP.
4. Check whether CPSec is enabled. This determines whether AP-to-controller communications are protected by an IPsec tunnel. Either way, the AP establishes a control plane with the MD, receives configuration, and establishes GRE tunnels to the controller.

Communication between AP and controller

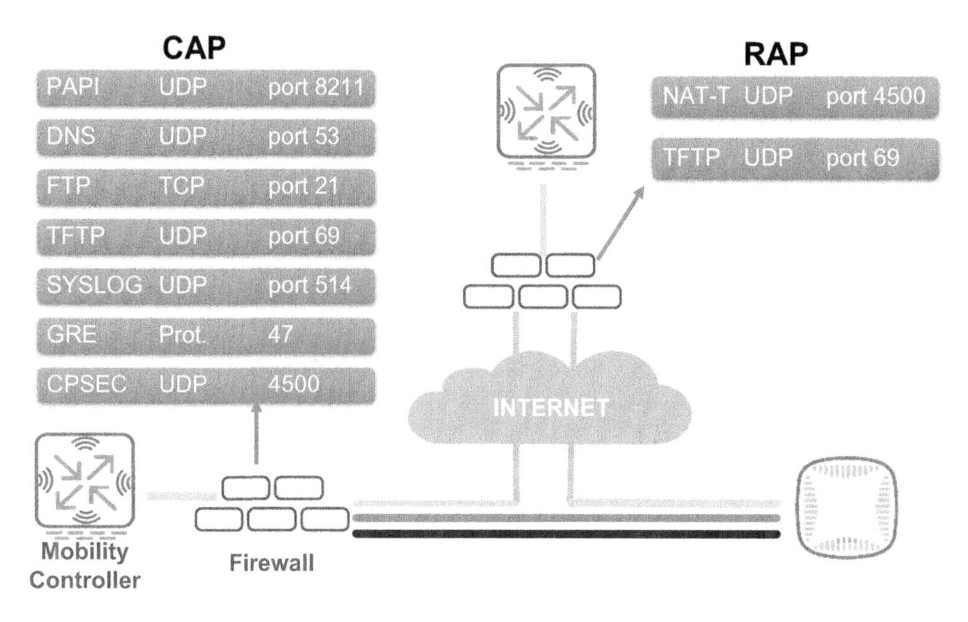

Figure 5-4 Communication Between AP and controller

Figure 5-4 shows a scenario where firewalls exist in the wired network between WLAN components. There is one firewall between APs and the MC, and another between an MC and the Internet. In this scenario, you must ensure that firewalls permit all required communication between a Campus AP and Mobility Controller as well as the RAP and MC. These services are shown in Figure 5-4 and described below:

- For CAPs:
  - PAPI—UDP port 8211. Air Monitor (AM) mode APs need a permanent PAPI connection to MM

- DNS—UDP port 53. APs that do LMS discovery via DNS—the first connection attempt is to MM
- FTP—TCP port 21
- Trivial File Transfer Protocol (TFTP)—UDP port 69. CAPs that have a missing or mismatched image use TFTP for retrieval
- SYSLOG—UDP port 514
- GRE (IP)—protocol 47
- CPSec—UDP port 4500

- For RAPs:
  - NAT-T—UDP 4500
  - TFTP—UDP 69. Used for image download

Learning check

1. Which ports or protocol must be enabled between CAP and MC, assuming they have the same image version?
   a. UDP 4500
   b. GRE Protocol 47
   c. TFTP UDP 69
   d. FTP TCP 21
   e. Syslog UDP 514

Answers to learning check

1. Which ports or protocol must be enabled between CAP and MC, assuming they have the same image version?
   a. **UDP 4500**
   b. **GRE Protocol 47**
   c. TFTP UDP 69
   d. FTP TCP 21
   e. Syslog UDP 514

Controller discovery mechanism

Enable Controller discovery

You can statically configure an AP with its IP address, mask, and DG, as well as the MC IP address, and other parameters. However, this could prove to be exhausting for larger installations.

It is easier to use dynamic discovery. The AP powers up and sends a DHCP request. The AP checks for Option 43 in the DHCP reply. The AP perceives the IP address provided in Option 43 as the MC address. However, other vendors may use Option 43, say for Voice over IP (VoIP) services. The AP might contact the VoIP service instead of the MC and continuously reboot. To avoid this problem, configure your DHCP server with Aruba's Option 60 vendor code—"ArubaAP."

If there is no Option 43 in the DHCP reply, the AP tries the ADP discovery method. It transmits L2 broadcasts to destination FF:FF:FF:FF:FF:FF, and L3 broadcasts to 255.255.255.255. It also multicasts to destination 239.0.82.11. A controller responds to both L2 and L3 requests if it shares the AP's subnet. If AP and controller are in different subnets, L2/L3 broadcasts do not reach the controller.

This is where the multicast request has value. However, your wired network infrastructure must be configured to support multicast. Otherwise, ADP multicast packets will not reach the controller.

The DNS lookup method is recommended, as it involves minimal changes to the network and provides flexibility in AP placement. During the DHCP process, APs acquire the DNS server's IP address and their DNS domain name. The AP adds the name "aruba-master" to the learned domain name to formulate a DNS lookup request, "Please resolve the name **aruba-master.domain.com** to an IP address." It sends this request to the DNS server's IP address. If you have properly configured the DNS server, it replies with the controller's IP address, typically the master controller.

The AP uses the DNS lookup method if the aruba-master setting is configured. However, it only uses this method if the Option 43 and ADP discovery methods fail.

Controller discovery caveat

Unlike AOS 6.x or prior, a Mobility Master (MM) cannot terminate APs. AP HELLO messages received by the Mobility Master are dropped. The following is an example of a log message that would be displayed from the CLI of the MM.

```
(MM-1) ^*[mynode] #show log system 10

Oct 23 05:34:51 :305061:  <5451> <WARN> |stm|  sapm_proc_hello_req
Ignoring the HELLO message from AP 6c:f3:7f:c2:59:02. MM doesn't
support AP redirect or termination
```

You can see that the MM ignores the message and reminds you that it lacks support for AP termination.

The APs learn their Mobility Controller IP address from the Master. The AP terminates to this MC to get its configuration.

Learning check

2. Arrange Column 2 in the correct order.

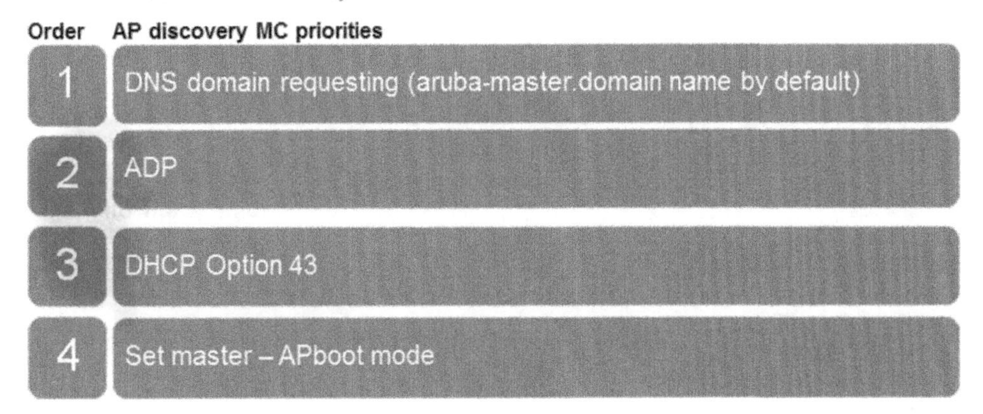

Answers to learning check

1. Set Master – AP boot mode.
2. DHCP Option 43.
3. ADP.
4. DNS domain requesting

CPSec

Introduction to CPSec

With CPSec, an IPsec tunnel is established to protect PAPI control plane communications between AP and controller. If CPSec is disabled, PAPI traffic is sent in the clear, unencrypted.

The CPSec feature relies on the standards-based IPsec protocol suite, for certificate-based, secure communications. CPSec can use factory-installed certificates, or self-signed certificates from the master controller.

Once CPSec is enabled, all controllers and APs should be registered and certified. CPSec is on by default, and it becomes effective for existing deployments after the AP's code is upgraded to match that of the controller. No special license is required for this feature.

How CPSec works

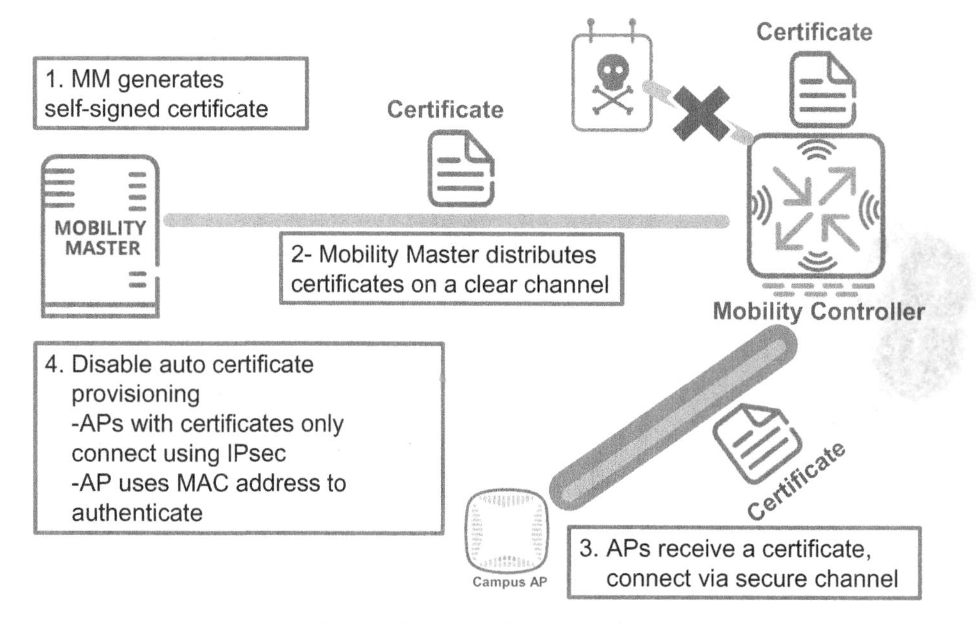

Figure 5-5 How CPSec works

Figure 5-5 highlights the major steps for CPSec operation. First, the Mobility Master generates its self-signed certificate. Next, the MM distributes these certificates, in the clear, to campus APs and managed devices. Soon all APs will receive a certificate and connect to the MC over a secure channel. Once this happens, you can access the **Control Plane Security** window and disable auto certificate provisioning. This prevents the controller from issuing certificates to any rogue APs that may appear on your network at a later time.

Configure CPSec Auto-Cert-Provisioning

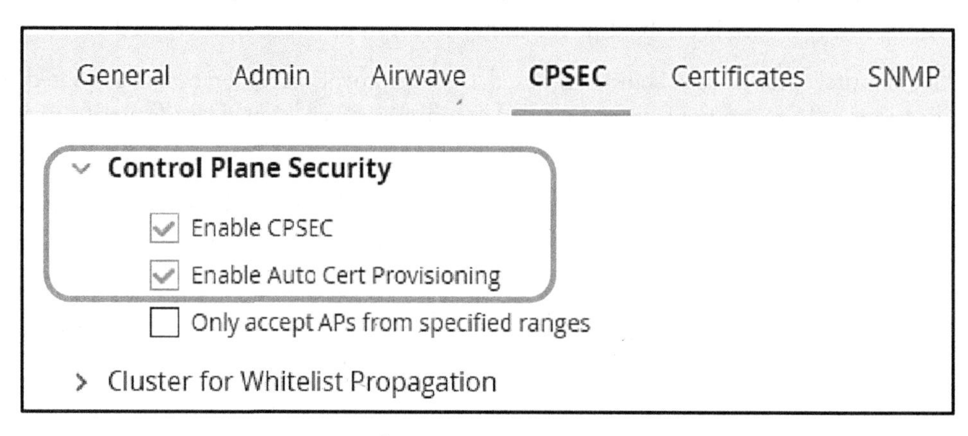

Figure 5-6 Configure CPSec Auto-Cert-Provisioning

Now that we understand how CPSec works, let us talk about configuration. To configure CPSec, navigate to **Managed Network > Configuration > System > CPSEC**. From there, expand the Control Plane Security section and check the Enable CPSEC check box. This is shown in Figure 5-6.

To control which APs are provisioned, you can create a whitelist of valid APs. One method to create this whitelist is using Auto-Cert-Provisioning. If you are confident that all currently attached APs are valid, you can use the initial setup wizard to configure automatic certificate provisioning on the controller. Thus, the controller sends the certificates to all CAPs within specific IP address ranges. These CAPs can now connect to the MC using a CPSec-secured channel.

Once this occurs, navigate to the Control Plane Security window, as shown in Figure 5-6 and disable Auto Cert Provisioning. This prevents the controller from issuing a certificate to rogue APs.
You can also configure this from the CLI, as shown in the example below:

```
(host)[mynode](config) #control-plane-security

(host)[mynode](Control Plane Security Profile)#auto-cert-allow-all

(host)[mynode](Control  Plane  Security  Profile)#auto-cert-allowed-addr
<start><end>

(host)[mynode](Control Plane Security Profile)#auto-cert-prov

(host)[mynode](Control Plane Security Profile)#cpsec-enable
```

CPSec manual whitelist

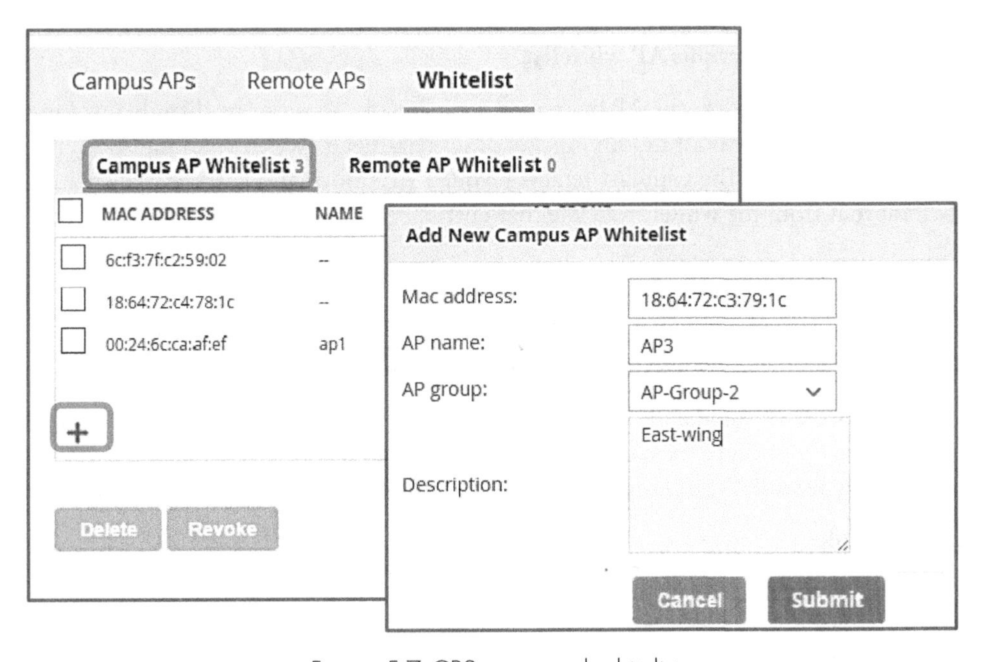

Figure 5-7 CPSec manual whitelist

By default automatic certificate provisioning is not enabled. Therefore, you must manually enter each campus AP's information into the AP whitelist, as described below.

WebUI

1. Access the Managed Network.
2. Navigate to **Configuration > Access Points > Whitelist** tab.
3. Click **Campus AP Whitelist** tab. Click +. This is shown in Figure 5-7.
4. Define the following parameters for each AP you want to add to the AP whitelist

Below, you see the same configuration performed from the CLI.

```
(host) #whitelist-db cpsec modify mac-address <name>

ap-group <ap_group>

ap-name <ap_name>

cert-type {switch-cert|factory-cert}

description <description>

mode {disable|enable}

revoke-text <revoke-text>

state {approved-ready-for-cert|certified-factory-cert}
```

Revoking APs from the Campus AP whitelist

You can revoke an invalid or a rogue AP by modifying its revoke status or by directly revoking it from the campus AP whitelist, without modifying any other parameter. When revoking APs, enter a brief description as your reason. The whitelist retains revoked AP information. To revoke an AP and permanently remove it from the whitelist, delete that entry.

Caveats of disabling CPSec

You can disable CPSec, as previously shown. Just be aware of the resulting caveats.

If you disable CPSec on a standalone controller or a managed device, all APs connected to that controller reboot then reconnect to the controller over a clear channel. The control plane is now unsecure and rogue APs can potentially join the controller.

Command example:

```
(MM-1) *[md] (config) #control-plane-security

(MM-1) *[md] (Control Plane Security Profile) #no cpsec-enable
```

Learning check

3. CPSec can encrypt both data and control plane.
    a. True
    b. False

Answers to learning check

3. CPSec can encrypt both data and control plane.
    a. True
    b. **False**

AP provisioning

AP provisioning overview

APs are provisioned into AP groups. The AP group has all the configuration the APs will need. This includes the SSIDs, RF channels, authentication requirements, firewall policies, radio usage (a/b/g/n/ac), and VLANs. Not all AP group configuration goes directly to the AP. The controller also uses some of the AP group information to apply firewall policies and to route and/or switch user traffic.

APs get a local IP address and then communicate with the controller to get its configuration. New APs are assigned to the default AP group.

APs can be staged prior to deployment and provisioned locally to prevent issues. This can help eliminate most IP problems and confirms the AP configuration. Once deployed, APs boot from their staged configurations.

AP provisioning manually via GUI

Campus APs | Remote APs | Whitelist

Campus APs 2

	AP NAME	AP GROUP	IPV4 ADDRESS	IPV6 ADDRESS	SWITCH IP	MAC ADDRESS	SERIAL #
☑	00:24:6c:ca:af:ef	default	10.1.60.150	--	10.1.60.100	00:24:6c:ca:af:ef	AL0217226
☐	18:64:72:c4:78:1c	default	10.1.60.152	--	10.1.60.100	18:64:72:c4:78:1c	CJ0166943

Provision

Figure 5-8 AP provisioning manually via GUI

To manually provision an AP, navigate to **Managed Networks > Configuration > Access Points**. This will provide a list of APs, as shown in Figure 5-8. It appears that these APs have not yet been provisioned—they have the default MAC-based name and are a member of the default AP group. Select an AP with the checkbox, as shown, and click Provision.

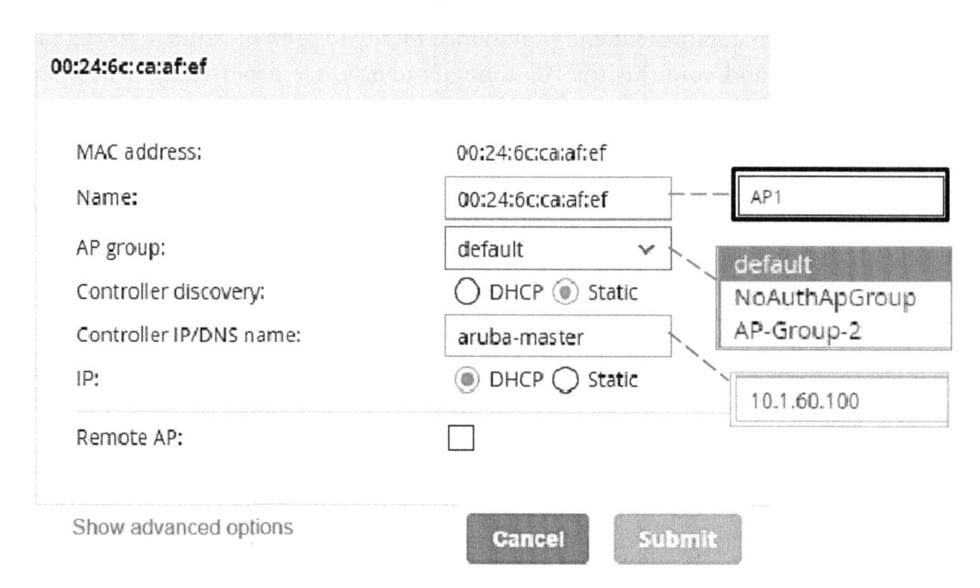

Figure 5-9 AP provisioning manually via GUI (Cont.)

You see a provisioning window, where you can enter the AP name, select an AP group, and specify the controller's IP address or DNS name (Figure 5-9). You should probably let the AP acquire its address via DHCP, but you can click the Static radio button and assign an IP address if desired.

AP provisioning manually via APboot

The following shows a scenario of directly provisioning the AP, from its own CLI. To do this, you connect your laptop directly to the AP's console port. Power up the AP. As it boots, press enter to stop the AP automatic boot sequence. You will get a prompt as shown below:

```
APBoot 1.4.0.5 (build 38142)

Built: 2013-04-21 at 22:03:44

Model: AP-11x

CPU:   QCA9550 revision: 1.0

..

Hit <Enter> to stop autoboot:  0

apboot>
```

Now it is simply a matter of knowing the provisioning syntax and entering this into the AP. Note that these commands are case-sensitive and must be entered in lower case. The most important commands are shown below and are fairly self-explanatory:

```
apboot> setenv ipaddr 10.1.60.150        configures AP IP address

apboot> setenv netmask 255.255.255.0     configures AP subnet mask

apboot> setenv gatewayip 10.1.60.1       configures AP default gateway

apboot> setenv group Ap-Group-2          configures AP group

apboot> setenv name AP2                  configures AP name

apboot> save                             saves the configuration

apboot> boot                             boots the AP
```

The additional commands are the following:

- **printenv**—List the environment variables and current settings
- **purgeenv**—Restore AP boot configuration to factory default, including environment variables
- **saveenv**— Save environment variables to persistent storage

- **reset**—Perform a RESET of the AP's CPU
- **ping**—Check network connectivity
- **factory_reset**—Reset the AP to factory default

AP provisioning via wizard

To access the AP provisioning wizard, navigate to **Managed Network > Configuration > Task > Provision new Campus APs**. This will lead you through an AP provisioning workflow.

AP operation

GRE tunnels and CLI commands

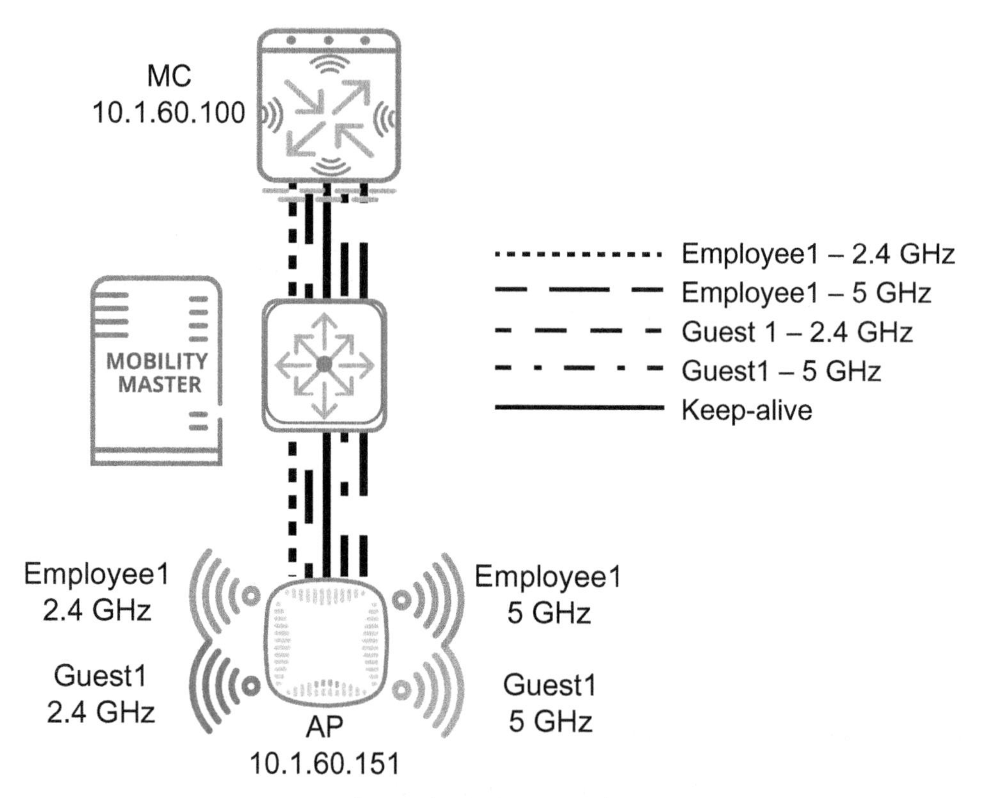

Figure 5-10 GRE Tunnels

Each AP's BSSID consumes one GRE tunnel. Each AP establishes one extra GRE tunnel with the Mobility Controller for keep alives. Figure 5-10 shows one physical AP with two SSIDs defined. This SSID is supported on both the 2.4 GHz and 5 GHz radios. That is a total of four SSIDs and so you see four lines representing GRE tunnels. The fifth one (a solid line) represents the tunnel for keep alives.

After you create the first SSID, named T-Employee1, the **show ap bss-table** command will show the BSSID (MAC address) assigned to the SSID named T-Employee1, for both the 2.4 GHz and 5 GHz radios, as seen in the following example:

```
(MD1) # show ap bss-table

Aruba AP BSS Table

bss                  ess          port   ip            phy

00:24:6c:2a:fe:f8    T-employee1  N/A    10.1.60.150   a-HT

00:24:6c:2a:fe:f0    T-employee1  N/A    10.1.60.150   g-HT
```

The top entry of the "phy" column is "a-HT." This tells you that the top entry is for a 5 GHz radio (the "a"), that supports 802.11n (HT, for High Throughput). As you can probably guess, the "g-HT" in the second entry means that this is for a 2.4 GHz radio that supports 802.11n. You can also tell that these BSSIDs exist on the same physical AP. The first five bytes of the BSSID are identical, as is the IP address.

Note

This ties in to our previous discussion from Chapter 1's WLAN Mobility section. Recall that each AP radio creates a unique BSSID per (E)SSID, derived from its base radio MAC address. This is why the first five bytes of the BSSID are identical.

The **show datapath tunnel table | include 10.1.60.150** command shows tunnel information on the MC. The addition of "| include 10.1.60.150" filters the output to show only tunnels for the AP with the specified address, as seen in the following example:

```
(MC1) # show datapath tunnel table | include 10.1.60.150

#    Source        Destination    Prt  Type  MTU

16   10.1.60.100   10.1.60.150    47   8200  1500

9    10.1.60.100   10.1.60.150    47   8300  1500

10   10.1.60.100   10.1.60.150    47   9000  1500
```

The first entry is for the T-Employee1 SSID 2.4 GHz radio, the second entry is for the T-Employee1 SSID 5 GHz radio, and the last entry is for keep alives.

Now if you define the second SSID, named T-Guest1, the **show ap bss-table** command will now show four BSSIDs: two for T-Employee1, and two for T-Guest1:

```
(MD1) # show ap bss-table

Aruba AP BSS Table

bss                  ess          port  ip            phy
00:24:6c:2a:fe:f1    T-Guest1     N/A   10.1.60.150   g-HT
00:24:6c:2a:fe:f8    T-employee1  N/A   10.1.60.150   a-HT
00:24:6c:2a:fe:f9    T-Guest1     N/A   10.1.60.150   a-HT
00:24:6c:2a:fe:f0    T-employee1  N/A   10.1.60.150   g-HT
```

And the **show datapath tunnel table | include 10.1.60.150** command will now show five tunnels on the MC:

```
(MC1) # show datapath tunnel table | include 10.1.60.150

#    Source         Destination    Prt   Type   MTU
16   10.1.60.100    10.1.60.150    47    8200   1500
13   10.1.60.100    10.1.60.150    47    8210   1500
9    10.1.60.100    10.1.60.150    47    8300   1500
10   10.1.60.100    10.1.60.150    47    9000   1500
11   10.1.60.100    10.1.60.150    47    8310   1500
```

In this example, the first line and the last line are for the T-Guest1 SSID 2.4 GHz radio and the T-Guest1 SSID 5 GHz radio, respectively.

Identify GRE tunnels with the CLI

All the various tunnels between a single AP and its Controller cannot be distinguished by IP addresses since they have the same source and destination IP. The way to distinguish between them is by their tunnel "type" listed in the **show datapath tunnel table** command.

In the following example, Type 8200 is for the first BSSID of the first SSID and 8210 is for that SSID's second BSSID. Type 8300 is for the first BSSID of the second SSID and 8310 is for the second BSSID.

```
(MC1) # show datapath tunnel table | include 10.1.60.152

#    Source         Destination    Prt   Type   MTU
10   10.1.60.100    10.1.60.152    47    8200   1500
18   10.1.60.100    10.1.60.152    47    8210   1500
9    10.1.60.100    10.1.60.152    47    9000   1500
19   10.1.60.100    10.1.60.152    47    8300   1500
20   10.1.60.100    10.1.60.152    47    8310   1500
```

GRE type 9000 is used as keep alive communications between the AP and the Controller. This always exists, regardless of any SSID broadcasting.

VLAN Tag in tunnel mode for an L2 deployment

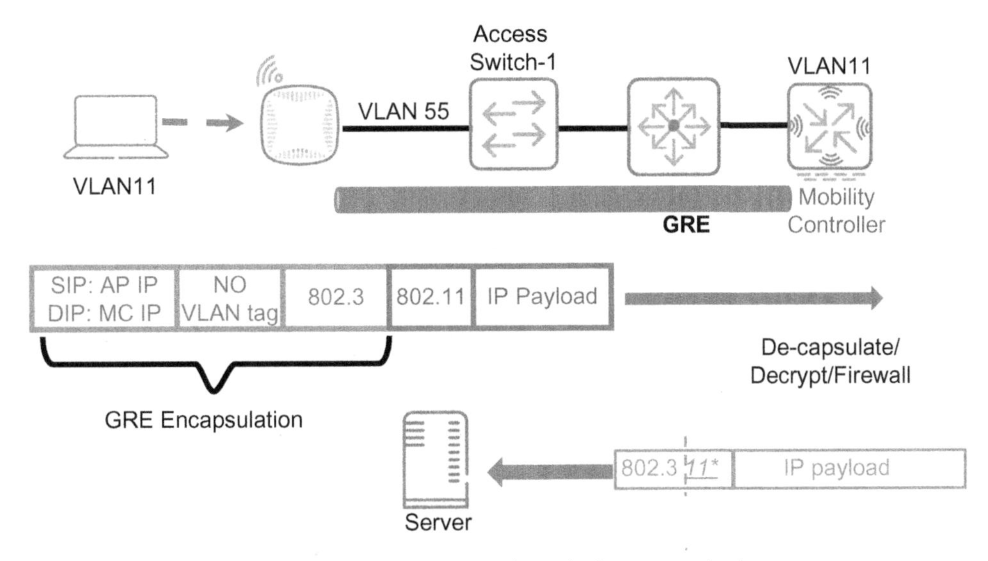

Figure 5-11 VLAN tag in tunnel mode for an L2 deployment

Figure 5-11 demonstrates how data flows through an L2 tunnel mode deployment, with a focus on how VLAN tags are used.

The client is connected to an SSID that maps to VLAN 11, so, of course, this client has a VLAN 11 IP address. It transmits an 802.11 frame, with its IP payload, to the AP.

Note that the GRE encapsulation shown in Figure 5-11 indicates "no VLAN tag" as opposed to VLAN tag 10 or 40. This can happen if these VLANs are configured as the "native VLAN" on the wired switches. Switches send native VLAN as a "native" Ethernet frame—in other words, without an 802.1Q VLAN tag.

The MC receives the L3 GRE packet in an L2 frame that has no VLAN tag (if 40 is the native VLAN) or with an L2 frame with 802.1Q tag = 40. It removes the tunnel headers to reveal the client's original 802.11 frame, converts this to an 802.3 frame, and sends it to its DG, as previously described.

Basic troubleshooting

Table 5-1 describes some basic troubleshooting commands, along with descriptions of each.

Table 5-1 Basic troubleshooting commands

Command	Descriptions
show ap active	See all APs with their UP/DOWN status
show ap database	See all APs with UP/DOWN status. If currently DOWN, the AP did communicate with the MC at one point
show datapath session table	
Show Datapath Tunnel Table	View GRE tunnels between controller and APs
show log errorlog 10	See the last 10 lines of error log output
show ap essid	See information for a specific SSID
show ap bss-table	Shows the BSS table for all APs. You can filter for a specific AP by including the AP name in the command.
show trunk	Validate the VLAN tagging and 802.1q trunking configuration of a port
show port status	Check the Layer 1 and 2 status of a port

Example—Troubleshooting AP provisioning

```
(MD1) #show ap database

AP Database
-----------
Name    Group        AP Type  IP Address    Status       Flags  Switch IP     Stand
----    -----        -------  ----------    ------       -----  ---------     -----
ap-115  controller9  115      10.1.90.150   Down         UEG    10.1.60.100   0.0.0
ap1     default      105      10.1.60.150   Up 45m:37s   2      10.1.60.100   0.0.0

Flags: U = Unprovisioned; N = Duplicate name; G = No such group; L = Unlicensed
       I = Inactive; D = Dirty or no config; E = Regulatory Domain Mismatch
       X = Maintenance Mode; P = PPPoE AP; B = Built-in AP; s = LACP striping
       R = Remote AP; R- = Remote AP requires Auth; C = Cellular RAP;
       c = CERT-based RAP; 1 = 802.1x authenticated AP; 2 = Using IKE version 2
       u = Custom-Cert RAP; S = Standby-mode AP; J = USB cert at AP
       i = Indoor; o = Outdoor
       M = Mesh node; Y = Mesh Recovery
       z = Datazone AP

Total APs:2
```

Figure 5-12 Example—Troubleshooting AP Provisioning

Figure 5-12 is focused on the "Flags" section of the show ap database output. In this case, the first APs status is down and flags are UEG. There is a legend at the bottom of the output to help you decode what these flags mean. This scenario is reviewed below:

- U this AP is **U**nprovisioned by the controller
- G the ap-**G**roup "controller9" does not exist on this controller
- E there is a mismatch between AP and controller regulatory domain settings. This can occur if an Instant AP with a specific country code setting is converted to AP mode.

The second AP is up normally. The "2" flag means the AP and controller have established IPsec tunnel using the IKEv**2** protocol.

Learning check

4. If one AP broadcasts three SSIDs in dual bands, how many GRE tunnels are there between the AP and MC?
   a. 5
   b. 6
   c. 3
   d. 7
   e. 4

5. Mobility Master (MM) can be used as an AP master and can terminate an AP tunnels.
   a. True
   b. False

Answers to learning check

4. If one AP broadcasts three SSIDs in dual bands, how many GRE tunnels are there between the AP and MC?
   a. 5
   b. 6
   c. 3
   d. **7**
   e. 4

5. Mobility Master (MM) can be used as an AP master and can terminate an AP tunnels.
   a. True
   b. **False**

6 WLAN Security

LEARNING OBJECTIVES

- ✓ The objective of this chapter is to prepare you to effectively deploy and maintain secure WLAN solutions. You will ultimately learn how to configure various security components. However, the focus here is on the concepts, processes, and protocols involved in WLAN security.
- ✓ We review 802.11 negotiation and authentication basics. Then you will engage more deeply with user authentication, machine authentication, and authentication servers. You will be able to compare and contrast data encryption options, and understand Wireless Intrusion Prevention Systems (WIPS).

802.11 Negotiation

Connecting to a secure WLAN

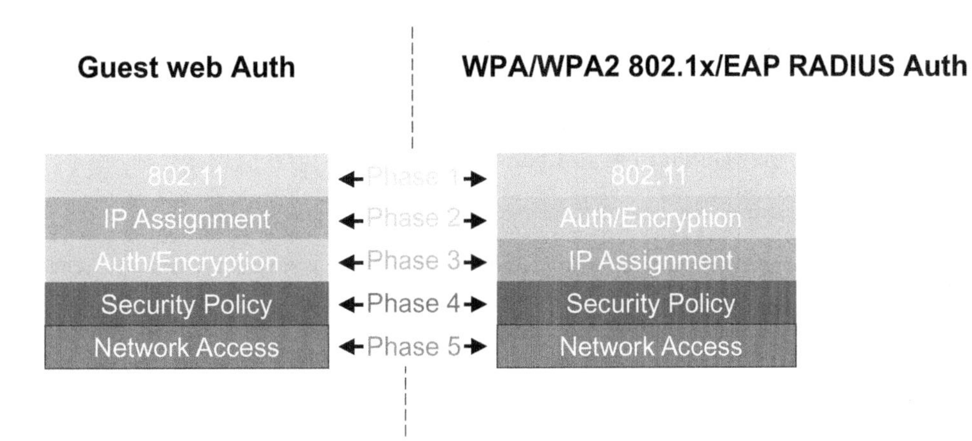

Figure 6-1 Connecting to a secure WLAN

The user connection to the WLAN happens in five phases. These vary depending on whether the client connects to a guest or an employee WLAN, as shown in Figure 6-1. What does not vary much is the initial 802.11 negotiation. The 802.11 negotiation process is defined in the 802.11 standard and, therefore, all devices must adhere to this process.

Phase 1 802.11 negotiation starts when a client discovers the WLAN. This discovery occurs through the use of various management frames—Beacon, Probe Request, and AP Probe Response. After the discovery phase is complete, the client sends an Authentication frame, and the AP sends an authentication frame in response. Next, the client sends an Association Request, and the AP sends an Association Response.

The so-called 802.11 "Authentication" is really just a handshake, not an actual authentication. The "real" authentication happens in a later phase. The reason "real" authentication takes place after the 802.11 Negotiation process because 802.11i is an amendment to the original 802.11 standard. So the original 802.11 negotiation process must occur first, which includes a step called authentication. Secure credential verification was not included in the original standard when it was published, so it has been added onto the process.

This original 802.11 negotiation phase is analogous to "plugging in the cable" for Ethernet. When you connect an Ethernet cable, the NIC senses "link up" and so can acquire an IP address, and perhaps perform 802.1X/EAP authentication, if that is configured.

Likewise, after the initial 802.11 request/responses for probe, authentication, and association, the WNIC is "connected," and can simply acquire an IP address (open guest WLAN), or perform the "real" L2 authentication (secure employee WLAN) and then acquire an IP address).

Guest WLAN

Phase 2 for the guest WLAN is to acquire an IP address, so the user can launch a web browser. This begins **Phase 3**. The web login captive portal page is displayed to facilitate guest access. This is called "Layer 3 authentication" because it occurs within a browser, which requires a Layer 3 connection.

In **Phase 4**, the security policy controls VLAN mapping and role assignment. This assignment can control user bandwidth, firewall filtering, and network access.

In Phase 5, now the guest user has network access.

This process may include some type of L3 authentication, but there is no encryption. Wireless frames are transmitted in the clear, but this is acceptable for guest users.

Secure employee WLAN

Like open guest WLANs, there are five phases to access a secure corporate WLAN. We know that **Phase 1** is the same—standard 802.11 negotiation. One main difference is that Phases 2 and 3 are reversed. Guests get an IP address and then authenticate. Employees authenticate and then get an IP address.

Note

802.11i authentication is a Layer 2 authentication, so it takes place before the client gets an IP address. Remember this fact! This can be very useful for troubleshooting connectivity issues.

Employee **Phase 2** involves WPA authentication, which is the same for both WPA and WPAv2. Either can use a PSK to authenticate (WPA-personal) or some 802.1X/EAP-based method (WPA-Enterprise).

Once the user's PSK or other credentials are validated, two things occur. The most obvious is that they gain network access and so can move on to Phase 3. The other result of successful authentication is *key exchange*.

Keying material is distributed to the AP and client for use in deriving session keys. What are these keys used for? Encryption and hashing functions, of course! This is the difference between WPA and WPAv2. WPA uses these keys to do RSA/TKIP-based encryption (and MIC hashing), while WPAv2 uses a more robust AES-based encryption and hashing.

In Phase 3, whether authentication occurs via PSK or 802.1x/EAP, the client and AP can now securely communicate, and so the endpoint can acquire an IP address via DHCP.

In **Phase 4**, the security policy is applied, again based on role assignment and security policies.

In **Phase 5**, the user is granted network access.

802.11 Negotiation

Protocol Summary	Summary
802.11 Beacon	FC=........,SN= 0,FN= 0,BI=100,SSID=CorpWLAN,DS=1
802.11 Beacon	FC=........,SN= 0,FN= 0,BI=100,SSID=GuestWLAN,DS=
802.11 Beacon	FC=........,SN= 0,FN= 0,BI=100,SSID=CorpWLAN,DS=1
802.11 Beacon	FC=........,SN= 0,FN= 0,BI=100,SSID=GuestWLAN,DS=
802.11 Probe Req	FC=........,SN= 28,FN= 0,SSID=CorpWLAN
802.11 Probe Rsp	FC=........,SN= 0,FN= 0,BI=100,SSID=CorpWLAN,DS=1
802.11 Ack	FC=........
802.11 Auth	FC=........,SN= 29,FN= 0 Algorithm=0 (Open System)
802.11 Auth	FC=........,SN= 29,FN= 0 Algorithm=0 (Open System)
802.11 Ack	FC=........
802.11 Assoc Req	FC=........,SN= 30,FN= 0,Listen=10,SSID=CorpWLAN
802.11 Assoc Rsp	FC=........,SN= 0,FN= 0 Status=0,AID=1
802.11 Ack	FC=........

Figure 6-2 802.11 Negotiation

Figure 6-2 shows the frames captured during client connection to an SSID. Each row represents a single frame. The first four frames are Beacons—sent by the AP. You can see that the AP is advertising SSID CorpWLAN and GuestWLAN.

The client hears these beacons and discovers the WLANs. This is how endpoints populate their WLAN list to be displayed for the user to choose. In this case, it appears the user selected CorpWLAN, as this is included in the fifth frame, a client Probe Request. Next, the AP received this request and responded with a Probe Response.

Note

Beacon frames and Probe Reply frames contain exactly the same information and fields. The difference is that beacons are broadcast frames and Probe Reply frames are unicast back to the sender of the probe request.

Next, the client sends an 802.11 Authentication frame. As typical, the algorithm number = 0, which means "Open System." In other words, if you ask to authenticate, the answer is always "yes," and so the AP responds accordingly. As previously explained, this is not "real" authentication—it is merely a simple handshake mechanism.

To complete the process, the client sends an 802.11 Association Request. The AP receives this and creates a unique Association ID (AID) for this client in its local RAM. This is how the AP tracks each connected endpoint's status. Next, the AP sends an Association Response, which includes "AID=1."

The 802.11 negotiation has been successfully completed. Again, this process is like "connecting the cable" for a wired Ethernet client. The link is up, and guest endpoints can do DHCP and L3 authentication. Employee endpoints can now perform L2 authentication, as described below.

Note

Since the AID = 1, this might be the first user attached to this AP. The next user attached may get AID=2, and so on. Please do not get too involved in the AID. It will rarely if ever be the focus for configuration or diagnostics. It is merely explained here so when you encounter it, you will know what it is.

Phase 2—WPA/WPA2 negotiation

Protocol	Summary
EAPOL-Start	FC=T.......,SN= 1,FN= 0
EAPOL-Key	FC=.F......,SN= 0,FN= 0
802.11 Ack	FC=........
EAPOL-Request	FC=.F......,SN= 0,FN= 0
802.11 Ack	FC=........
EAPOL-Response	FC=T.......,SN= 2,FN= 0
EAPOL-Request	FC=.F......,SN= 0,FN= 0
802.11 Ack	FC=........
EAPOL-Response	FC=T.......,SN= 3,FN= 0
EAPOL-Request	FC=.F......,SN= 0,FN= 0
802.11 Ack	FC=........
EAPOL-Response	FC=T.......,SN= 4,FN= 0
EAPOL-Request	FC=.F......,SN= 0,FN= 0
802.11 Ack	FC=........
EAPOL-Response	FC=T.......,SN= 5,FN= 0
EAPOL-Request	FC=.F......,SN= 0,FN= 0

Figure 6-3 Phase 2—WPA/WPA2 Negotiation

WPA or WPA2 negotiation occurs after successful 802.11 association. Remember, Figure 6-2 showed this 802.11 association, with request/response frames for probe, authentication, and association. This is a continuation of the previous discussion.

Figure 6-3 shows several "EAPOL" frames—EAP-Over-LAN. Yes, a WLAN is still a type of LAN, so the standard works for both wired and wireless. Also, "EAP-Over-LAN" frames will also be seen with WPA-Personal. While it is true that WPA-Personal does not use EAP, it uses PSKs. Even so, EAPOL frames are in play.

We would not dive into the details of the EAP process here. The specific order, format, and process of EAPOL exchange vary depending on whether you are using PSKs or 802.1.X/EAP. It will vary based on which "flavor" of EAP you use (PEAP, EAP-TLS, and so forth). For now, it is enough to know that these negotiations serve to authenticate users and to distribute keys for encryption and hashing functions.

Let us move on to user authentication.

Authentication basics

User authentication

The authentication process validates user and/or device identity. Typically, you only grant access for authorized users with proper credentials, but what credentials might be used? All typical options are described below:

MAC Address—Endpoints with a MAC address on the configured whitelist are authorized. Others are not. Warning—WLAN MAC headers are not encrypted. Attackers can see authorized MAC addresses, and spoof those addresses. For this reason, the use of MAC addresses for authentication can *augment* the overall security of your systems. However, it should not be considered an effective, stand-alone method of security.

If you must rely solely on MAC-based authentication, consider isolating these users, so they can ONLY access the specific resources and applications required for the job function. The use of access-lists and firewalls are appropriate for this.

Passphrase or PSK— Pre-Shared Keys are configured on each endpoint. Essentially, everyone on the same SSID has the same authentication credential. This can lead to inadvertent or blatant sharing of the WPA-PSK to unauthorized users. Also, this is only "one-factor" authentication—based only on something you *have*—and is usually preconfigured on end devices. Thus, if an intruder gains access to such an end device, they can easily connect to the corporate WLAN.

Username/password—the endpoint associates, and the end user is prompted to supply a username and password. This information arrives at a RADIUS server, which validates these credentials against some database. Appropriate levels of access are granted.

PEAP-MSCHAPv2 is a popular authentication method. PEAP must be configured on the RADIUS server, and on each endpoint. Now, the RADIUS server wants your user credentials. But should you reveal this to an untrusted network? Perhaps this is not your actual corporate server. What if an attacker is spoofing your corporate infrastructure?

The use of **x.509 certificates** can mitigate this risk. To properly enforce certificate validation, administrators install a copy of the corporate RADIUS Server Certificate on each endpoint. The endpoint can use this to validate the server's identity.

The authentication server submits its x.509 certificate to the client. The client compares this submitted certificate with the valid certificate it has (stored locally, and signed by the corporate Certificate Authority [CA] server). If the submitted certificate is determined to be genuine, then it is safe for the user to pass user credentials to the server.

A hacker attempting to spoof a RADIUS server will not have the same signature. Both sides authenticate to each other, and network access is granted. In this manner, use of x.509 certificates can be used to enable two-way authentication.

Use of certificates will be covered in more detail later in this chapter.

Note

The country where you live defines a standard format for your passport or your license to operate a motor vehicle. Your government is a central authority that issues these standardized passports and driver permits.

When interacting with government officials or police officers, you can prove your identity by showing your passport or driver's license. To validate your identity, officials compare your identification against known standards. Once your passport is deemed valid, you are allowed access to another country.

Similarly, x.509 defines the standard format for a digital certificate. This certificate is digitally signed by a central, trusted authority—your corporate CA. Since everyone agrees that the CA is trustworthy, devices can prove their identity by submitting an x.509 certificate. This certificate is validated against a known, properly signed CA certificate. Once this x.509 certificate is validated, a device or a user is allowed network access.

So an x.509 certificate can be compared to a passport or a driver's license, and a Certificate Authority can be compared to the government that creates and distributes certificates.

This concept will be explained in more detail soon.

One-way authentication

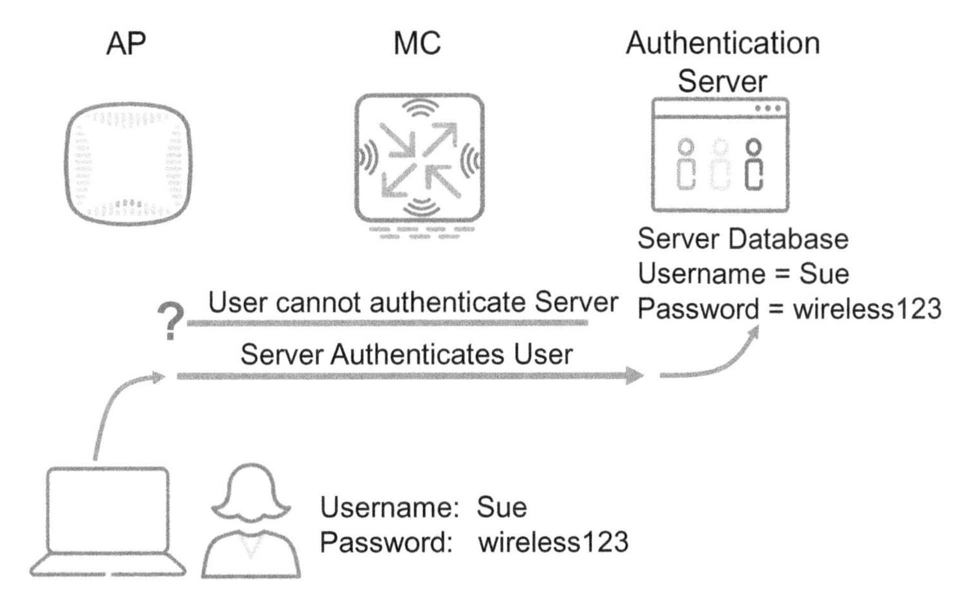

Figure 6-4 One-way authentication

With one-way authentication, the user or the device provides a credential to the Authentication Server, via the AP and Mobility Controller. In this example, the credential is a username and a password (Figure 6-4). The Authentication Server validates these credentials against a database and returns a message to the AP—either "access granted" or "access denied."

Poor Sue. She could be revealing her secret credentials to a hacker. One-way authentication cannot mitigate bogus Server spoofing attacks.

One example of one-way authentication is WPA-Personal, which uses simple PSK authentication. With PSK, there is no mechanism for the user to validate the network-side devices. PEAP-MSCHAPv2 is a secure, two-way authentication method. However, the "Server Validation" option can be unwisely disabled. This makes PEAP-MSCHAPv2 a one-way authentication method.

 **Note**

In some situations, it may be appropriate to uncheck the "Validate Server Certificate" option on a client. This is primarily to test the configurations or for troubleshooting the EAP/RADIUS setup. You might also disable this option in a simple lab environment, to simplify your systems. However, this option should not be disabled permanently in a corporate environment.

Two-way authentication

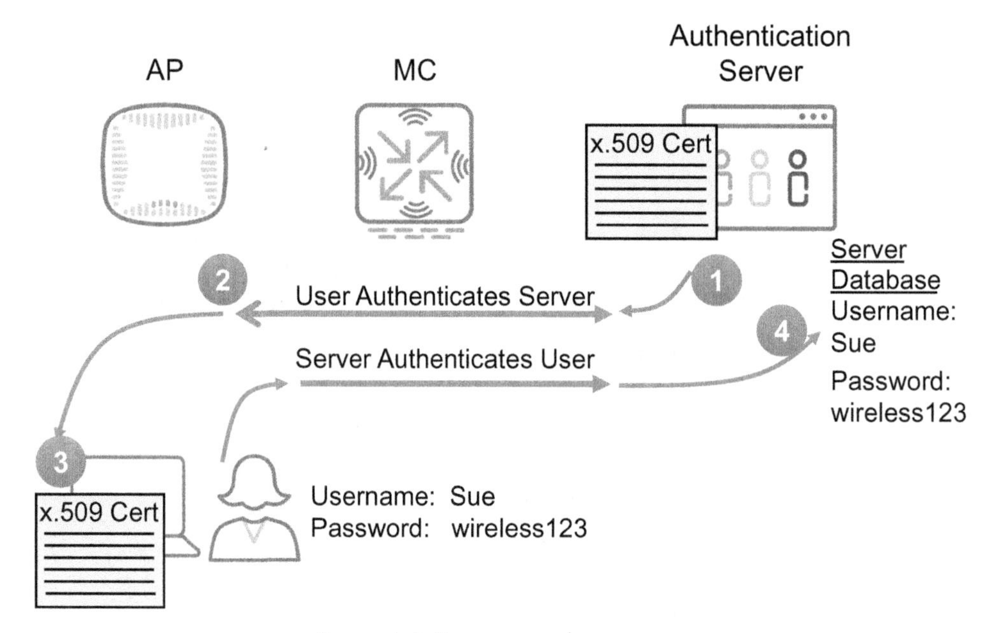

Figure 6-5 Two-way authentication

Two-way authentication mitigates risk, since the Authentication Server is validated by the user, as described below (Figure 6-5).

1. The Server provides an x.509 certificate credential to the MC/AP, which is sent to the client.
2. The endpoint validates this certificate against a locally stored copy of the Server's Certificate, which was signed by a CA or trusted CA chain (The CA name, identity, and signature are stored in a field in the certificate as well as the validity date).
3. The client validates the Server certificate and passes the user credential to the server. The client user credential is protected using the public key of the server, which is also included in a field in the certificate. This process will be covered in more detail later in the chapter.
4. The Authentication Server verifies the username and password in the authentication database and returns a success or failure message to the MC/AP.

An example of bidirectional authentication is when entering a secure building you would be asking the security guard for identification prior to providing your own ID.

The scenario shown in the graphic depicts a common WLAN scenario, using PEAP-MSCHAPv2. As depicted, the network proves its identity with a certificate, and users prove their identity with username/password.

EAP-TLS is another commonly used method, in which both servers AND clients prove their identity with certificates.

Learning check

1. In bidirectional authentication, only the server's certificate is validated.
   a. True
   a. False

2. What is used to validate the server's identity in bi-directional authentication?
   a. The server's certificate
   b. The client's certificate
   c. The username and password for the user
   d. The username and password for the server

Answers to learning check

1. In bidirectional authentication, only the server's certificate is validated.
   a. True
   b. **False**

2. What is used to validate the server's identity in bi-directional authentication?
   a. **The server's certificate**
   b. The client's certificate
   c. The username and password for the user
   d. The username and password for the server

SSID hiding

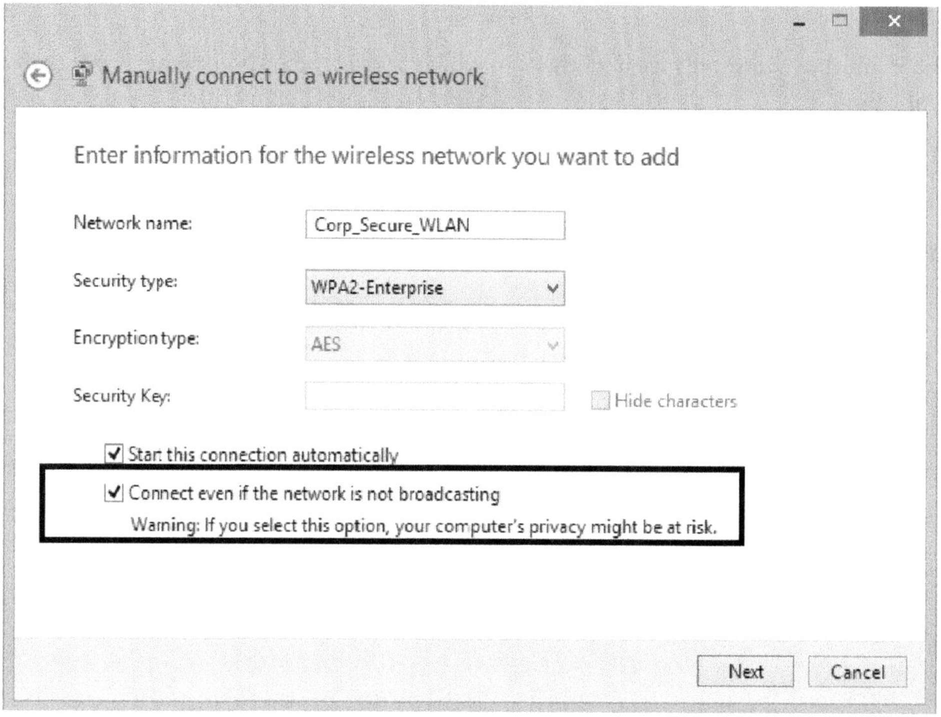

Figure 6-6 SSID Hiding

The SSID names are listed in the SSID field in Beacon frames, which are sent every 100 ms or so. A so-called "hidden" SSID is one that has been excluded from this list. Wireless LAN clients will not automatically discover it. Only clients who already know the SSID can connect.

Figure 6-6 shows a Microsoft Windows-based system that has been explicitly configured with and SSID named "Corp_Secure_WLAN." The option "Connect even if the network is not broadcasting" has been selected. Thus only the corporate devices that are preconfigured can automatically connect to this SSID.

This makes it a bit more secure, but not much more. Hidden SSIDs can be easily discovered with free software tools, such as inSSIDer and Wireshark. This is because clients that already know the SSID send a Probe Request and get AP Probe Replies. These frames include the "hidden" SSID name in the SSID field, which is not encrypted.

Still, this is often considered a best practice for employee SSIDs—at least you are not blatantly advertising your private SSIDs to the general public! However, guest SSIDs need to be broadcast, so visitors can easily access your WLAN. This is because guests will not have the SSID preconfigured in their client software.

How you configure an endpoint to connect to a network with a hidden SSID depends on the endpoint client. You will need to enter the SSID name, which is case-sensitive. You will also need to configure other appropriate security configurations.

MAC filtering

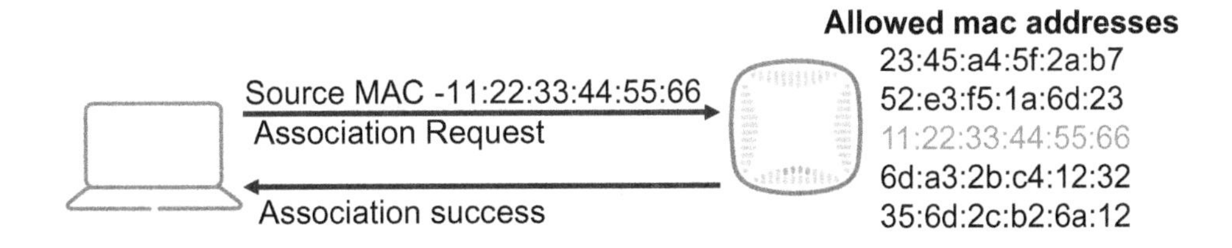

Figure 6-7 MAC filtering

To do MAC filtering, you configure the HPE Aruba system with a "whitelist" of authorized MAC addresses. WLAN clients with a MAC address on the configured whitelist are authorized (Figure 6-7). Others are not. The normal 802.11 Probe request/response and Authentication request/response take place. Then the normal Association request is sent by the client. Before responding, the AP checks the list. If your endpoint's MAC address is on the list, you are granted access. A normal Association Success message is sent to the client.

This is sometimes used as one component of a good security architecture, with other features like Role-Based Access Control (RBAC). However, MAC filtering is not very strong security on its own. WLAN MAC headers are not encrypted. Attackers can see authorized MAC addresses, using tools like Wireshark.

Then attackers can change their client MAC address to that of an authorized client. How is this possible? Well, when a client boots, what it reads is MAC address from the physical hardware NIC. This physical NIC address cannot be changed. However, this address is stored in RAM on the machine, and THIS is what is used for network communications. You can modify this value to your liking, by changing a registry setting in Windows, for instance.

Public Key Infrastructure (PKI)

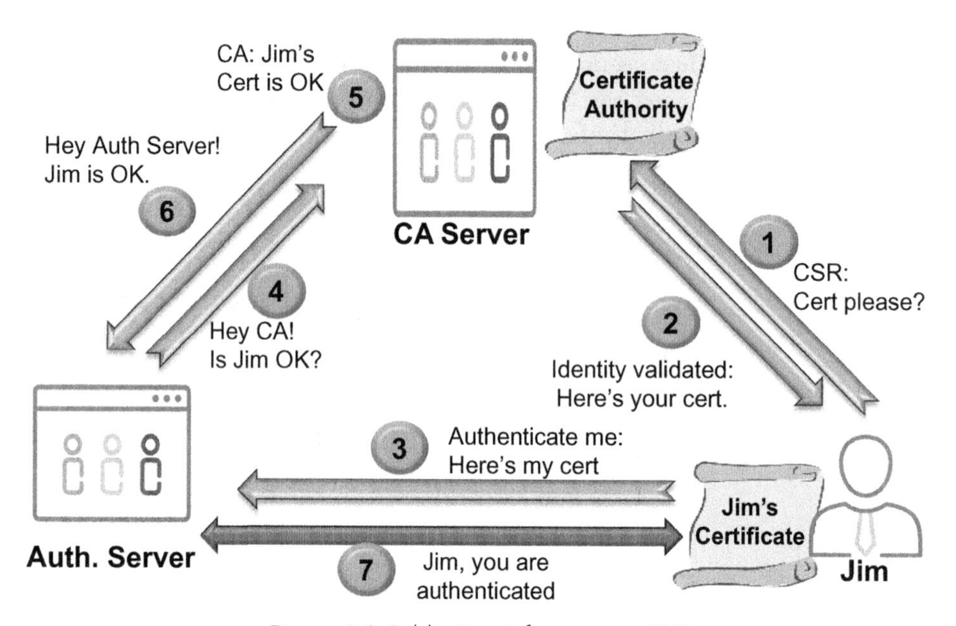

Figure 6-8 Public Key Infrastructure (PKI)

The Public Key Infrastructure (PKI) is a system to create, distribute, and validate certificates. Certificate credentials provide the most secure authentication, but are also more complex, and require more setup and configuration.

Certificate Authorities (CA) issue certificates to users, computers, and servers as a form of credential. Certificates presented for authentication are validated by the receiver against a local copy that has been signed by the CA. This "signature" is a highly secure, one-way hash value. If the received certificate is properly signed, the device is authenticated.

Essentially, you do not trust each device, unless they have a signed certificate from a trusted authority—the CA. This is analogous to airline travel. The gate agent does not allow you to enter, unless you have a passport or a driver's license. The passport is generated and distributed to you by a trusted, central authority—some government agency. The agent checks passport validity against known standards.

You request a passport from your government by filling out a form and providing certain information. Likewise, users can request a certificate from the CA by creating a Certificate Signing Request (CSR)—an electronic file the user creates and sends to the CA.

As time passes, your driver's license may expire. If you misbehave, your license may be revoked. Likewise, PKIs use Certificate Revocation to revoke x.509 certificates. During authentication, the certificate credential is checked against a Certificate Revocation List (CRL). If the provided certificate is on the CRL, you are denied access.

Figure 6-8 shows a scenario where Jim uses a certificate to authenticate. This is described below:

1. Jim asks the CA to issue a certificate in his name. This is usually done with a CSR.
2. The CA validates Jim's identity, then issues him a certificate.
3. Jim presents his certificate to an Authentication Server or other system.
4. The Server does not know Jim, and so asks the CA to check the authenticity of his certificate.
5. The CA checks the certificate. If its hash signature value is correct, and it is not on the CRL, the certificate is valid.
6. The CA informs the Authentication Server that Jim's certificate is valid.
7. The authentication server informs Jim that he has successfully authenticated.

Certificates

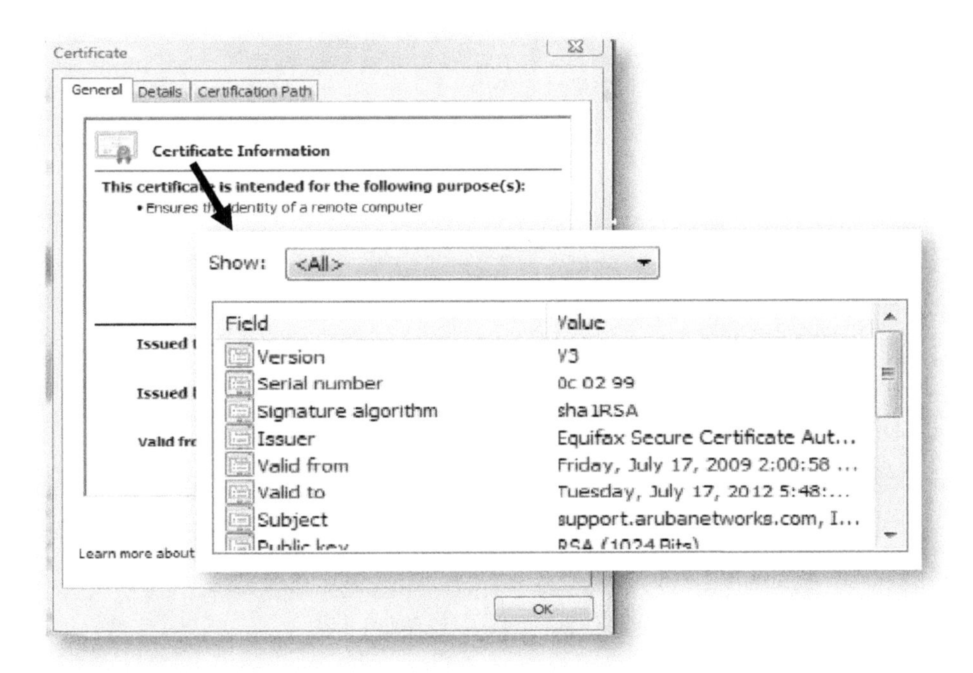

Figure 6-9 Certificates

You can easily view the server certificate used for an SSL connection, which is the basis for HTTPS—the most common type of secure web browser communication (Figure 6-9). When you connect to a web server via HTTPS, that server must prove its validity to you.

The server sends its Certificate to your endpoint, which checks it against a local store of valid, trusted certificates. If it is valid, the HTTPS connection is formed.

You can view the server's certificate from your browser—just click the lock symbol in the browser address bar, then choose to view the certificate. This can help you understand certificate structure. You can see the certificate version, serial number, Issuer (CA) and validity dates, and much more, all under the "General" tab. There is also good information under the "Certification Path" tab.

Certificate authorities

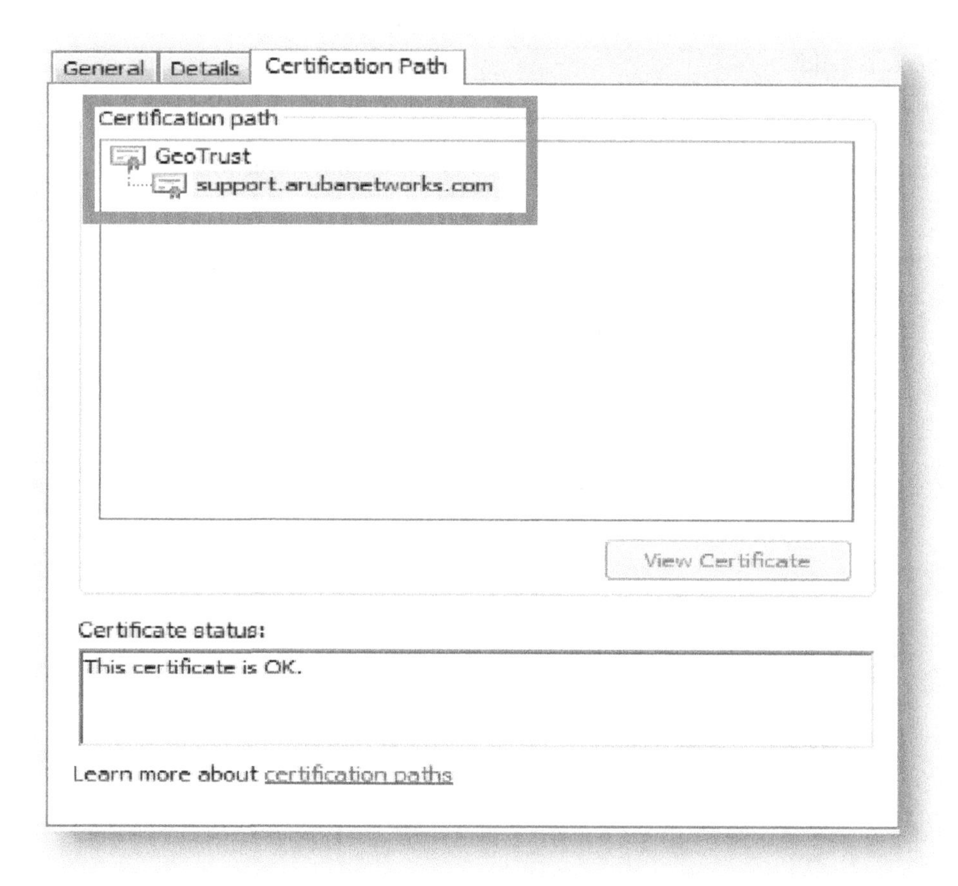

Figure 6-10 Certificate authorities

We now move from the General tab to the Certification Path tab (Figure 6-10). This reveals the CA that issued the certificate. Often, special "root CAs" will issue certificates to other "secondary CAs"—allowing them to issue certificates. Verisign, Entrust, GeoTrust, and others act as root CAs. A certificate signed by a root CA is inherently trusted by most endpoints. This is because they ship with root CAs pre-installed and configured to be trusted.

For a fee, Verisign and others will generate a root certificate for your organization's local PKI/CA services. Since your CA creates certificates based on a root certificate, they will be inherently trusted by Microsoft, Android, Apple, and other devices.

The certification path establishes a trust chain, so certificates issued by a corporate server will be accepted. The example shows a root CA issued by GeoTrust. Based on this root certificate, support.arubanetworks.com issued your machine a certificate. Since your machine inherently trusts GeoTrust, it trusts Aruba, and the HTTPS connection is successful.

Note

SSL and 802.1X/EAP use a slightly different implementation of Certificates, but the concepts of PKI and Certificate issuing and use are the same.

User authentication

WPA-Personal PSK

The simplest Authentication mechanism used by WPA or WPA2 is referred to as WPA(v2)-Personal, which uses PSK. PSKs are an ASCII string, similar in use to the old WEP key, except the WEP key is a hexadecimal value.

We know that WEP is not secure, and one of the reasons is that every user uses the same key for authentication, and this key is also used for encryption. If one WEP user is hacked, all users are hacked.

WPA-PSK is *significantly* more secure than WEP, for several reasons. WPA and WEP both use the same RC4 encryption algorithm. However, WPA adds a kind of "wrapper" called the Temporal Key Integrity Protocol (TKIP).

Think about the words in the TKIP acronym. "Temporal" means time...every **Time** a frame is encrypted, a unique **Key** ensures frame **Integrity,** and each user has a unique set of keys. With WEP, ALL user frames are encrypted with the same set of keys.

Also, WPA uses an actual HASH algorithm, which mitigates against Man-In-The-Middle (MITM) or anti-replay attacks. WEP lacks any hashing protection. WPA uses a much larger key than WEP and offers "broadcast key rotation" and other features that further improve its security.

Still, PSK is primarily used in residential applications or in simpler, lower-cost business applications where a RADIUS server is not used. For the most secure employee access, you should use WPA-Enterprise (802.1X/EAP), instead of WPA-Personal (PSK). The names provide a good hint, yes?

Authentication methods review

You previously learned that Layer 3 Authentication is often used for guests, typically with a Captive Web Portal, and perhaps with a VPN. Layer 2 authentication is typically used for employees and use the following security methods:

You know that **hiding SSIDs** is a very weak form of "security" but is still a best practice for employee WLANs. You typically want guests to easily attach, so hidden guest SSIDs are not appropriate.

MAC Address—Endpoints with a MAC address on the configured whitelist are authorized. Others are not. Warning: WLAN MAC headers are not encrypted. Attackers can see authorized MAC addresses, and spoof those addresses.

WPA (v2)-Personal uses PSKs, with TKIP and MIC. This offers significantly higher security than WEP, but is still only one-factor authentication. If a thief steals a machine with a PSK, they automatically have access to the network—no unique user credentials are used.

WPA (v2)-Enterprise uses 802.1X/EAP methods, in which unique credentials are validated against a RADIUS, LDAP, or CA server. Microsoft Active Directory and Aruba ClearPass can offer such services.

802.1X enables authentication and key management in wired and wireless networks, such as 802.3 Ethernet and 802.11 Wireless. EAP is an authentication framework, with common flavors such as PEAP, EAP-TLS, and EAP-TTLS. EAP messages are encapsulated directly in 802.1X messages. This is why we often use the term "802.1X/EAP." Let us talk about 802.1X/EAP in more detail.

Authentication with 802.1X/EAP

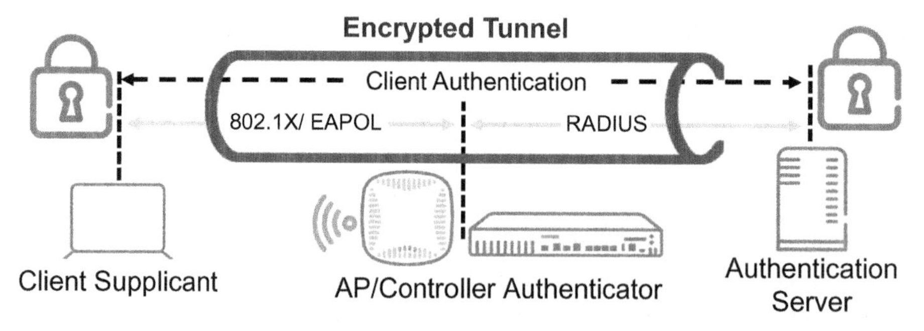

Figure 6-11 Authentication with 802.1X/EAP

Figure 6-11 shows a typical 802.1X/EAP deployment. To support this deployment, clients use 802.1X/EAP-capable software called a "supplicant." The supplicant requests network access from the Authenticator. The Authenticator can be the AP, the controller, or even a switch or firewall in a wired network. It depends on the situation.

To request this access, clients and authenticators use 802.1X/EAP—"sister protocols," united to provide secure WLAN communications. During authentication, 802.1x carries EAP-over LAN (EAPoL) messages between Supplicant and Authenticator.

The Authenticator harvests user credentials from these EAPoL messages, places them in a RADIUS message, and sends it to the Authentication Server. The Authentication Server checks the credentials against a local or external directory and sends a RADIUS Access-Accept (or Access-Reject) message to the Authenticator. The Authenticator receives this, creates an 802.1X/EAPoL message, and sends it to the Supplicant.

Notice that the actual client authentication is between Supplicant and *Authentication Server*. This means the Supplicant and Authentication Server must use matching EAP flavors—EAP-TLS, PEAP, or EAP-TTLS.

You do not configure EAP types on the Authenticator—it does not care! The Authenticator is like a mediator or a translator, speaking "802.1X/EAPoL" to the supplicant, and speaking "RADIUS" to the Authentication Server. This is a good thing to remember during troubleshooting scenarios.

As shown in Figure 6-11, the communication is protected within an encrypted tunnel. Early in the process, keys are exchanged—just enough to create an SSL tunnel. Now secret credentials and other information can be securely sent via this encrypted tunnel.

The 802.1X/EAP authentication process

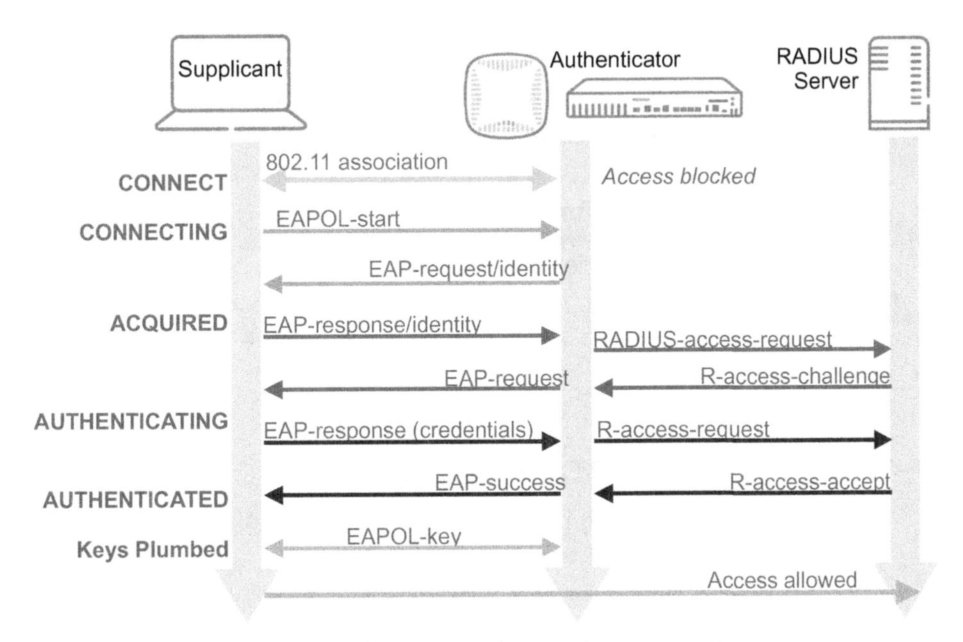

Figure 6-12 The 802.1X/EAP Authentication Process

The 802.1x authentication process starts after successful 802.11 association (Figure 6-12). Next, the client sends an EAPOL Start frame to the Authenticator. Since the client has not yet successfully authenticated, only authentication-related frames are accepted—actual network access is blocked.

The Authenticator sends an EAP-request/identity frame, and the client responds as shown. Depending on the EAP type, this response contains either a real or an anonymous username. The Authenticator sends a RADIUS Access-Request to the RADIUS Authentication Server, which responds with a RADIUS-access-challenge. The Authenticator creates an EAP-Request to convey this challenge to the supplicant.

The supplicant checks the RADIUS server's certificate to ensure authenticity. Satisfied of this authenticity, the user sends its username/password or certificate credentials to the server. This should not happen in the clear, so the supplicant uses the server's public key to encrypt this response, which ultimately arrives at the RADIUS server.

The server thus receives a message, encrypted using its own *Public* key. This message can only be decrypted with the server's *Private* key.

Note
Public/Private Key technology will be explained in more detail later in this chapter.

The RADIUS server checks the credential against the user database. If valid, a RADIUS-access-accept is sent to the authenticator. This is converted to an EAP-Success message and sent to the supplicant.

Now that authentication is successful, keys are generated. These unique keys are used as a basis to encrypt/decrypt and hash/check-hash every client frame.

Extensible Authentication Protocol (EAP)

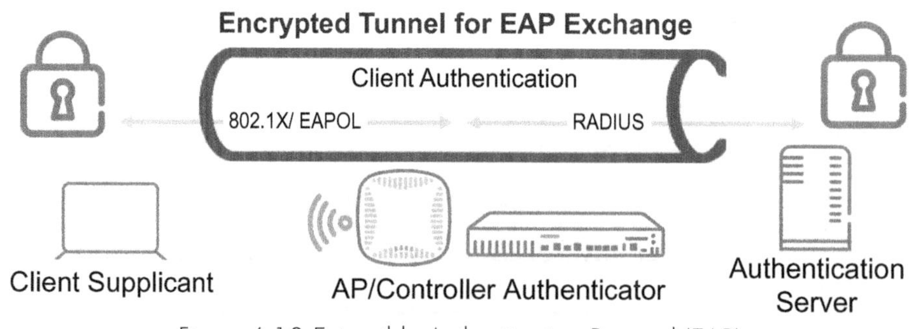

Figure 6-13 Extensible Authentication Protocol (EAP)

EAP is an authentication framework, defined in RFC 3748. EAP provides a way for supplicants to authenticate, usually against a back-end RADIUS server. EAP authentication messages are carried inside the 802.1x payload field (Figure 6-13).

Remember, the actual authentication is between the Supplicant and the Authentication Server. The Authenticator (AP or Controller) merely acts as a mediator/a translator and is not aware of EAP types. A trusted or secured link is created between the Authenticator and the RADIUS server through the use of a shared secret.

Some common EAP types or "flavors" are described below:

- EAP-PEAP (Protected EAP) is commonly supported on the Microsoft Windows operating system.
- EAP-Transport Layer Security (TLS) is the strongest form of EAP and uses Client-side Certificates as the client credential.
- EAP-Tunneled Transport Layer Security (TTLS) was originally created by Funk software (acquired by Juniper) to support legacy EAP types in a secure tunnel.
- Cisco lightweight extensible authentication protocol (LEAP) and EAP-flexible authentication via a secured tunnel (FAST) are proprietary to Cisco Systems.

Of the EAP types listed above EAP-PEAP and EAP-TLS are the most common. For many years, EAP-TLS was not as popular as PEAP. This was because PEAP only requires server-side certificates, while EAP-TLS requires both server and client-side certificates. This meant IT staff needed to manually and securely distribute thousands of certificates—one for each employee device.

However, there are now secure, automated methods of distributing certificates, and so the load on a well-trained IT staff is no greater for EAP-TLS than it is for other methods. Thus, EAP-TLS continues to gain in popularity. The use of certificates for both client and server-side authentication makes EAP-TLS seem more secure than other methods.

EAP termination

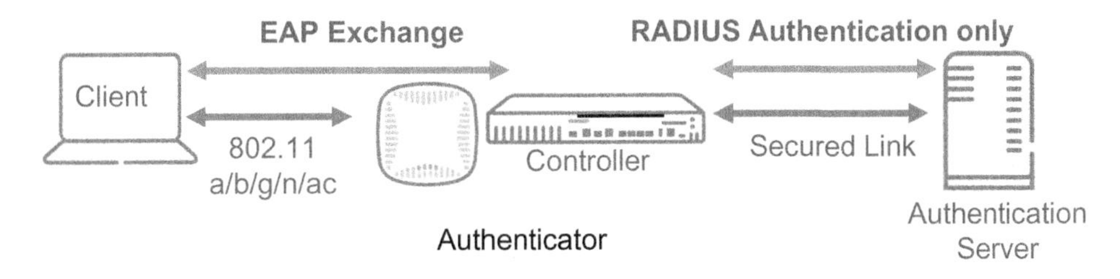

Figure 6-14 EAP termination

EAP termination is also known as AAA-Fast Connect or EAP Offload. EAP termination allows the controller to terminate the EAP session with the client supplicant instead of passing the EAP messages back to the RADIUS server (Figure 6-14). Only the RADIUS request is sent to the server for user authentication against the RADIUS database.

In some cases, EAP termination can help resolve authentication failures caused by EAP timeouts. These timeouts can happen when the authenticator and RADIUS server are separated by slow WAN links.

Learning check

3. What are some reasons to enable EAP termination?
   a. To speed up client roaming between controllers
   b. To speed up the EAP process when the Server is across a WAN link from the controller
   c. To speed up network access for the client when authenticating
   d. To bypass network firewall restrictions

4. Which EAP types are supported by an Aruba Controller in nontermination mode?
   a. PEAP
   b. EAP-TLS
   c. EAP-TTLS
   d. EAP-FAST

Answers to learning check

3. What are some reasons to enable EAP termination?
   a. **To speed up client roaming between controllers**
   b. **To speed up the EAP process when the Server is across a WAN link from the controller**
   c. To speed up network access for the client when authenticating
   d. To bypass network firewall restrictions

4. Which EAP types are supported by an Aruba Controller in nontermination mode?
   a. **PEAP**
   b. **EAP-TLS**
   c. **EAP-TTLS**
   d. **EAP-FAST**

Machine authentication

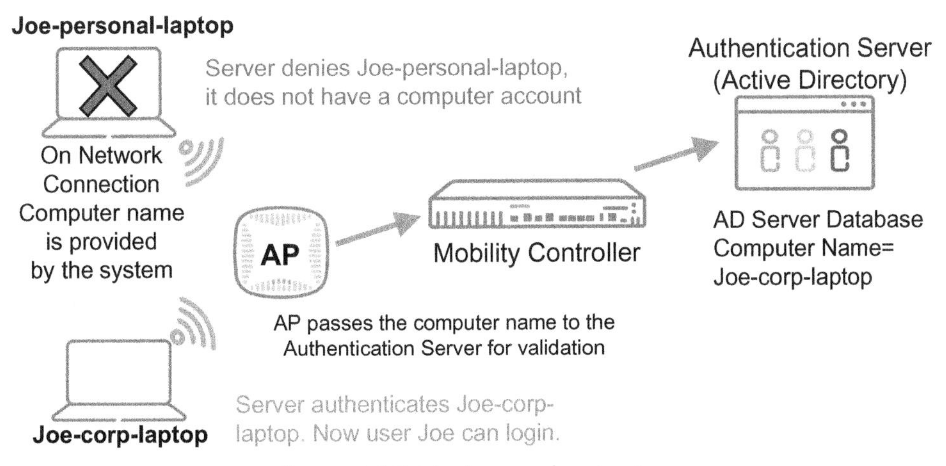

Figure 6-15 Machine authentication

As the name implies, machine authentication validates the endpoint itself, rather than the user. You can use machine authentication to ensure that only authorized computers may be used to access the network.

In Figure 6-15, Joe can use the WLAN from his corp-laptop but not his personal-laptop. The reason for enforcing machine authentication is to ensure only systems that are under the administrative control of the organization are used on the network. Often, the motivation is to ensure that users have up-to-date virus definitions, service packs, approved software, and patches.

Authentication servers

Common Authentication servers

RADIUS and LDAP are the primary authentication server types. HPE Aruba controllers have an internal database to support limited authentication functions but is not a full RADIUS server. Aruba offers complete RADIUS services and much more in the ClearPass product.

A comparison between RADIUS and LDAP is summarized below.

- RADIUS
  - Carries EAP securely, by using the public key in the Server certificate
  - Needs a User Account Database
  - Communicates with NAS clients (authenticators), such as a controller
  - Role-Based Security has configuration attributes to control user access levels or VLAN assignment
- LDAP
  - Main type of database used by Microsoft Windows servers running Active Directory
  - Hierarchical database that can be used directly for authentication

Active directory

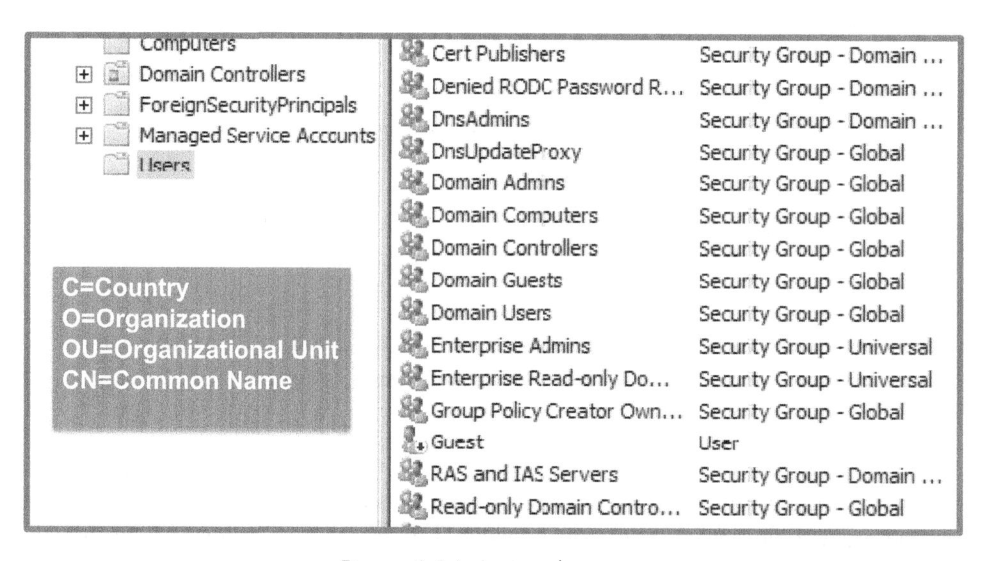

Figure 6-16 Active directory

Active directory is a hierarchical database of user, computer, and group accounts. Large numbers of users or computers can be organized by Country, Organization (O), and Organizational Units (OU) (Figure 6-16). The user or computer is identified by a Common Name (CN).

Microsoft Windows Servers are common in WLAN deployments. They can act as a Certificate Authority (CA), a RADIUS server (in the form of the Network Policy Service [NPS]), and they have this hierarchical database of users, computers, and certificates. This is a nice solution to authenticate users.

Remember, the result of authentication is twofold. One result is that network access is granted. The other result is that keys are generated and distributed. These keys are used for Encryption and hashing, our next topic.

Learning check

5. Which of the following are RADIUS Servers?
   a. ClearPass
   b. NPS
   c. Controller Internal Database
   d. Free RADIUS

Answers to learning check

5. Which of the following are RADIUS Servers?
   a. **ClearPass**
   b. **NPS**
   c. Controller Internal Database
   d. **Free RADIUS**

Encryption

Confidentiality

Encryption Suites ensure privacy for over-the-air transmissions. This ensures that "man-in-the-middle" hackers cannot capture and read your data. Hashing functions ensure the integrity of your data. In other words, if a "man-in-the-middle" attacker captures your data, modifies it, and then resend it, there will be a hash value mismatch, thus revealing the attack. The chart below summarizes L2 WLAN security usage:

Protocol	Encryption	Hashing
WEP	RC4	none
WPAv1	RC4/TKIP	MIC
WPAv2	AES	AES

Encryption/Decryption can be performed with either Symmetric or Asymmetric Keys.

Symmetric key encryption

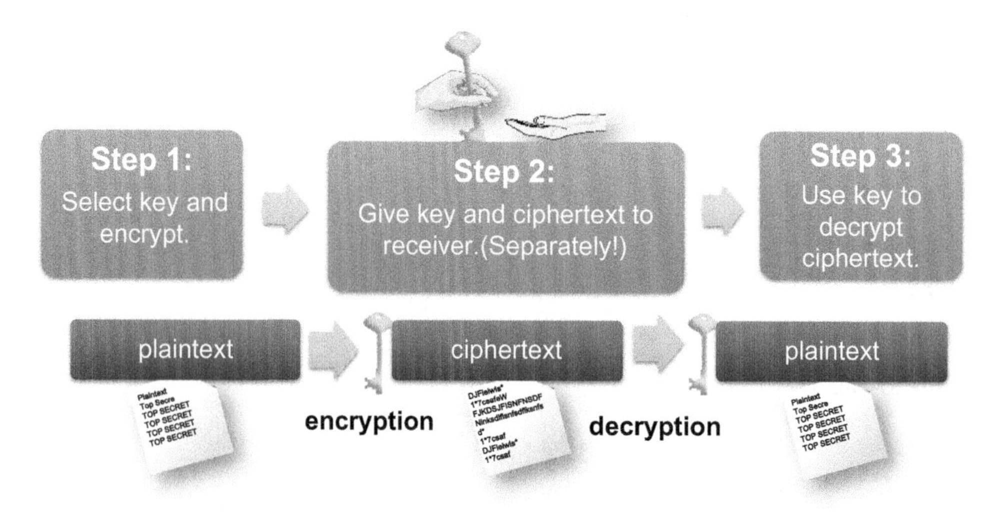

Figure 6-17 Symmetric key encryption

In Symmetric key encryption, the same key is used for both encryption and decryption. An advantage of symmetric key encryption is speed (Figure 6-17). This is why symmetric key encryption is used to encrypt every single wireless data frame.

And a key is used to encrypt these frames. How do we securely get these keys over the air? …to a wireless endpoint?

Symmetric key encryption has a weakness in distributing encryption keys. This is why we have asymmetric key encryption.

Asymmetric key encryption

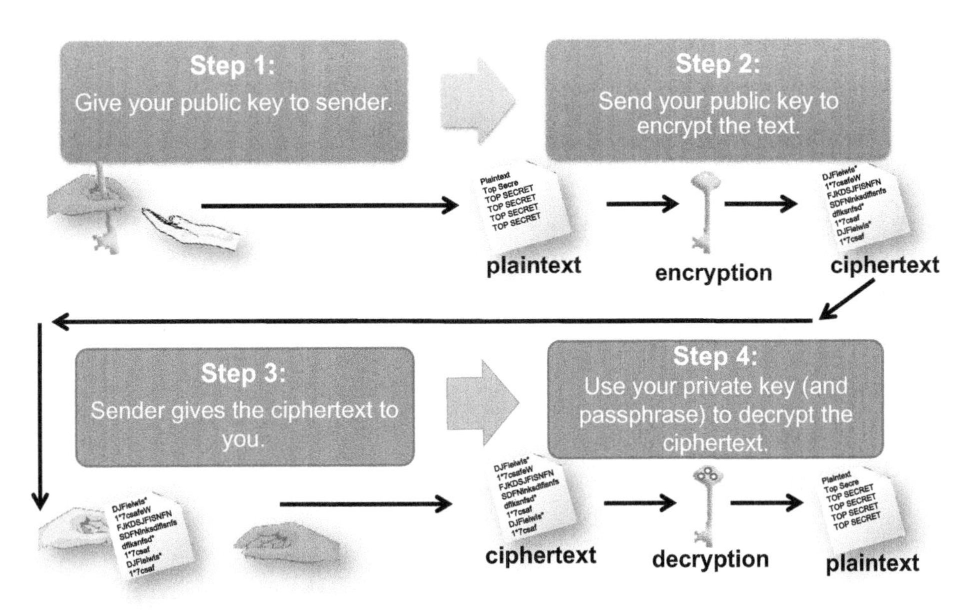

Figure 6-18 Asymmetric key encryption

Asymmetric key encryption is far more secure for key distribution, but slower than symmetric key encryption in performing encryption/decryption processes. Since key exchange is not frequently done this is a perfect match.

With Asymmetric key encryption, a unique pair of keys are generated with a complex mathematical relationship—a public key and a private key. It does not matter who gets the public key—let anyone use this key to encrypt frames if they want. However, no device can decrypt these frames without the private key—a closely guarded entity on the server.

Figure 6-18 shows asymmetric keys in action. Each side has a public/private key pair. Client sends its public key to the server, and server sends its public key to the client.

The client uses the server's public key to encrypt data. This data can only be decrypted with the server's private key.

The server uses the client's public key to encrypt data, which can only be decrypted with the client's private key.

So Asymmetric key encryption is used to get keys between devices. The devices then use these symmetric keys as a basis for every transmitted frame's encryption.

Learning check

6. What type of encryption is the fastest for bulk data encryption?
   a. Symmetric key encryption
   b. Asymmetric key encryption

Answers to learning check

6. What type of encryption is the fastest for bulk data encryption?
   a. **Symmetric key encryption**
   b. Asymmetric key encryption

Wireless threat overview

Uncontrolled wireless devices

Properly deployed WLANs are usually quite secure. Even so, threats remain, and the more you understand them, the better equipped you are to protect against them. We describe the most common threat types below.

Often WLAN security is compromised by internal employees. This can be malicious but is often just because they are unaware of how their actions create risk. For example, an employee might bring an unsecured wireless AP to work and plug it into the corporate wired network. The best way to mitigate this type of risk is through employee education.

A rogue AP is one that does not belong in your environment—they are unmanaged APs, of unknown origin, and/or used for unknown purposes, many of which could be malicious. This represents one of the single largest threats to network security. A rogue AP in an office, with default configuration parameters, poses a similar risk to having an Ethernet jack on the outside of your building.

Ad-Hoc WLANs do not use APs. This is direct endpoint-to-endpoint wireless communication. Most corporations have a policy against these unsecured Ad-Hoc networks, as they could turn a PC into a wireless-wired bridge. An employee device could have an Ethernet connection to corporate systems. With Ad-Hoc mode enabled on their WNIC, an attacker could attach to the employee device, and then use it to gain access to the wired corporate system.

WLAN infrastructure attacks

Malicious hackers could create Denial-of-Service (DoS) attacks, perhaps by flooding the network with Probe requests, or any of several other types of WLAN frames. They could spoof a corporate AP and send deauthentication or disassociation frames to clients, knocking them off the network.

In Man-in-the-Middle attacks, attackers capture WLAN frames, modify them, and retransmit them. This is mitigated by hashing algorithms, like WPA's MIC and WPAv2's AES-based hashing.

Denial of service attacks

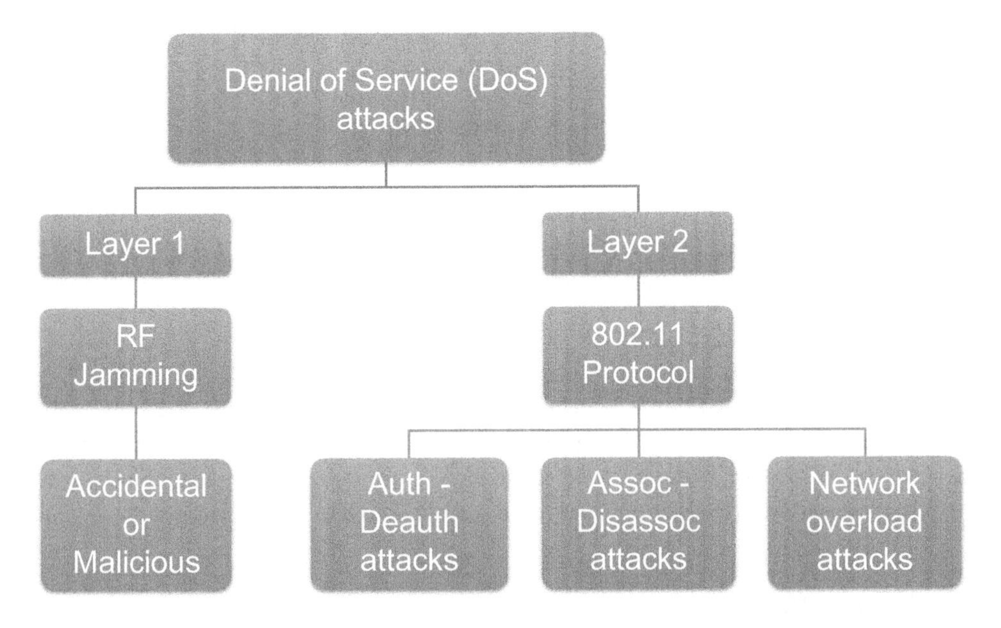

Figure 6-19 Denial of service attacks

RF DoS attacks can occur at Layer 1 (physical RF) or Layer 2 (mac layer). A layer 1 RF DoS attack involves sending sufficient RF noise to drown out any 802.11 communication (Figure 6-19). This can be done either maliciously or accidentally. Malicious RF jamming is Illegal in most countries, but difficult to enforce. This is because the jammer can be hard to find.

Layer 2 MAC-based DoS Attacks work within the 802.11 protocol framework. Example of these types of attacks are:

- associate/disassociate attacks
- authenticate/de-authenticate attacks
- network overload attacks and NIC firmware flaws

An attacker may try to disconnect clients from an AP by sending de-authenticate or disassociate frames with a spoofed source address. This often signals an attempted man-in-the-middle attack.

In the US, these attacks are not prohibited by the FCC, but are covered by other information security laws (Communications Act of 1934).

Note

Poorly written or outdated WNIC drivers can appear to be sources of DoS and other issues. Many an experienced WLAN engineer has spent too long troubleshooting an issue that simply required driver updates, or some WLAN utility, or some software "shim" between the driver and the software. Keep your drivers and related software up to date!

Access Points, Air Monitors, Spectrum Monitors

Wireless Intrusion Detection Systems (WIDS) are used to *detect* potential WLAN threats. Wireless Intrusion Protection Systems (WIPS) serve to both detect and *prevent* WLAN threats.

APs are primarily responsible for serving WLAN clients, passing user data to the controller. They can supplement WIDS and WIPS on a time-allowed basis, but this is their secondary responsibility.

HPE Aruba APs can be configured to serve as Air Monitors (**AM**) or Wireless Sensors. Thus, they are used to scan the air and detect Rogue APs or other IDS events. AMs are dedicated to WIDS and WIPS. They do not transmit RF energy/data and do not service clients. They are passive scanning devices.

HPE Aruba APs can also be configured to serve as Spectrum Monitors (**SM**). Like AMs, SMs do not service clients. However, they differ in function. AMs look mostly at 802.11 L2 frames that could indicate malicious WLAN activity. SMs capture Layer 1 RF data, which is sent to the controller for spectrum analysis. This is used to detect RF noise and interference. Examples include non-802.11 devices like microwave ovens, baby monitors, video cameras, Bluetooth, and more.

Wireless IPS process

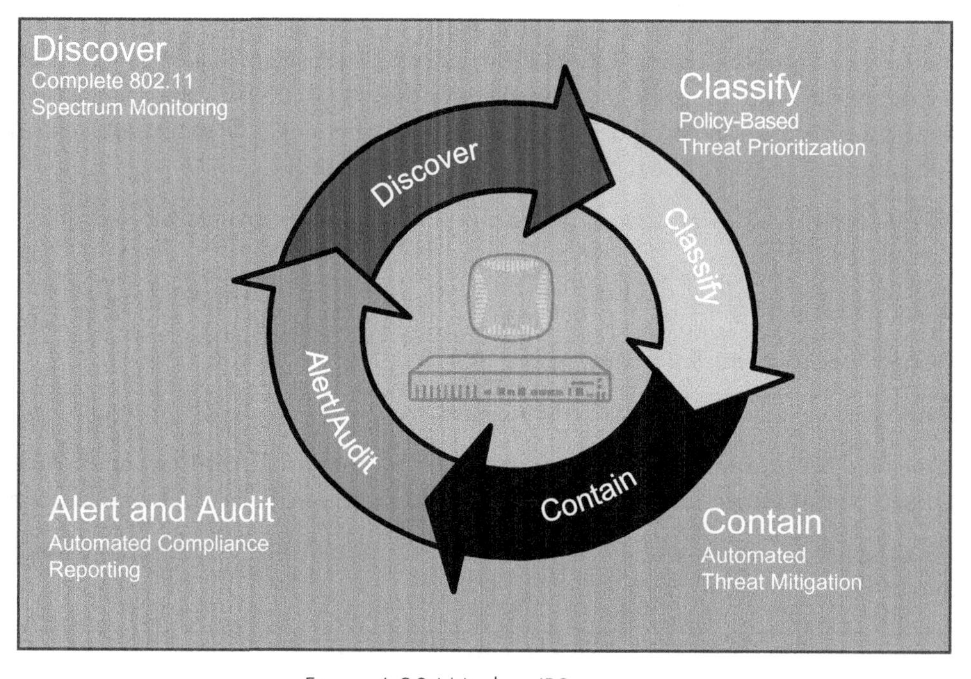

Figure 6-20 Wireless IPS process

The Wireless IPS process includes four major phases (Figure 6-20). It discovers threats, classifies them, and then uses threat containment to mitigate them. WIPS includes alerting and auditing functions to help track and managed WLAN security.

The **discovery phase** relies on AMs for continuous monitoring, across all 802.11 channels. It learns about wireless devices, activity, and configuration, so you know what is out there in the RF spectrum.

Next, WIPS automatically **classifies** everything into various threats and nonthreats. This increases your situational awareness, informing you of threat severity and potential risk.

For Intrusion *Detection* Systems (IDS), the story ends here—it is now up to you to analyze the data, strategize, and take appropriate action. But we are talking about Intrusion *Protection* Systems (IPS).

The next step involves automated threat **containment**. This includes blocking anyone from using a rogue AP or denying intruder access.

Finally, **alert and auditing** functions enable ongoing analysis, reporting, and report distribution. The can help you with compliance reporting to validate that the WLAN conforms to corporate regulations and compliance requirements.

Locating rogue APs

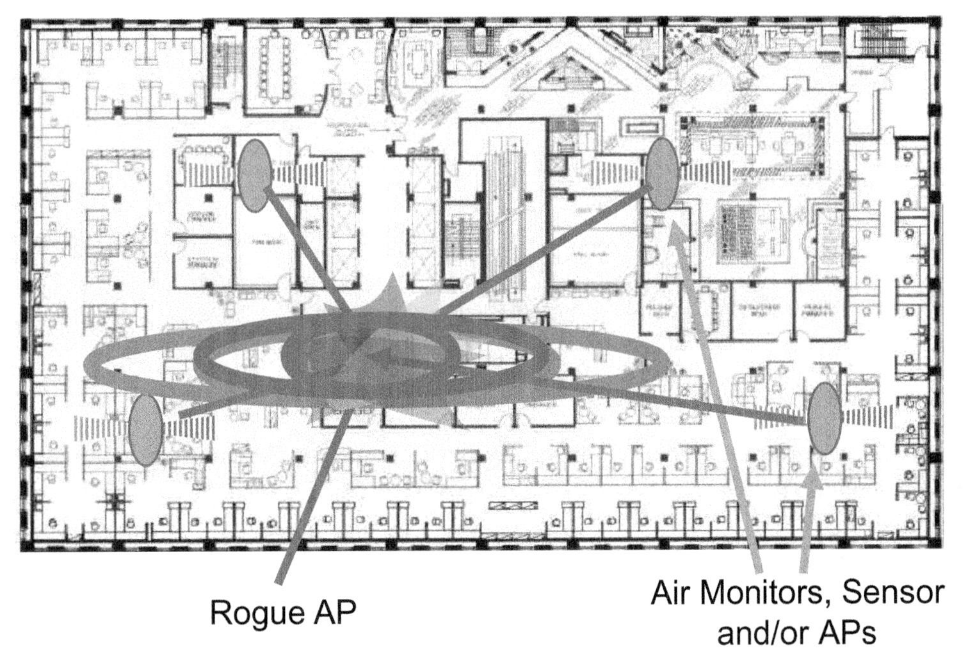

Figure 6-21 Locating rogue APs

One of the most dangerous types of WLAN attacks is a rogue AP (Figure 6-21). This is often a simple, inexpensive AP meant for Small-Office Home-Office (SOHO) applications. However, it has been brought to your facility and connected to your wired network.

Unsecured rogue APs allow anyone to access your wired network. HPE Aruba APs and AMs collect information about rogue APs and send it to the controller. This includes signal strength information. In Figure 6-21, a rogue is located between four valid, corporate APs. Each AP reports rogue AP signal strength. This allows the controller to estimate rogue AP location using triangulation.

If you use the HPE Aruba AirWave management platform, this information can be displayed on floor plan maps in your campus. You can easily visualize rogue location and send a team out to find and eliminate the rogue.

Client Tarpit containment

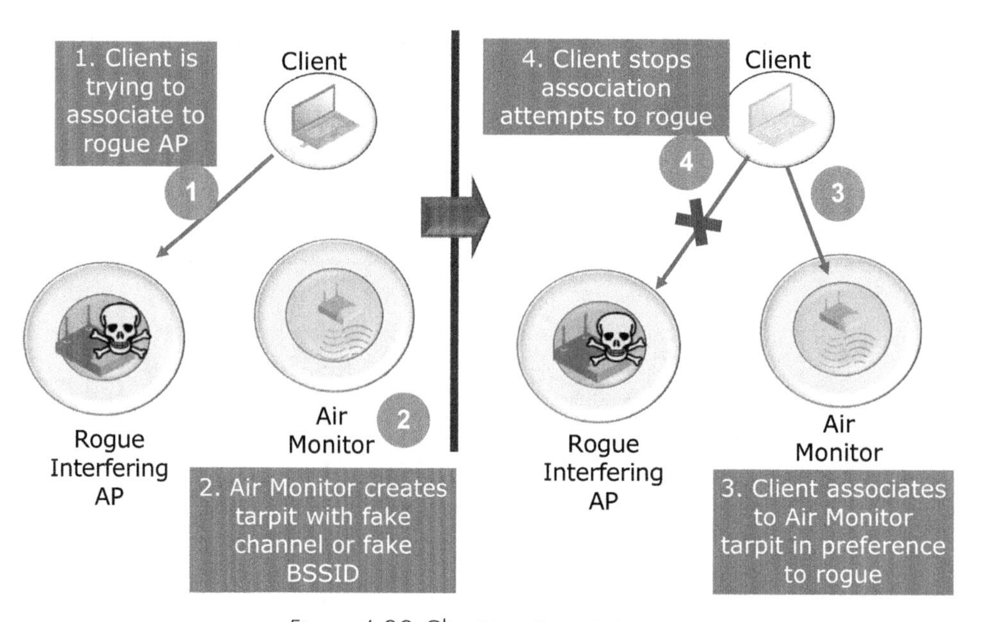

Figure 6-22 Client tarpit containment

Tarpit containment prevents clients from connecting to rogue APs (Figure 6-22). It takes time for you to discover and eliminate rogue APs. You do not want corporate clients connecting to a rogue during this time.

Client tarpit containment is off by default. You enable it to automate the response to a rogue AP. Client tarpit containment does not waste air-time during threat mitigation and it works against any brand and type of wireless device, as described below:

1. Client attempts association to a rogue AP. The rogue AP has been identified by the AM and APs.
2. The AM spoofs the rogue AP's SSID. The AM transmits frames as if it was the rogue AP.
3. Client associate to the AM tarpit or fake AP instead of the actual rogue AP
4. Client stops association attempts to the rogue.

Learning check

7. What is an example of an RF DoS attack?
   a. RF Jamming
   b. RF Obstruction
   c. RF Tarpit
   d. RF Containment

Answers to learning check

7. What is an example of an RF DoS attack?
   a. **RF Jamming**
   b. RF Obstruction
   c. RF Tarpit
   d. RF Containment

7 Firewall Roles and Policies

LEARNING OBJECTIVES

- ✓ In previous chapters, you have learned about end-user connectivity and security. You have learned that HPE Aruba WLAN systems can control user access with firewall functions, policies, and rules. This is all based on the roles assigned to connected users.
- ✓ You will now learn how these features operate, the relationship between them, and how to configure them.

Aruba integrated firewall introduction

Aruba firewall

With HPE Aruba networks and identity-based firewall, each client is associated with a user role. This role determines client access and network privileges.

A stateful firewall tracks client sessions and is aware of bidirectional traffic. No rules are needed for the returning traffic. Stateful firewalls understand that if you allow certain client traffic to exit the firewall, you want to allow the appropriate return traffic.

The firewall uses deep packet inspection—it is "application-aware." It can permit or deny traffic based on the application in use. Based on certain actions, the system can blacklist users, or apply QoS to limit their bandwidth or traffic priority.

The HPE Aruba firewall is dynamic. Address information in the policy rules can change as policies are applied to users.

Users will get the same role and firewall policies, even while moving from subnet to subnet. Users may get role assignment directly from the Mobility Controller, or server-derived roles from a RADIUS server. The server can apply both standard RADIUS attributes and Aruba's Vendor Specific Attributes (VSA).

ClearPass has full integration with the Aruba controller to facilitate more involved Role derivation, Change-of-Authorization (CoA), and downloadable Roles.

Aruba firewall role

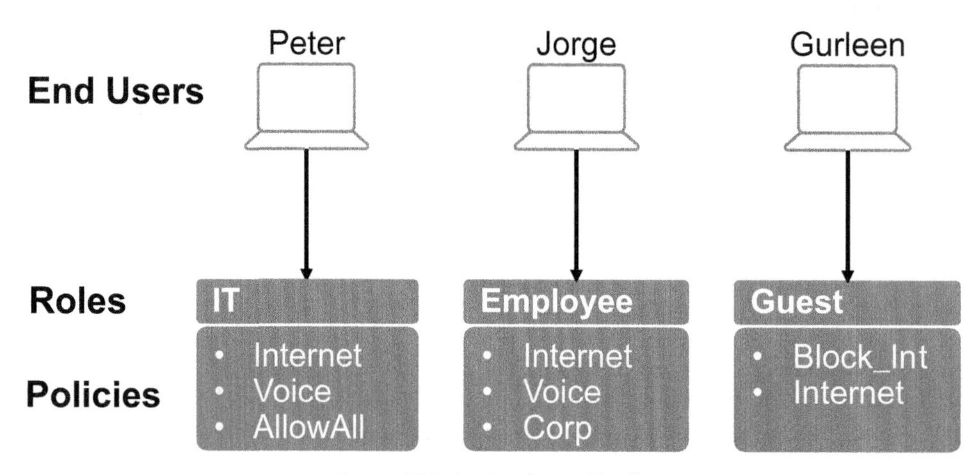

Figure 7-1 Aruba firewall role

As users connect, they are assigned a role. In Figure 7-1, Peter and his peers are assigned to the IT role. Jorge is assigned to the Employee role, and Gurleen is assigned to the Guest role, as are all visitors.

Each role may have one or several policies. Figure 7-1 reveals that the IT role has three policies, named Internet, Voice, and AllowAll. The Employee role also has the Internet and Voice policies assigned, along with a policy named Corp. The Guest role has a policy that blocks access to all Internal Corporate subnets (Block_Int). This role also includes the Internet Policy, which allows access to the Internet. These policies control what each connected user can and cannot do.

Aruba firewall policies

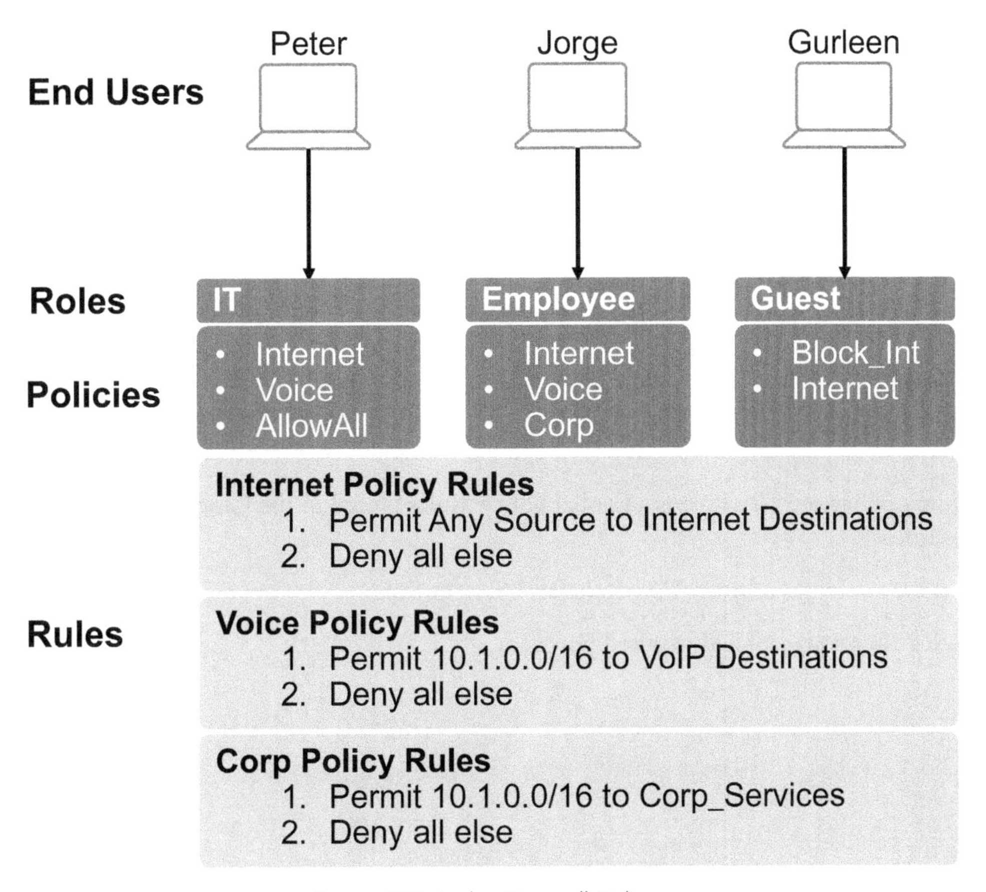

Figure 7-2 Aruba Firewall Policies

Each policy is merely an ordered set of rules, which define what users can and cannot do. When a user packet arrives, it is associated with that user's role. Then the first rule of the first policy is examined. The rule analyzes packet type, source/destination addresses, ports, and services. If these packet criteria match those specified in a rule, some action is taken. Typical actions are to permit the packet, deny the packet, and to apply QoS to the packet. If there is no match then the next rule is compared. This is similar to the top–down processing of a typical Access Control List (ACL).

If the packet matches no rules in the first policy, the next policy's rules are checked, top-down. If there is still no match, then the implicit deny all is applied and traffic is blocked.

Figure 7-2 does not attempt to represent actual CLI rule syntax. Here, the objective is to convey rule concepts. For example, the Guest role has two policies—Block_Int and Internet. The Block_Int policy is there to prevent guest users from internal corporate access. Although not shown, you can imagine that this rule likely blocks any guest-assigned source subnet from accessing any internal corporate subnet destinations.

If a guest attempts to access internal corporate resources, the packet will match a rule in this first policy and be discarded (denied). If the guest is attempting Internet access, their packet will not match any rule in the Block_Int policy. Thus, the next policy (Internet) is assessed.

The Internet policy has two rules. The first rule permits any source IP address to access any Internet destination. All other traffic is denied. This "deny all" rule is implicit. Meaning you do not actually configure it, nor do you see it. However, it is there. Regardless of how many rules are in a list, the last rule is always this implicit "deny all."

Below are a few other facts about policies and roles:

- A role can also invoke a specific Captive Portal page, typically used for guest access. You can impose bandwidth limits to all users of this role. The Role can also assign a specific VLAN to the users.
- The first policy of every Role is a Global policy. Rules placed in the Global policy are applied to ALL Roles.
- There are two types of roles, User Roles and System Roles.
- Policies can also be applied to wired ports.

Roles, policies, and role derivation

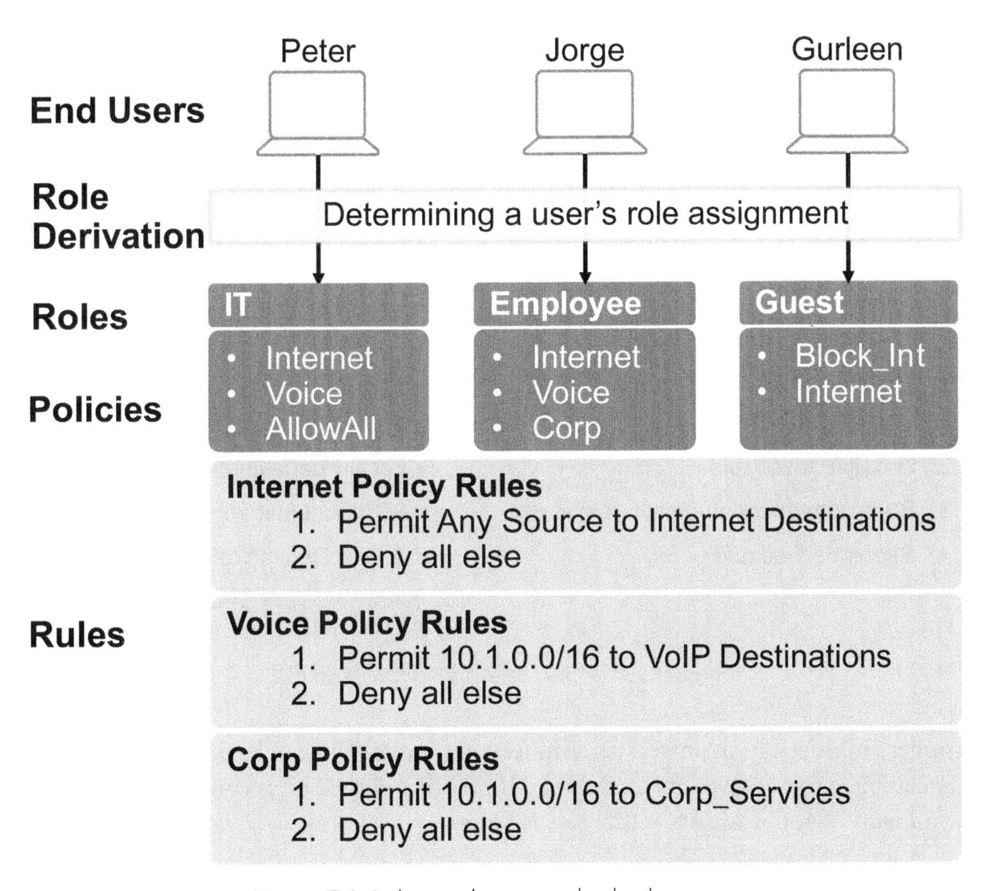

Figure 7-3 Roles, policies, and role derivation

You know that roles are created for groups with similar access needs (Figure 7-3). These access needs are controlled by a policy. In this example, the administrator wisely used intuitive policy names, so you can easily see that users assigned the IT role will have access to the Internet, VoIP, and everything. Users assigned to the Guest role have only Internet access.

Firewall policies are stateful, bi-directional, and dynamic. Policies are made up of rules. If traffic matches a rule, then an action is taken. No other policy rules are examined.

Note

Once the first few frames have been inspected and a match is found, then the action is applied. This action information is cached. Therefore, no lookup is required for subsequent frames—the action is simply applied.

When users attach, the system must determine what roles to apply. This is called role derivation. This can be performed based on local configuration, or indicated by an external server. If neither occurs, then the default authentication role is applied.

Aruba role derivation example

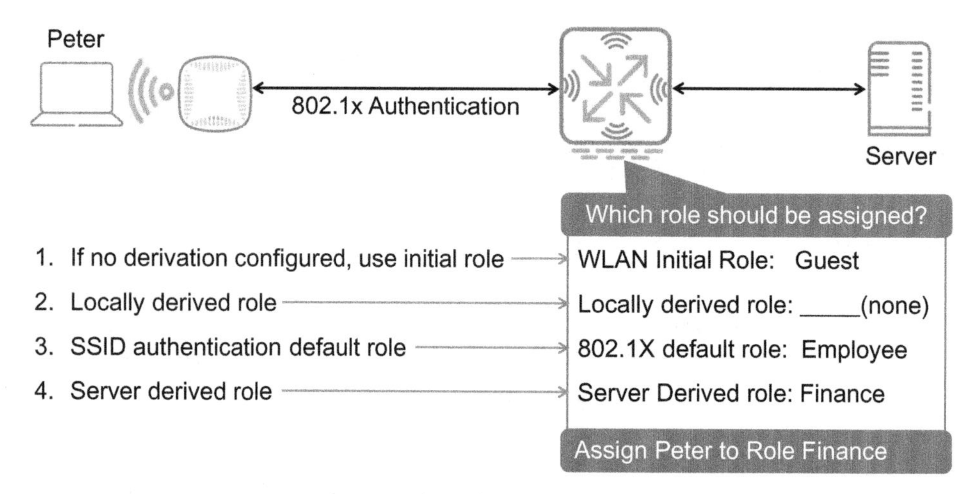

Figure 7-4 Aruba role derivation example

The controller's role derivation process determines a user's role. This can be done in many ways. Upon initial association, a user is given the initial role. If no other derivation is configured, then that role is applied. In Figure 7-4, the WLAN initial role is Guest.

There can be locally derived roles, which take precedence over the initial role. In this case, there is no locally derived role. Even if there were, this would be overridden by any role configured for an authentication method. In the example, Peter used an 802.1X-based authentication method to connect to an SSID. The default role assigned to this SSID's authentication method is Employee.

So far, it looks like Peter will be assigned to the Employee role. However, this Company's 802.1X authentication system uses an external RADIUS server, which has access to some external store of user names, passwords, and group memberships (often a Microsoft Active Directory (AD) server). The RADIUS server has been configured to override the 802.1X default role for any user in the Finance group. Since Peter is a member of the Finance AD group, he is ultimately assigned to the Finance role.

This is a very general description of role derivation. This advanced topic will be covered more thoroughly in the HPE Aruba advanced mobility course.

Identity-based Aruba firewall

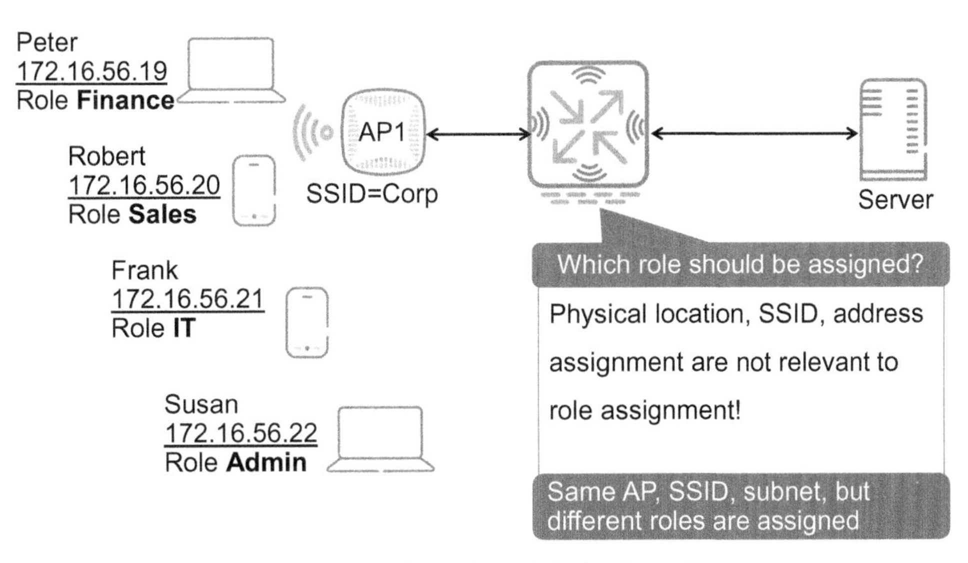

Figure 7-5 Identity-based Aruba firewall

Regardless of a user's connected AP, SSID, or subnet, role derivation determines their Role, and therefore, their network access levels.

In Figure 7-5, Peter, Robert, Frank, and Susan are on the same AP, SSID, and subnet. However, each is assigned to a unique role. This is assigned locally by the controller, or from an external server.

If there is no locally derived Role and no Server-derived Role, then users fall into the Authentication default Role.

Your location, attached AP, and IP address is usually irrelevant to your role assignment. However, there are rules available to base role assignment on connected AP or VLAN assignment.

Centralized and consistent security

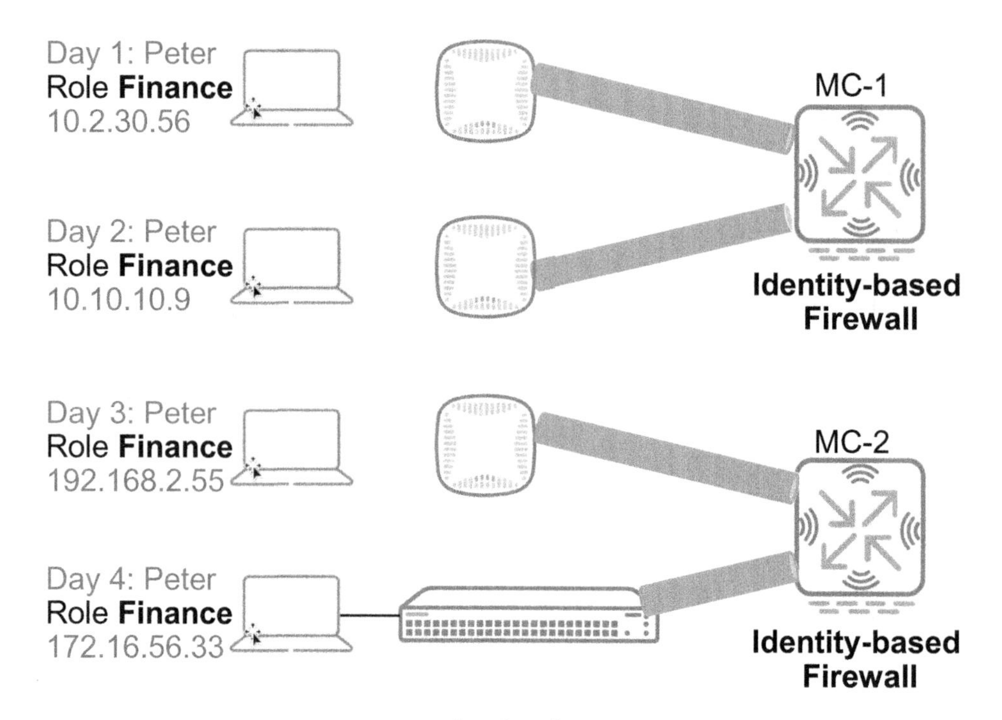

Figure 7-6 Centralized and consistent security

In this scenario, Peter associates with the top-most AP in Figure 7-6, is bridged to Subnet 10.2.30.0, and is assigned to the Finance role.

As Peter moves from one AP to another, he may be mapped to a different subnet, but his assignment to the Finance role is unaffected.

This can also apply to wired connections. Tunnel node is an HPE Aruba switch that creates GRE tunnels to a dedicated MC. The wired user is treated like any other wireless user.

This ensures that you have a central control over a consistent security policy.

Learning check

1. Most commonly, user role assignment is based on AP or SSID association.
   a. True
   b. False

2. A policy may be used by many roles.
    a. True
    b. False

Answers to learning check

1. Most commonly, user role assignment is based on AP or SSID association.
    a. True
    b. **False**
2. A policy may be used by many roles.
    a. **True**
    b. False

Policies and rules

Policy rules

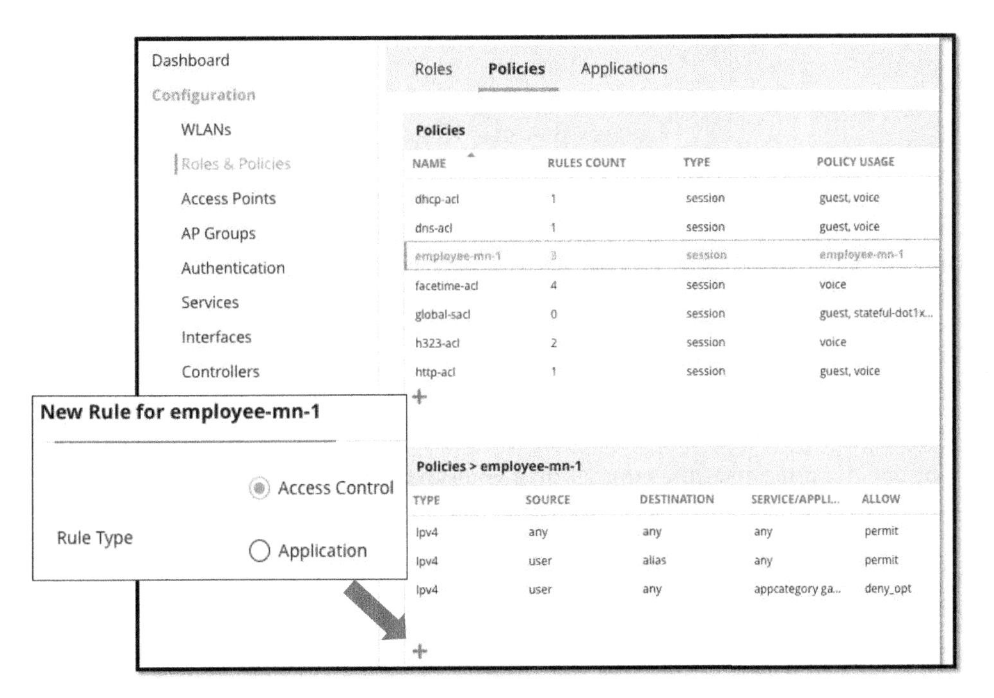

Figure 7-7 Policy rules

To add a new policy, navigate to **Configuration > Roles & Policies > Policies**. Always choose a name that has meaning (Figure 7-7). Click on a policy to edit or click on the "+" to add a new policy.

There are two types of policies—Access Control and Application.

The rules you add to a policy are processed in a top-down fashion, as previously described. You should be aware of the hierarchy. Policies can only be modified from the group or the device where it was created.

Access Control rule

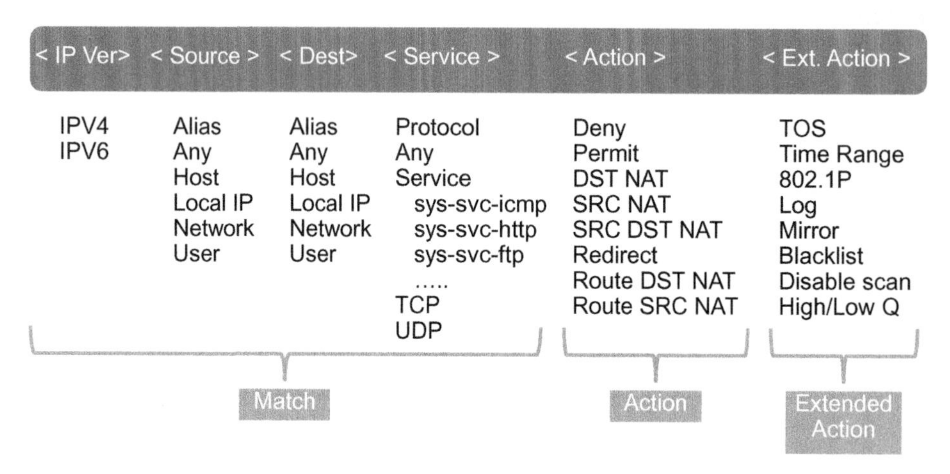

Figure 7-8 Access Control rule

Access Control rules can be created for both IPv4 and IPv6 traffic. The source and destination can be for an individual user, a subnet or network, any, Local IP, or an alias (Figure 7-8). An alias is a meaningful name you apply to an entity. The service can be selected from predefined services like DHCP, a specific port like TCP 21.

The various actions include permit, drop, NAT, and route. Then the extended actions such as log, mirror, and time range will be executed along with the main action. Add the rules as required.

For example, Figure 7-8 shows a rule that permits ANY source address, to access ANY destination address, but only for a specific time range, which you can define.

Configuring service rule in policies

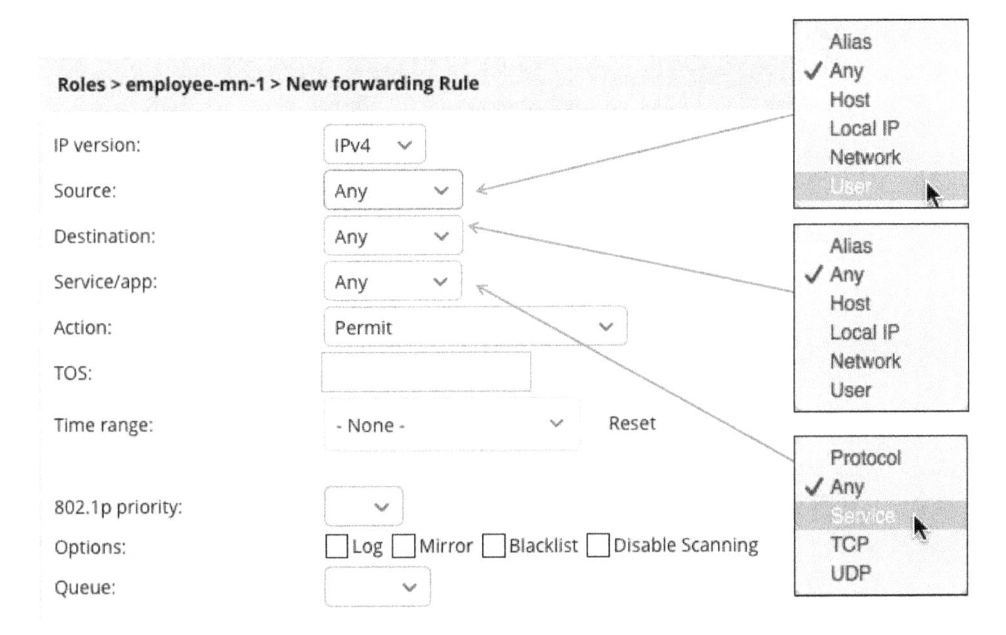

Figure 7-9 Configuring service rule in policies

The Figure 7-9 shows rule configuration via the GUI. A rule is being added to the policy employee-mn-1.

Use the drop-downs to select the parameters shown, as previously described.

Application rule

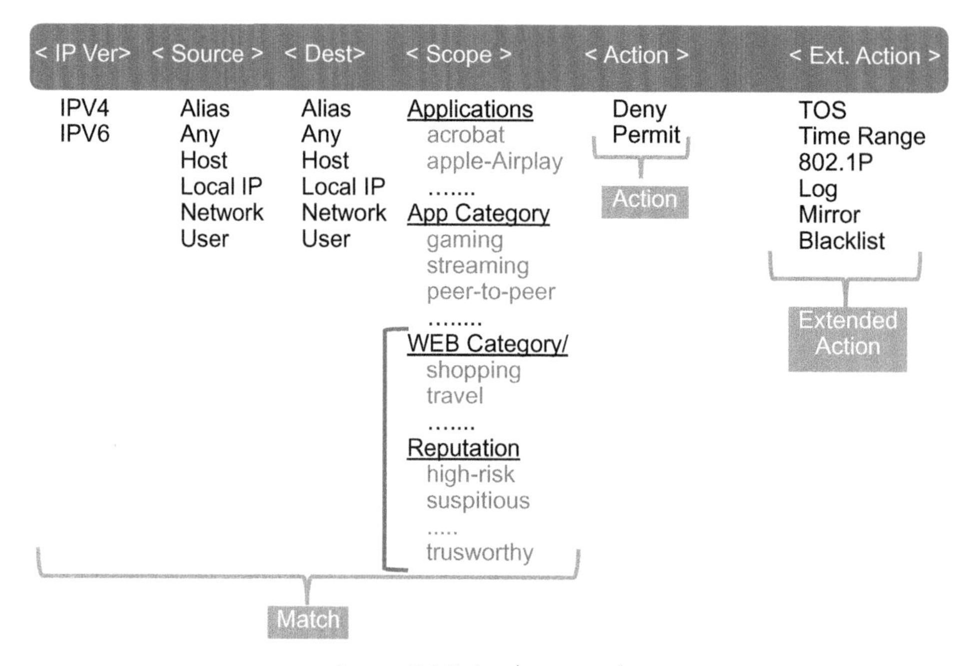

Figure 7-10 Application rule

You just learned about Access Control rules, which are based on specific IP protocols and ports. As the name implies, Application rules are based on applications (Figure 7-10). You specify source and destination, just like an Access Control rule.

Then you configure service scopes, which fall into three categories.

- Application scopes deal with applications such as acrobat, apple-airplay, and many more.
- The Application Category breaks down into gaming, streaming, and many others.
- Web Category and Reputation are grouped together. A category, like shopping or travel, is grouped with several reputation settings, from "high-risk" up to "trustworthy."

Only two actions can be configured—deny or permit. Then the extended action can be applied.

Configuring application rule in policies

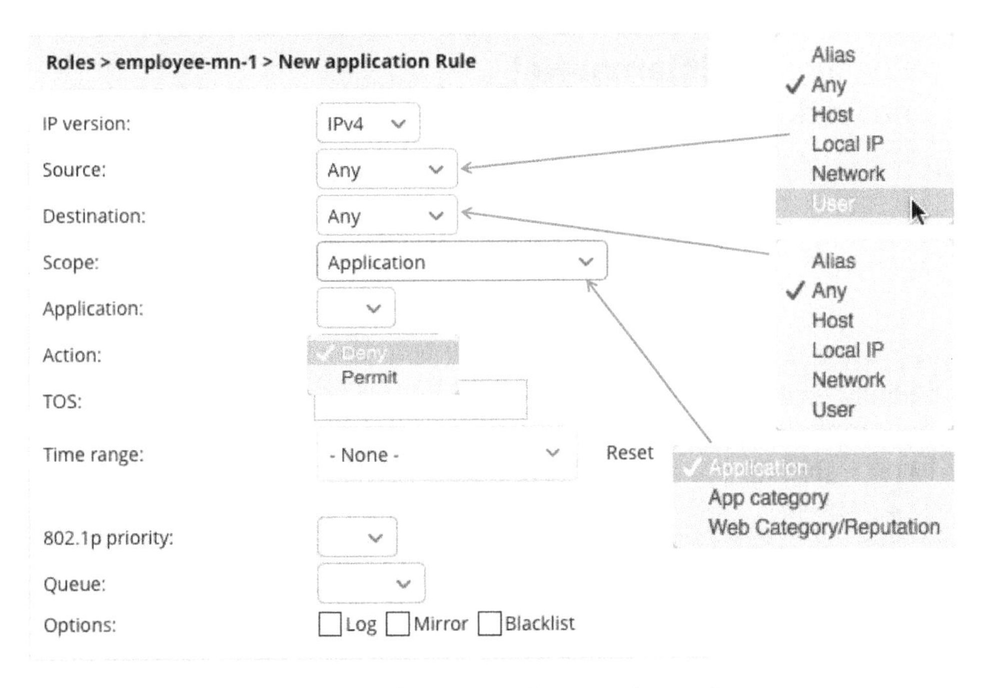

Figure 7-11 Configuring application rule in policies

The Figure 7-11 shows the application rule GUI page. The source and destination have the same options as Access control rules—alias, any host, Local IP, network, and User. The scope is one of the three categories, as previously explained.

Deny and permit are the only two actions allowed.

Aliases

Aliases add a self-documenting facet to your rules and increase their readability. For example, you can allow access to your email server by referencing its IP address—10.254.2.50. Or you can create an Alias named "Email-Server" that references the 10.254.2.50 address.

So the rule would be "ipv4 user Email-Server permit." The intent of this rule is quite obvious.

Aliases also ease and scale your configuration efforts. If you change the IP address of your email server, or add another one, you do not need to modify the multiple policy rules where it is referenced. You simply update the Alias. All rules using this Alias are automatically updated.

There are Destination Aliases and Service Aliases. A few predefined destination aliases exist, and you can create your own.

Service Aliases are predefined with all the common protocols, such as HTTP, HTTPS, DHCP, ICMP, and more. Again, you can create your own.

Aliases improve workflow scalability

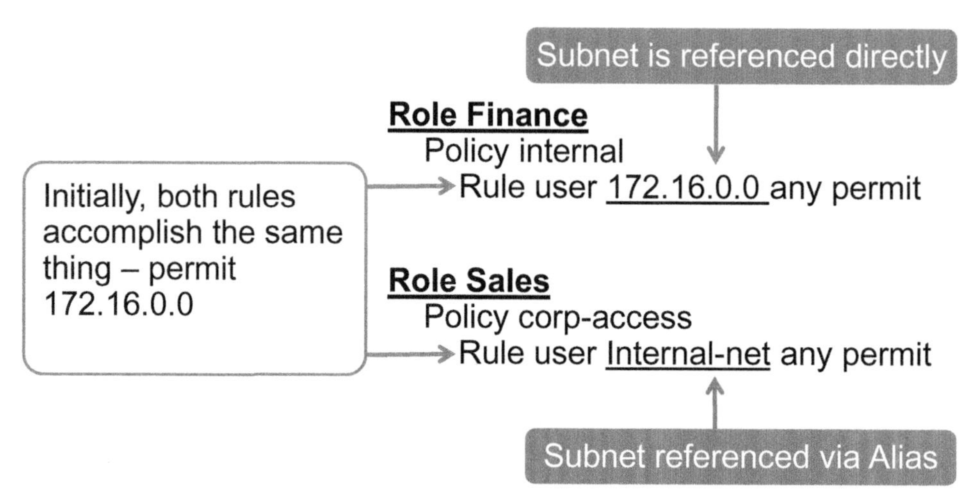

Figure 7-12 Aliases improve workflow scalability

Aliases ease the process of creating and maintaining policy, making it more scalable.

One reason for this is that they are self-documenting, as previously described. An alias of "Internal Network" is more intuitive than "172.16.0.0."

In Figure 7-12, two roles are shown, and each references the internal network. Role Sales reference subnet 172.16.0.0 via the Destination Alias. However, role Finance directly references the subnet. This creates an inconsistency in how the subnet is referenced and is not a best practice. To see why, look at Figure 7-13.

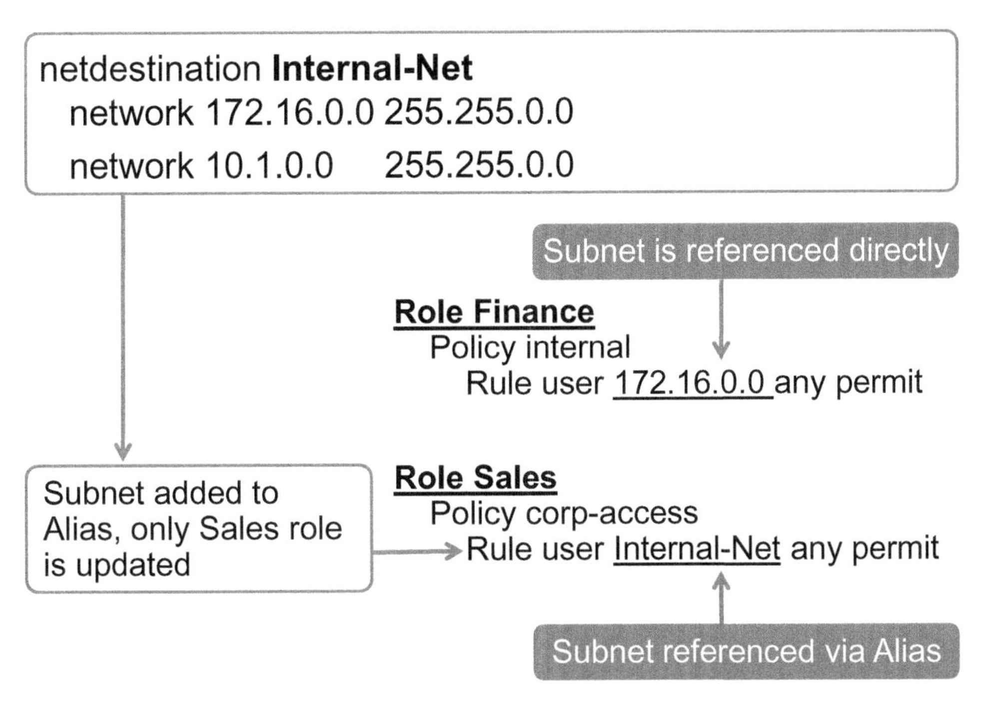

Figure 7-13 Aliases improve workflow scalability

Figure 7-13 shows a scenario where several months have passed, and you have subnet 10.1.0.0/16 to your internal networks. When you add this entry to the Alias, all roles that reference this alias are automatically updated with the new network. This is the big scalability advantage of an alias. This means that employees who need access to the new subnet will automatically get it. Guests' roles that reference this alias will automatically be denied access.

However, users assigned to the finance role will not have access to the new subnet, since the alias was not used for this role.

If you use an alias, use that alias everywhere it is needed. Otherwise, you may update the alias, and falsely assume the internal network is updated in every policy rule.

Predefined destination aliases

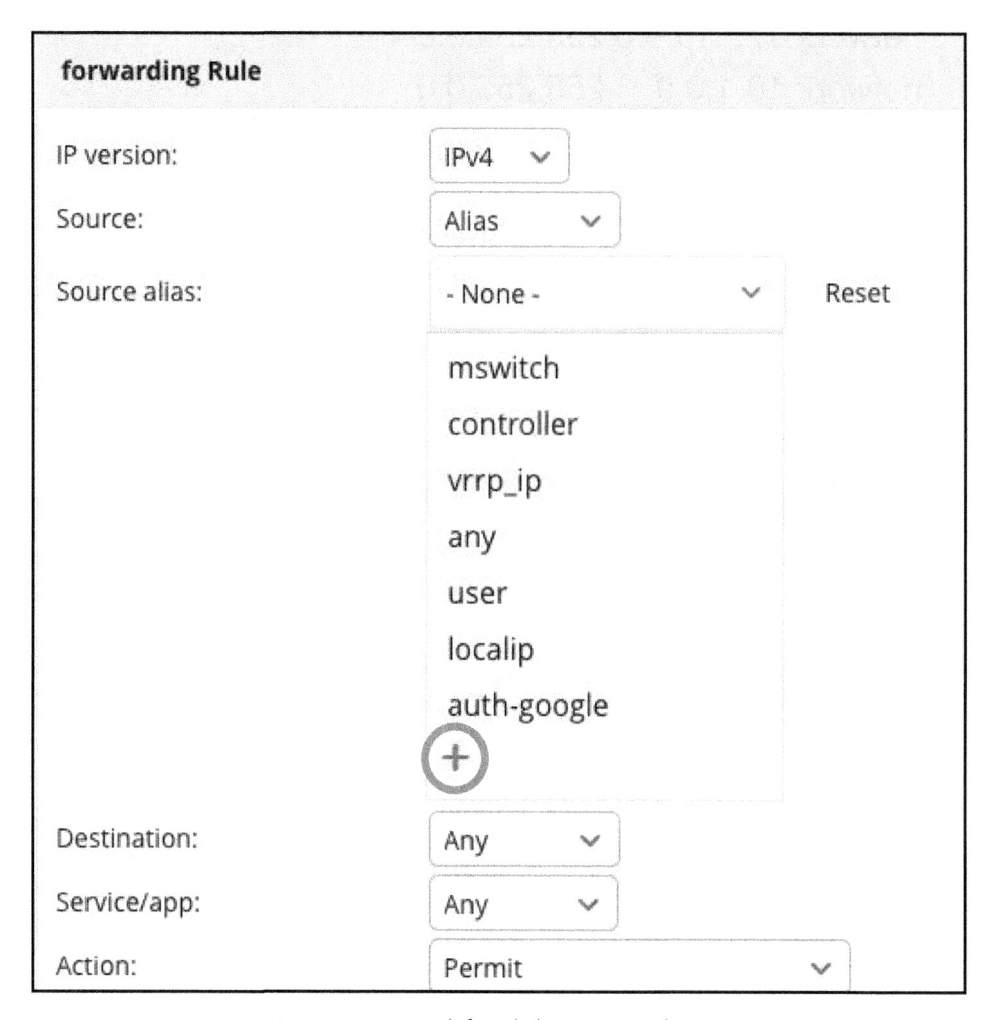

Figure 7-14 Predefined destination aliases

The following is a list of *predefined* source/destination aliases:

- **User**—This is a very important alias and is explained in detail next.
- **Mswitch**—This translates to the MM primary IP address.
- **localIP**—This translates to the RAP local IP address.
- **Controller**—This translates to the Controller IP address.
- **Auth-google**—This is used for Google authentication in certain cases.

- **Auth-facebook**—This is used for Facebook authentication in certain cases.
- **Wificalling-block**—This is used to block voice calls.
- **Vrrp_Ip**—This references the IP address used for the Virtual Router Redundancy Protocol (VRRP)
- **Any**—This translates to match any address.

To add your own administratively defined alias, click the "+" sign at the bottom Aliases drop-down menu, as shown in the figure.

When creating a forwarding rule, you can select some destination, a service or application, and Action, as shown in Figure 7-14.

Alias USER

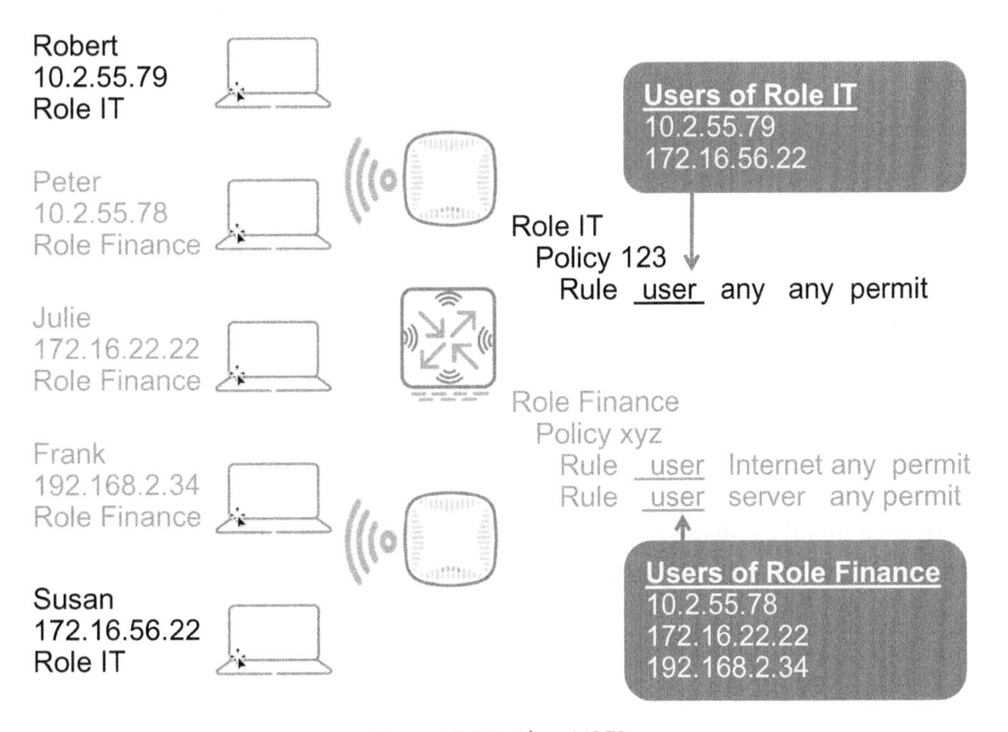

Figure 7-15 Alias USER

Figure 7-15 shows a scenario where Robert and Susan are on different subnets. They are both users assigned to the IT role. All of their traffic is processed through the policies and rules of the IT role.

Peter, Julie, and Frank are also on different subnets and assigned to the Finance role. Of course, all of their traffic is processed through the policies and rules of the Finance role.

Peter and Robert are on the same subnet but are assigned to different roles, and you know that their traffic is processed based on role assignment, not on AP or subnet connectivity.

User versus any

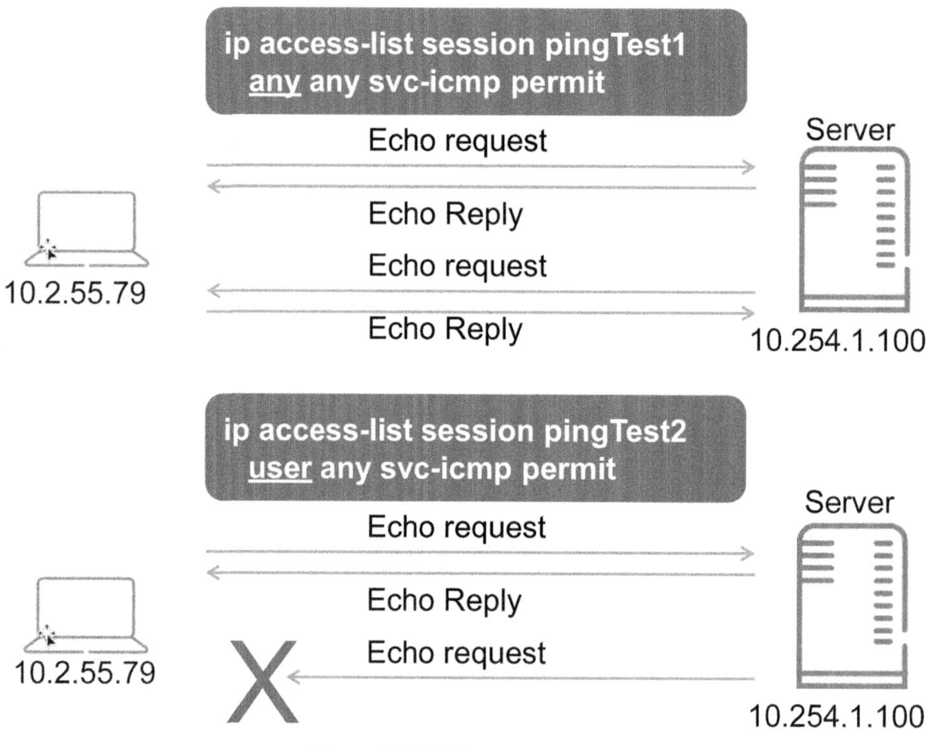

Figure 7-16 User versus any

Figure 7-16 illustrates the difference between alias Any and alias User. The top rule has an "any" for source and therefore, the user can ping the server and the server replies. But if a ping is generated from the server, this will also be accepted.

The bottom rule uses the alias User. The User can ping and get a reply since this is a stateful firewall. However, the server is not a user, within the Mobility Controller user database, and therefore not allowed to originate an echo request.

Service alias

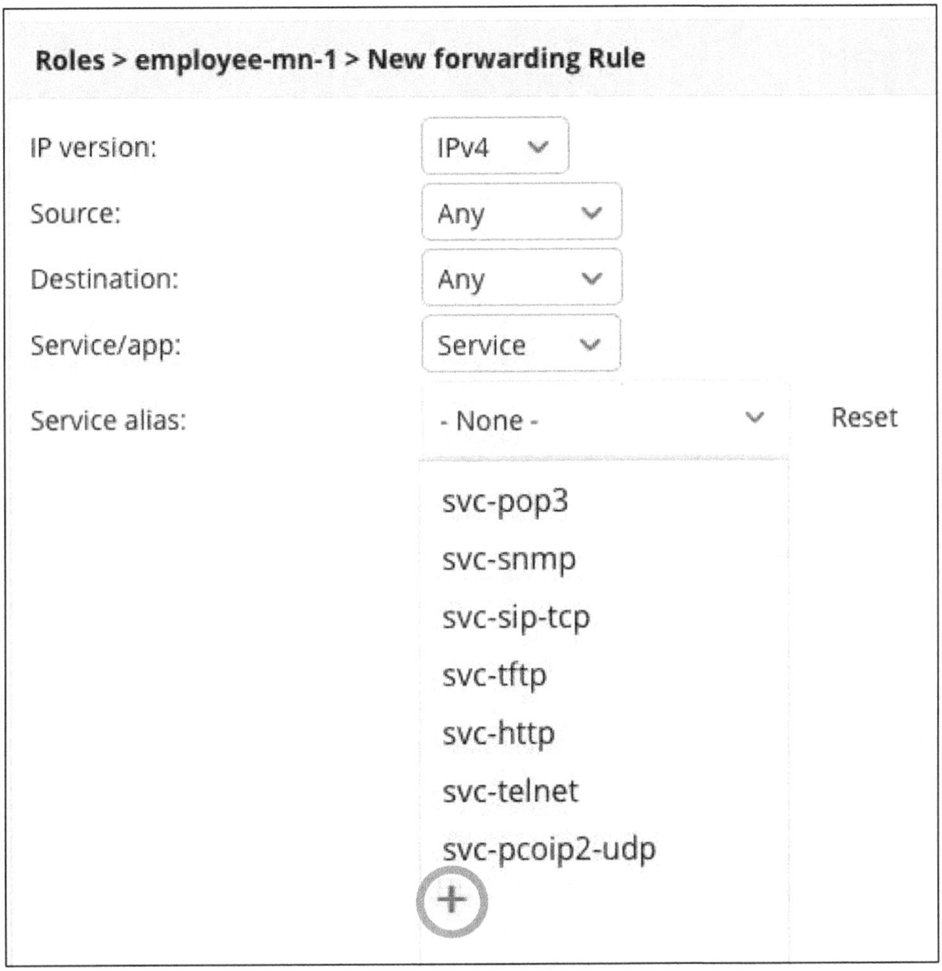

Figure 7-17 Service alias

There are a large number of predefined services (Figure 7-17). Most common protocols are predefined; however, you may wish to create your own service alias. To add a new service alias, click the "+" sign at the bottom of the Service Alias drop-down menu, as previously shown in Figure 7-14. A pop up window lets you define your own services based on TCP, UDP, or specific protocols.

Learning check

3. What are the four match criteria for Access Control rules?
   a. IP version
   b. Source
   c. Destination
   d. Scope
   e. Service

Answers to learning check

3. What are the four match criteria for Access Control rules?
   a. IP version
   b. Source
   c. Destination
   d. Scope
   e. Service

Practice: What is wrong with these firewall policies?

Example 1: Wireless users are not getting IP addresses

```
ip access-list session DHCP_rules
  user any udp 68    deny
  user any svc-dhcp permit
```

Example 2: Wireless Guest users cannot access the internet

```
netdestination Internal-Network
        network 172.16.0.0 255.255.0.0

ip access-list session Guest_Access
        user host 172.16.15.2 svc-dns permit
        user host 172.16.16.2 svc-dns permit
        any alias Internal-Network deny
        any  user any permit
```

Figure 7-18 Practice—What is wrong with these firewall policies?

Take a few moments to analyze the policies shown in Figure 7-18. Can you describe how these policy examples should be improved? Write your answers below, or jot them down in a notepad.

____

____

____

____

____

____

____

____

Solution to policy examples

Example 1: Wireless users are not getting IP addresses

ip access-list session DHCP_rules
user any udp 68 deny
~~user any svc-dhcp permit~~ Wrong
any any svc-dhcp permit

Example 2: Wireless Guest users cannot access the internet

netdestination Internal-Network
network 172.16.0.0 255.255.0.0

ip access-list session Guest_Access
user host 172.16.15.2 svc-dns permit
user host 172.16.16.2 svc-dns permit
any alias Internal-Network deny
~~any user any permit~~ Wrong
user any any permit

Figure 7-19 Solution to policy examples

Example 1

The first rule is to stop rogue DHCP servers (Figure 7-19). Example—Wireless Client laptops with DHCP server configured.

The second rule is the problem. The reason users fail to get an IP address is because the server is not a user. A DHCP request is a broadcast. So when the server sends a DHCP reply, its a new packet from the server to the user. The Server is not a user in the network. Rectify this as shown in Figure 7-16.

Example 2

The fourth rule is incorrect. This rule status that the source "any" is permitted to access destination "user," using any service. Therefore guest-to-guest traffic is allowed. This should not be the case, as a guest could potentially attack another guest.

Also, there is no rule to allow guest Internet access. This is typically the main reason for creating a guest network. To rectify this, the statement at the bottom of Figure 7-19 was added "user any any permit."

Global and WLAN policies

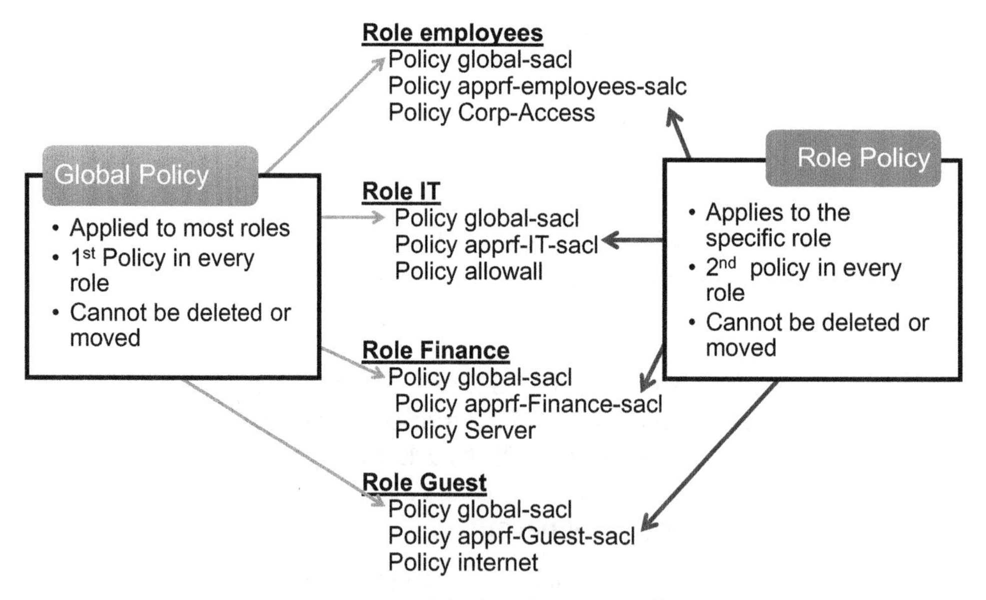

Figure 7-20 Global and WLAN policies

By default, two Session ACLs (SACL) are added to every role—the global-sacl and the apprf-<role>-sacl. These session ACLs cannot be deleted (Figure 7-20). The global-sacl is a policy that applies to all Roles, as shown in Figure 7-17. For example, if you created a global policy that denied the application "gaming," then all Roles will deny gaming.

The apprf-<role>-sacl is an application limitation placed on a specific user role.

In the global and apprf role session, you can define APP priorities, bandwidth limitations, or even blocking rules.

If there are no rules in these two policies, then the next policy in the role is examined for a match.

Global rule configuration

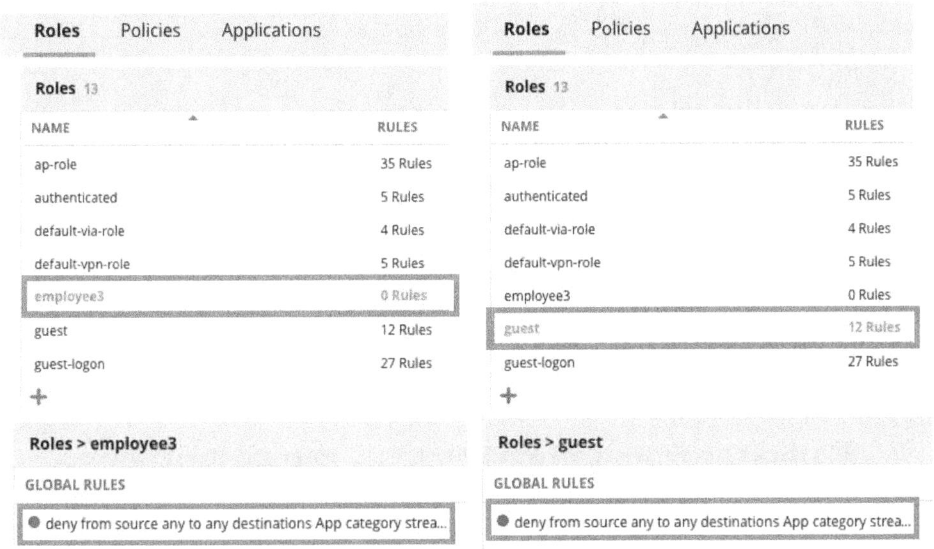

Figure 7-21 Global rule configuration

The global rule can be created in any role. Once a global rule has been added to a role, it is propagated to all roles.

Figure 7-21 shows how the Role employee3 and the role Guest have the same global rule.

Rule for this role

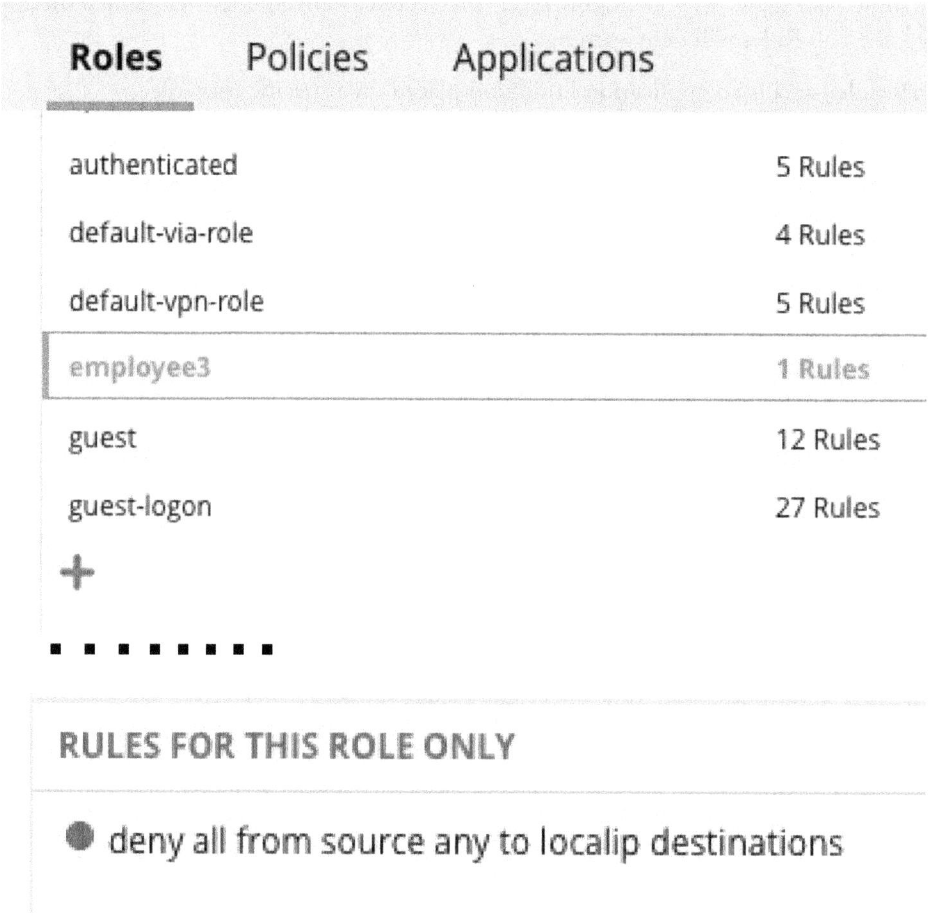

Roles Policies Applications

authenticated	5 Rules
default-via-role	4 Rules
default-vpn-role	5 Rules
employee3	1 Rules
guest	12 Rules
guest-logon	27 Rules

RULES FOR THIS ROLE ONLY

- deny all from source any to localip destinations

Figure 7-22 Rule for this role

Under Roles you have the area "Rules for this Role only." All rules added will be placed in a policy with the same name as the role (Figure 7-22). Unlike other policies that can be shared between roles, this Role policy is only applicable to this role. This is normally found in the third position, but can be moved.

Learning check

4. A global rule will affect what roles
   a. The default role
   b. The role assigned to that policy with that rule
   c. All roles in the network
   d. All roles in an AP-group
   e. All roles in an MD

Answers to learning check

4. A global rule will affect what roles
   a. The default role
   b. The role assigned to that policy with that rule
   c. **All roles in the network**
   d. All roles in an AP-group
   e. All roles in an MD

Roles

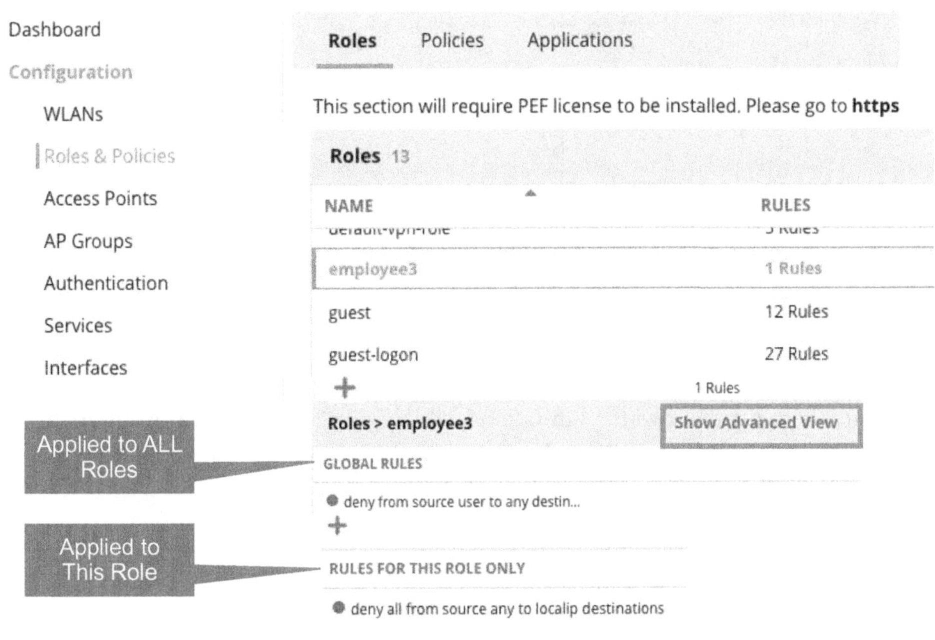

Figure 7-23 Rule for this role

Roles can be defined in the system group or any subgroup Figure 7-23. The role is propagated down the hierarchy. In the Role GUI page, add new roles by clicking on the "+" sign. You can also add global rules, which will be propagated to all roles. Rules for this role will create a policy with the same name as the role name. All defined rules are placed in this policy.

To see more options, click **Show Advanced View**.

Role advanced view

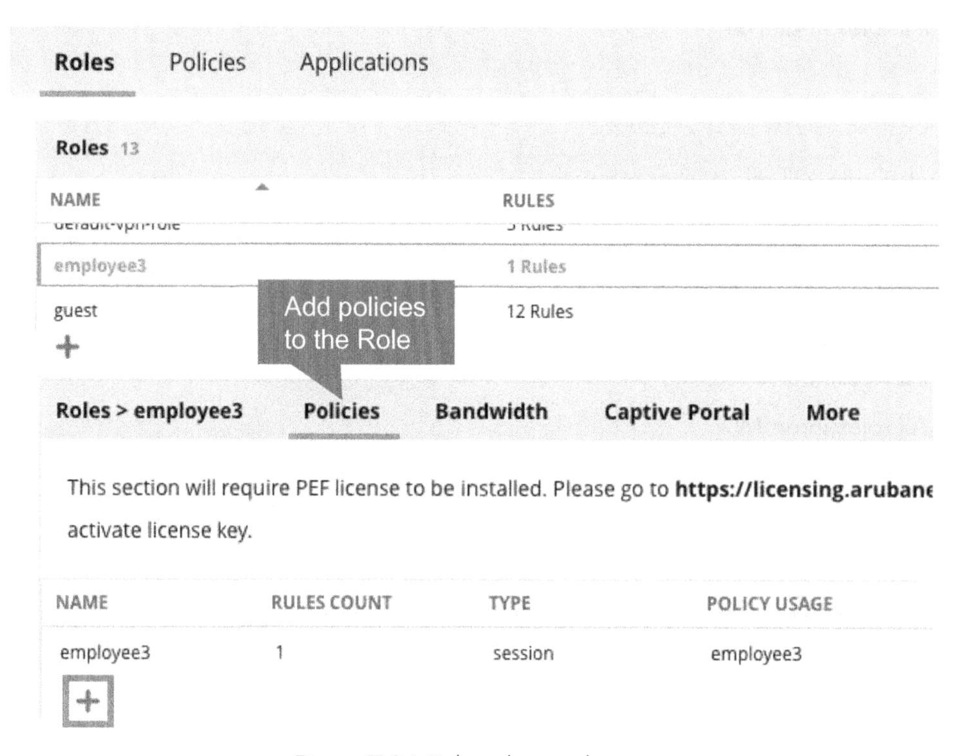

Figure 7-24 Role advanced view

The Roles advanced view lets you add new policies to the selected role. In the example shown, there is already a policy named employee3 for the employee3 role (Figure 7-24). This was created in the Basic View under "Rule for this Role only" section. Click the "+" sign to add or create other policies for this Role.

Another option is to setup Bandwidth limitations, which create per-application limits for the Role. You can also have an exception list to create exceptions to defined rules. This is not shown in Figure 7-24.

If you are editing the guest role, you can also select the desired captive portal type. This is explained in more detail in the Captive Portal chapter.

Adding policies to roles

Add Policy

Add an existing policy
Create a new policy
Policy name:
Policy type: Session
Position:
Cancel
Submit

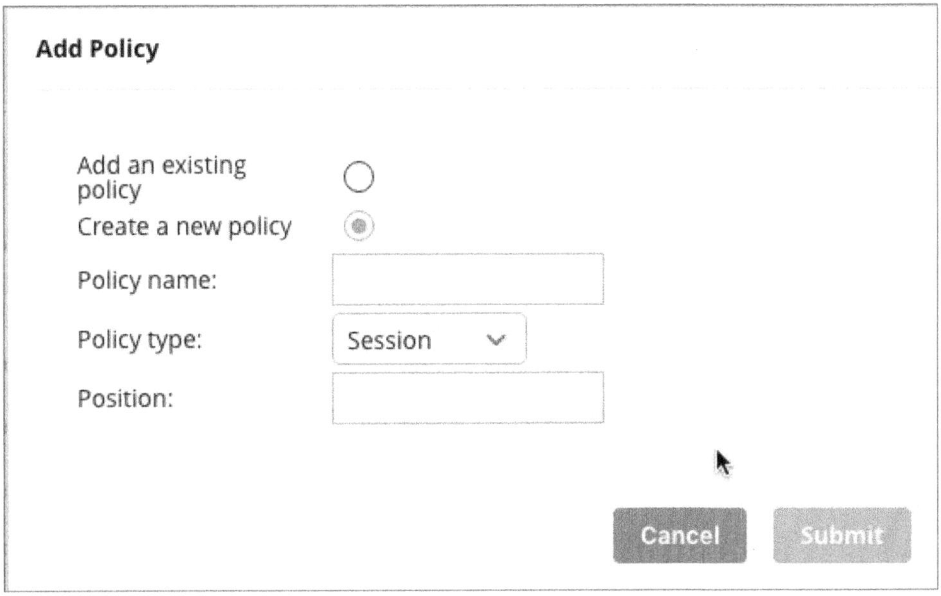

Figure 7-25 Adding policies to roles

When you click on the "+" symbol to add a new policy to a Role, the Add Policy window appears, as shown in Figure 7-25. You can add to an existing policy or create a new one. The policy type should always be a session. All other selections create standard-type access lists that are not stateful, which is not recommended.

When creating a new policy, you can also select its position in the Role.

WLAN default role assignment

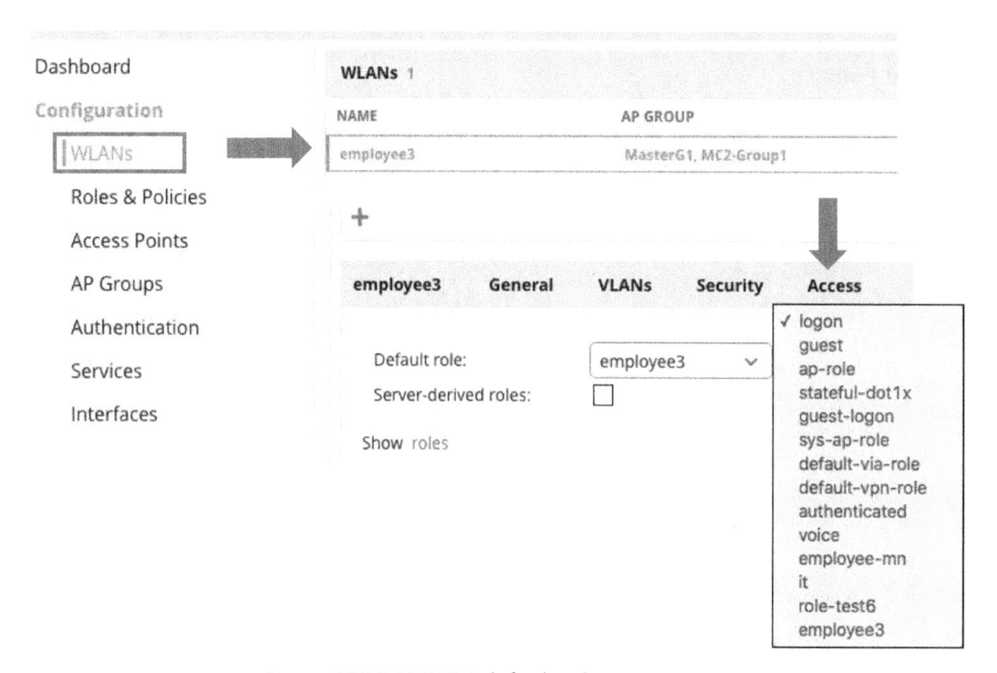

Figure 7-26 WLAN default role assignment

When you create the WLAN, you can assign roles (Figure 7-26). To assign a role after the WLAN is created, select the WLAN, go to Access, and modify the Default Role, as shown. The default role will depend on the authentication method of the WLAN. The default role is assigned to the user on a WLAN, when no role is assigned by an external server.

AAA profile

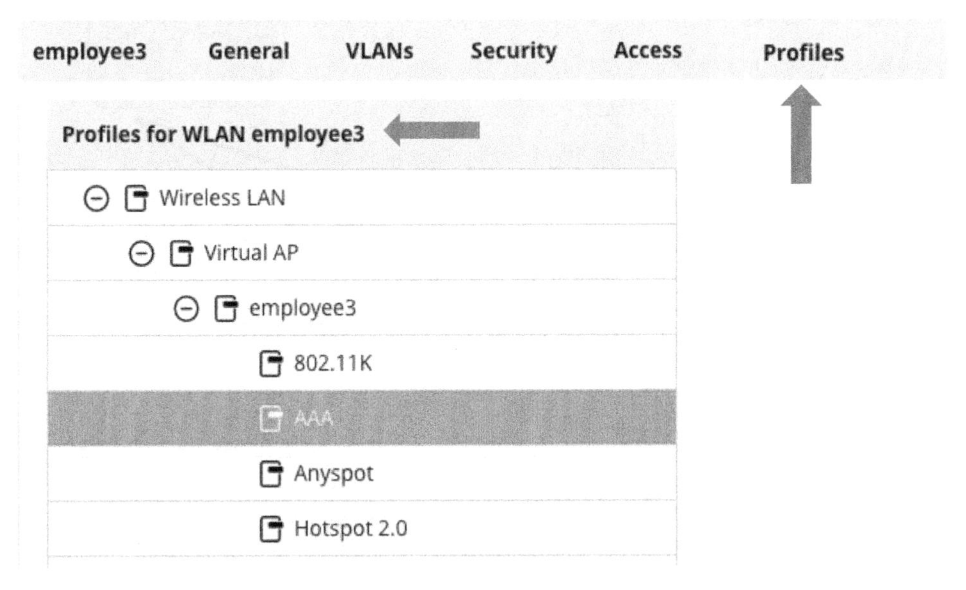

Figure 7-27 WLAN profiles

The Aruba WLAN configuration structure consists of profiles. Every WLAN has a VAP profile, which references an SSID profile and a AAA profile. You can see these profiles by clicking on the profile tab. Figure 7-27 shows the empoyee3 Role has been selected.

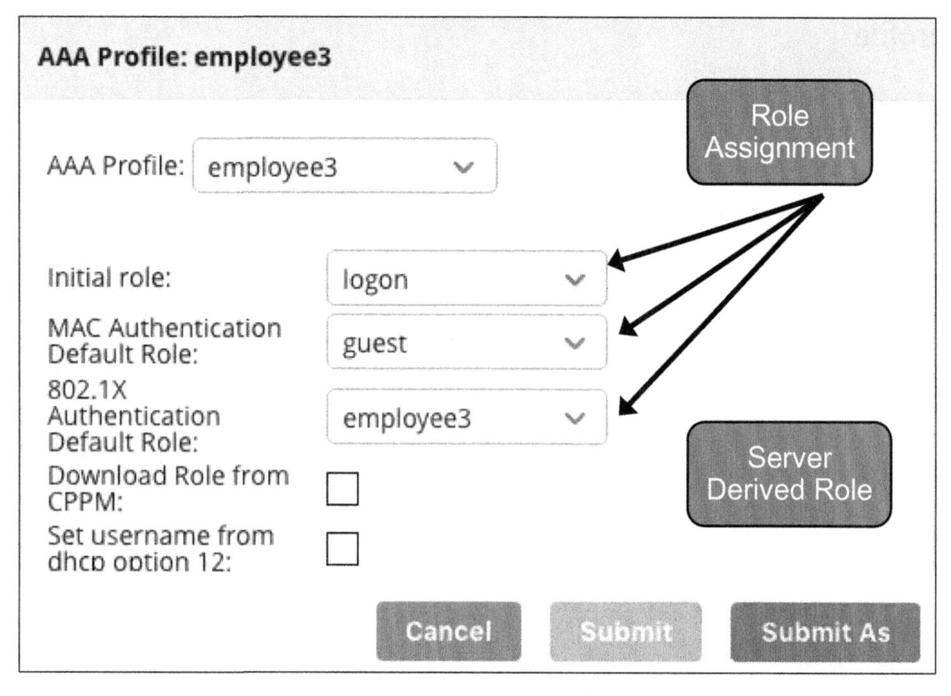

Figure 7-28 WLAN profiles

Figure 7-28 shows the details of the AAA profile for employee3. The employee3 role uses the 802.1x Authentication Default Role because the WLAN uses 802.1X-based security methods.

Remember, any role returned by an external RADIUS server overrides other, locally derived roles. If the RADIUS server simply returns an access-accept message, then the user is placed in the 802.1x Default role.

Learning check

5. What four methods can be used to assign a Role to a user?
    a. Initial role
    b. Locally derived
    c. Authentication default role
    d. Server derived role
    e. AP derived role

Answers Learning check

5. What four methods can be used to assign a Role to a user?
    a. **Initial role**
    b. **Locally derived**
    c. **Authentication default role**
    d. **Server derived role**
    e. AP derived role

8 Dynamic RF Management

LEARNING OBJECTIVES

✓ This chapter is focused on HPE Aruba's Adaptive Radio Management (ARM). You will learn how ARM automatically manages and optimizes the RF environment, under ever-changing, dynamic conditions. This includes a discussion of key ARM features, such as AirMatch and Legacy ClientMatch. You will also learn how ClientMatch works in AOS8, which uses an MM/MD solution.

Dynamic RF management introduction

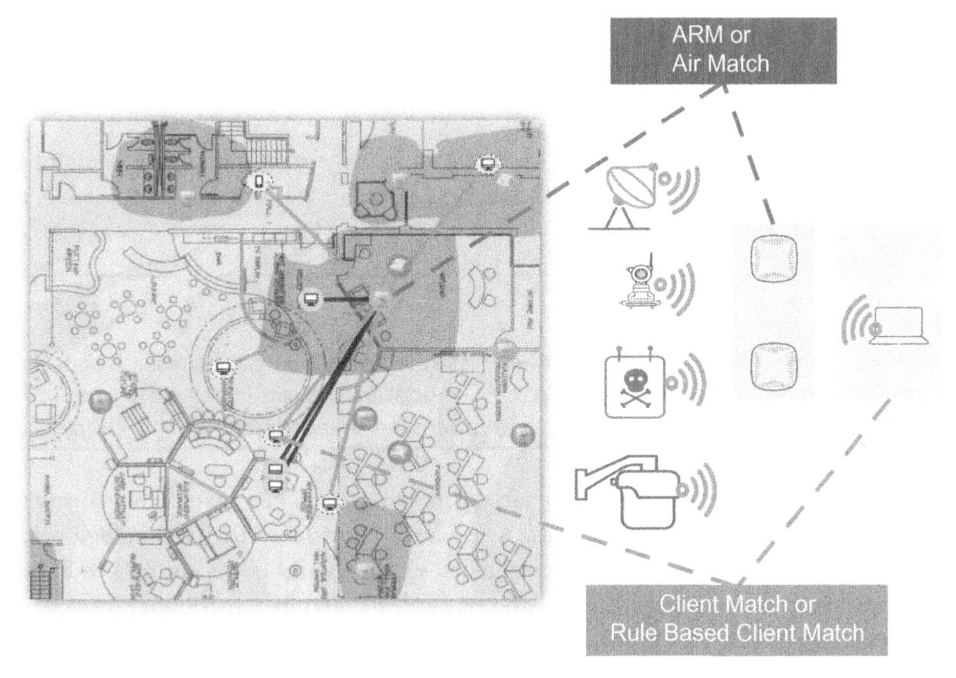

Figure 8-1 How ARM works

In the real world, all kinds of devices are in the RF environment. Some of these devices constantly transmit RF energy, while others emanate RF energy in bursts. Either way, this can negatively impact WLAN performance. Dynamic RF management helps to ensure that each AP is configured with optimal channel and power settings. Dynamic RF management enhances a client's ability to find and connect to the best available AP.

Figure 8-1 shows how HPE Aruba's AirWave management product provides insight into AP coverage patterns. Each round AP icon is surrounded by areas of red, orange, and green. These colors represent areas of strong, good, weak, and unacceptable coverage.

This coverage can be affected by interferers, such as video cameras, rogue APs, radar systems, and more. ARM features can adjust AP power and channels in response to this and other changes in the RF field. AirMatch and ClientMatch can help client devices find and attach to the best AP.

Adaptive Radio Management (ARM)

ARM dynamically manages the RF spectrum to choose the best 802.11 channel and transmit power for each Aruba AP. As you add or remove APs, build or tear down walls, ARM automatically adapts. ARM analyzes everything in the RF environment—the Aruba APs, rogue APs, and any other sources of interference. Aruba APs coordinate these changes with the controller and can use Over-the-Air updates to help each other.

Over the Air (OTA) Updates allow APs to get RF environment information from neighbor APs. This works even when the AP cannot scan. If you enable this feature, APs send OTA updates every time they do an off-channel scan.

These updates are sent toward neighbor APs in an 802.11 management frame. This contains information about that AP's home channel, the home channel's current transmission EIRP value, and one-hop neighbors seen by the AP.

Note

As of AOS 8.0, ARM is only supported in standalone mode or a Master/Local deployment. ARM in AOS 8.x using the MM/MC model utilizes an improved Dynamic Radio Management technology, AirMatch, to be discussed in more detail in this chapter.

ARM technology addresses the challenges of large deployments and dense deployments with standalone and Master/Local deployment features. As capacity increases, ARM technology provides the ability to keep 802.11n and 802.11ac APs on the best channels and power settings. It continuously learns and implements an optimized channel and power level plan in a distributed network. It is also continuously aware of the needs of Voice and Video services.

How ARM works

APs scan their current (home) channel and report their channel and power information to the controller to help create an Interference Index (sum of SNRs) and Coverage Index (a number representing

RF overlap). The Interference Index and Coverage Index are compared against a configurable Ideal Index. Should certain thresholds be met, the AP seeks to change channels and/or power levels. Under normal scanning, APs could find that other channels are better than its existing channels, by a configurable factor.

Specific aspects of ARM operation are configured in ARM profiles. For example, the **scan-interval** parameter controls how often APs scan. If the AP has no associated clients (or if most clients are inactive), ARM will dynamically readjust this default scan interval. This allows the AP to obtain better information about its RF neighborhood by scanning nonhome channels more frequently.

If an AP makes an unsuccessful attempt to scan a nonhome channel, it makes additional rescan attempts before skipping it and continuing on to other channels.

Indices in ARM profile

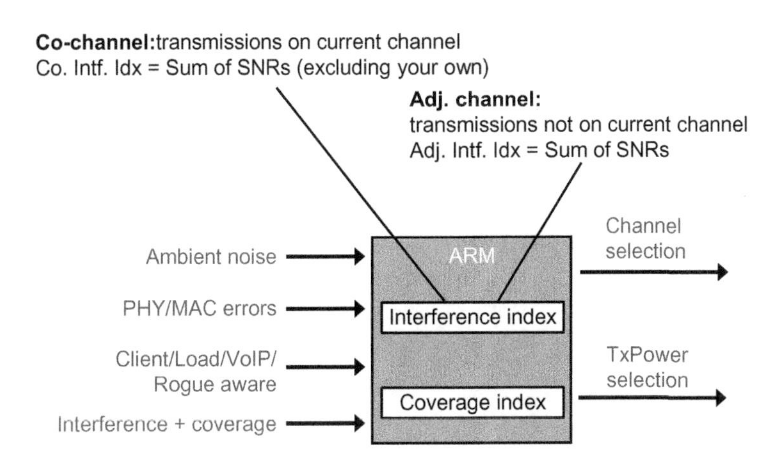

Figure 8-2 Indices in ARM profile

ARM uses the Interference Index and Coverage Index to select optimal parameters (Figure 8-2). APs use the Interference Index to measure co-channel interference, including third-party APs on specific channels, which is then calculated and weighted for all APs. APs use the Coverage Index to measure RF coverage, calculated and weighted for all Aruba APs seen on a specified channel.

ARM compares the co-channel interference index on current and adjacent channels. If an adjacent channel has significantly less interference than the current channel, the AP can change to a new channel. The system must prevent an AP from wildly bouncing between channels because the client traffic can only be associated with an AP and sending traffic while the AP remains on a fixed channel. To help prevent the AP from bouncing between channels, eight complete scans of the new channel are completed before an AP can change channels. The AP can also update its power setting, if it detects very low co-channel interference, from other APs on the same channel.

If the Client Aware option is enabled, the AP would not change channels while active clients are associated. "Activity" is defined by the "inactivity-time" parameter, as specified in the IDS general profile. By default, active clients are those that have sent or received traffic within the last 60 seconds. If Client Aware is disabled, the AP may change to a more optimal channel. However, this change may disrupt current client traffic. This feature is enabled by default.

If you have enabled both the Scanning and Rogue AP options, Aruba APs may change channels to contain off-channel rogue APs with active clients. This security feature allows APs to change channels even if the Client Aware setting is enabled.

Note

This setting is disabled by default and should only be enabled in high-security environments, where security needs take precedence over the desire for low network overhead.

You may prefer to receive Rogue AP alerts via SNMP traps or syslog events. This feature is disabled by default.

ARM optimizing channel and power

```
(Aruba7205) [mynode] #show ap arm history ap-name AP115

Interface :wifi0
ARM History
-----------
Time of Change       Old Channel  New Channel  Old Power  New Power  Reason  Result
--------------       -----------  -----------  ---------  ---------  ------  ------
2016-11-12 06:05:00  157+         157+         15         18         P+      Configured

Interface :wifi1
ARM History
-----------
Time of Change       Old Channel  New Channel  Old Power  New Power  Reason  Result
--------------       -----------  -----------  ---------  ---------  ------  ------
2016-11-12 05:54:22  11           11           6          9          P+      Configured
2016-11-12 05:52:32  1            11           6          6          I       Configured

I: Interference, R: Radar detection, N: Noise exceeded, Q: Bad Channel Quality E: Error tl
 exceeded, INV: Invalid Channel, G: Rogue AP Containment, M: Empty Channel, P+: Increase I
-: Decrease Power, 40INT: 40MHZ intol detected on 2.4G, NO40INT: 40MHz intol cleared on 2
: Turn off Radio, ON: Turn on Radio, D: Dynamic Bandwidth Switch, I*: CCA Interference
(Aruba7205) [mynode] #
```

Figure 8-3 ARM optimizing channel and power

The **show ap arm history** CLI command is a good way to see changes in AP channel and power settings.

In Figure 8-3, the AP's 2.4 GHz radio has changed channels, from 1 to 11. This is because it detected interference. Once the AP was on channel 11, it experienced less co-channel interference, likely because other APs on the same channel are farther away. Thus, the AP increased power setting.

General ARM profile configuration

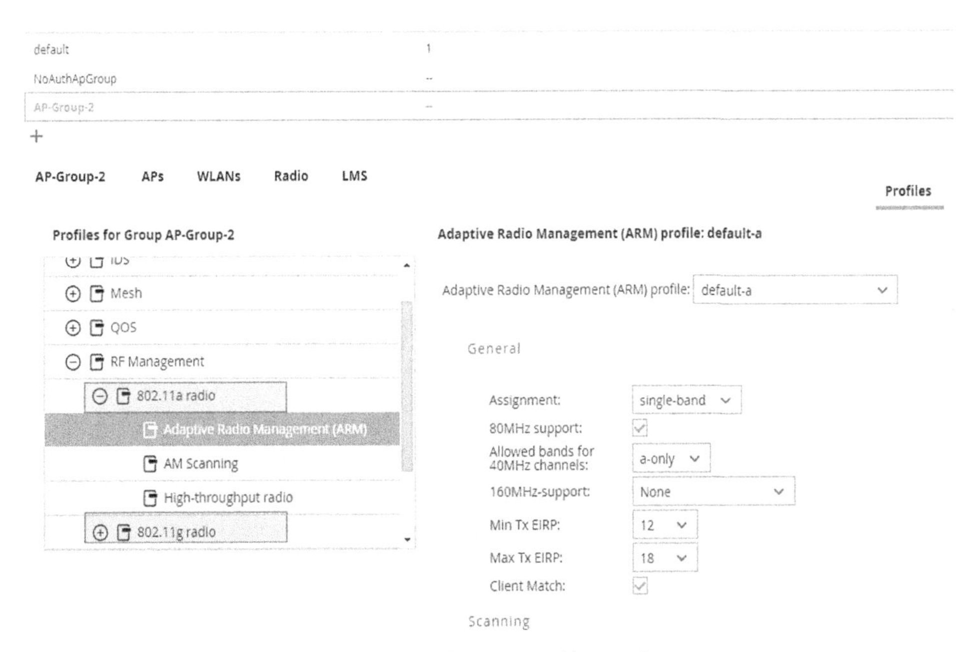

Figure 8-4 General ARM profile configuration

In each AP group, there is one ARM profile for the 802.11a (5 GHz) radio and another for the 802.11g (2.4 GHz) radio. This gives you great configuration flexibility. The different profiles can accommodate the unique characteristics of each band. You can also configure different ARM settings for different AP groups.

To view the ARM profiles, choose **Configuration > AP Groups** and choose an AP group to view. From the AP Group window, choose **Profiles > RF Management** and expand either the 802.11a radio or 802.11g radio. From there, click ARM to see which ARM profile is being used, and what its settings are. The Max Tx EIRP and the Min Tx EIGRP can be a value between 3 and 33 dBm, in 3 dBm increments. The number 127 has a special meaning; it indicates a regulatory maximum. Figure 8-4 shows part of this configuration.

The default ARM profile is best for most networks, and so it is recommended, as is. Only those with extensive radio knowledge should modify this profile. Both scanning and client match are on by default.

Advanced ARM profile configuration

The following provides a brief introduction to some of the advanced settings for the ARM profile. It is typically best to leave these at their default values. To see these settings you would navigate exactly as described above, for Figure 8-4. Instead of expanding the "General" section, you simply expand the "Advanced" section.

- **Client Aware**—Shows if the client aware feature is enabled or disabled. When enabled, the AP does not change channels when there are active clients.
- **ARM over the Air Updates**—The feature allows an AP to get information about its RF environment from its neighbors, even if the AP cannot scan. If you enable this feature, when an AP on the network scans a foreign (nonhome) channel, it sends an Over-the-Air (OTA) update in an 802.11 management frame. This frame contains information about that AP's home channel, the current power setting of the home channel, and one-hop neighbors seen by that AP.
- **Backoff Time**—Time in seconds that an AP backs off after requesting a new channel or power level. When the AP changes channel or power settings, it waits for the backoff time interval, and then requests a new channel or a power setting.
- **Mode Aware Arm**—If enabled, ARM will turn APs into Air Monitors (AMs) if it detects higher coverage levels than necessary. This helps avoid higher levels of interference on the WLAN. Although this setting is disabled by default, you may want to enable this feature if your APs are deployed in close proximity (for example, less than 60 feet apart).

ARM caveat

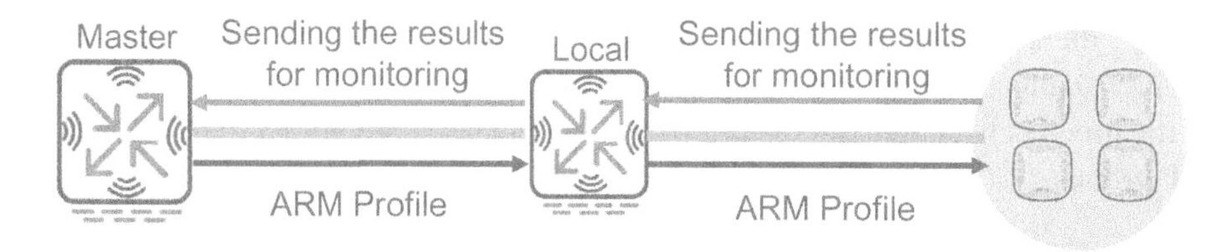

Figure 8-5 ARM caveat

ARM works fine in many cases, but there are caveats (Figure 8-5):

- ARM is responsible for AP channel and power-level assignment.
- Each AP calculates its own channel and power level based on RF information and its proximate radio neighbors.
- The RF info used for channel and power-level calculation is an instantaneous RF snapshot.

- Frequent channel and power-level changes might happen in an AOS 6.x network. This results in client disconnection and RF instability.
- Without a global view of the RF environment, the decentralized ARM mechanism also could cause uneven use of channels.

Learning check

1. In ARM who decides channel and power of an AP?
   a. Master Controller
   b. Local Controller
   c. Wireless Client
   d. AP self

Answers to learning check

1. In ARM who decides channel and power of an AP?
   a. Master Controller
   b. Local Controller
   c. Wireless Client
   d. **AP self**

AirMatch

What is AirMatch?

AirMatch is Aruba's next generation automatic RF planning service. This can replace the legacy solution—ARM's automatic channel and power assignment.

AirMatch provides unprecedented quality of RF network resource allocation. It consumes the past twenty-four hours of RF network statistics and proactively optimizes the network for the next day. RF changes are applied at 5 AM by default, but can be user-defined.

AirMatch reacts to detrimental RF events such as radar and high noise and minimizes channel and power changes. Network throughput is also predictively maximized.

AirMatch only runs on the MM, running AOS 8.0 or higher. For a standalone MD, or Controllers, running in Master/Local mode ARM will perform the RF optimization.

How AirMatch works

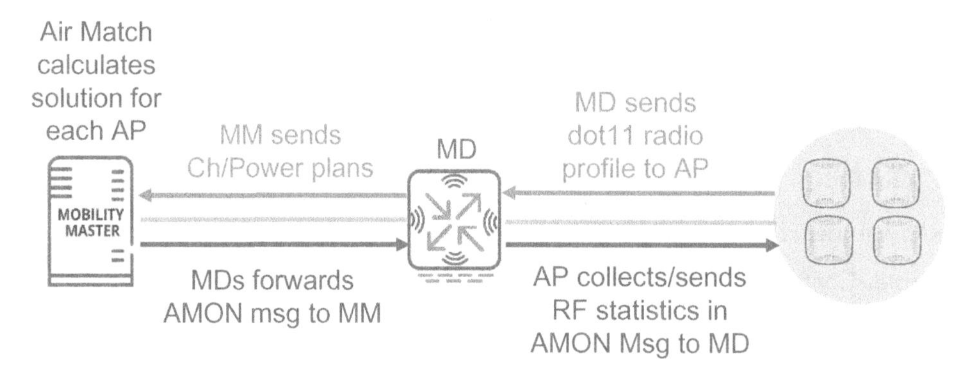

Figure 8-6 How AirMatch works

As of AOS 8.0, AirMatch replaces ARM's decentralized RF planning mechanism. AirMatch runs on the Mobility Master, as a centralized RF planning and optimization service. It models the network as a whole to devise better RF channel and power plans (Figure 8-6). It is one of the Loadable Service Modules (LSM) in the Mobility Master and so can be upgraded independently.

Let us review how it works:

1. APs periodically collect RF information and statistics about their RF neighborhood. This info is sent to the mobility controller/managed device via Application Monitoring (AMON) messages.
2. The MC forwards AMON messages to the MM.
3. AirMatch consumes this RF information and generates an RF solution to optimize channels, bandwidth, power, and mode of operation—per radio.
4. The MM sends the solution as CLI commands down to the MD.
5. The MD receives these configurations, creates a dot11 radio profile override, and sends it to the AP.
6. The AP receives this configuration and applies the changes to its radio configuration.
7. AirMatch calculates RF channel and power settings based on the past 24 hours RF information. Channel and power changes occur daily at 5AM—based on MM time zone.
8. AirMatch also minimizes channel coupling, where adjacent radios are assigned to the same channel.

AirMatch and ARM comparison

AirMatch channel and power optimization is superior to that of legacy ARM. AirMatch is only supported on the Mobility Master and, therefore, is only supported on AOS 8.0 and higher. Legacy ARM is still used by standalone controllers running ArubaOS 8.0.

The discussion below compares ARM and AirMatch. The discussion uses an analogy of how the companies might operate: ARM company and AirMatch company.

ARM company

ARM is like a boss (master or standalone controller) who makes an ARM policy (ARM profile). The employee (AP) makes its own work list and completes it under the policy, then reports their working process and results for monitoring.

AirMatch company

The AirMatch company has several robots (APs), which are more dependent on management. They listen for RF information and report it to the manager (MD). Then the manager reports to the CEO (MM). The CEO guides the AP's channel and power settings. However, if a radar or noise event occurs, the robots (APs) do have consciousness to avoid those. This new information would again be sent up the chain of command and appropriate configuration changes would be sent back down.

Table 8-1 provides a summary of the differences between AirMatch and ARM.

Table 8-1 AirMatch and ARM comparison

Features	AirMatch	ARM
Supported topology	Only on Mobility Master	Legacy controller
Computation	Centralized	Decentralized
RF information used	Past 24 hours	Instantaneous snapshot
Run period	Once per 24 hours by default	As often as 5 minutes
Outcome quality	Superb	Good
Deploy time	5 AM by default	Any second necessary
Optimization scope	The entire RF network	Each individual AP
Radar avoid	Yes	Yes
High noise avoid	Yes	Yes

AirMatch optimization

When the Mobility Master first boots, the AirMatch database is empty. The MM detects APs on the network and enters its initial optimization phase. It collects data from all APs and generates an incremental solution, every 30 minutes, for the next 8 hours. After this initial 8-hour period, a new RF configuration is pushed out every 24 hours.

During the initial 8-hour optimization phase, newly deployed APs join the network with its preassigned channel and power values. The AirMatch service detects the newly deployed AP, restarts its RF computations, and sends an incremental RF configuration to the new AP 30 minutes later. APs added after the initial 8-hour optimization period do not receive an update until the next scheduled update period.

Users may invoke AirMatch computation on-demand, instead of waiting for the next scheduled deployment, with the following commands:

- **AirMatch runnow full**—This command performs the same quality optimization that would be done in a periodically scheduled optimization.
- **AirMatch runnow quick**—This command performs a quick-and-dirty scan of random channels and computes correct EIRP for cell sizes.
- **AirMatch runnow incremental**—This command only considers new APs added while keeping existing AP configuration intact.

For example, suppose AirMatch has already optimized the ten APs you have deployed. You add one more AP, x, to the network. You want AP x to be optimized without disrupting the other ten. An incremental optimization leaves the existing APs alone. It assigns the best settings possible to new AP x, *considering the current, unchangeable configuration of the ten existing APs.*

Note

Dot11a/g radio profile channel/power parameters are ignored in AOS 8.x MM.

When you want to assign a particular channel/power plan for select AP radios, use the "AirMatch ap freeze" command. For example:

```
(mobility-master) *[/] #AirMatch ap freeze ap-name my_ap_name band
5ghz channel 149 eirp 16
```

You can use the "AirMatch ap unfreeze" command to remove the static assignment. For example:

```
(mobility-master) *[/] #AirMatch ap unfreeze ap-name my ap_name band
5ghz channel EIRP
```

View the AirMatch RF plan

```
(MM-1) *[mynode] #show airmatch solution
Seq  Time               APs  [5GHz] Radios Cost  Conflict Deploy Radios Cost  Conflict Deploy    Type
---- ------------------ ---         ------ ----- -------- ------ ------ ----- -------- ------    ---------
#44  20161112_05:03:05  1              1   2.0        0.0 No        1   4.0        0.0 No        Scheduled
#43  20161111_05:00:13  1              1   2.0        0.0 No        1   4.0        0.0 No        Scheduled
#42  20161110_05:03:33  1              1   2.0        0.0 No        1   4.0        0.0 No        Scheduled
#41  20161109_05:03:32  1              1   2.0        0.0 No        1   4.0        0.0 No        Scheduled
#40  20161108_05:03:32  1              1   2.0        0.0 No        1   4.0        0.0 No        Scheduled
#39  20161107_05:04:03  1              1   2.0        0.0 No        1   4.0        0.0 No        Scheduled
#38  20161106_05:00:36  1              1   2.0        0.0 No        1   4.0        0.0 No        Scheduled
#37  20161106_00:59:09  1              1   2.0        0.0 Yes       1   4.0        0.0 Yes       Quick
#36  20161105_05:03:36  1              1   2.0        0.0 No        1   1.0        0.0 No        Scheduled
#35  20161104_05:03:36  1              1   2.0        0.0 No        1   1.0        0.0 No        Scheduled
#34  20161103_05:03:36  1              1   2.0        0.0 No        1   1.0        0.0 No        Scheduled
#33  20161102_05:03:36  2              1   2.0        0.0 No        2   2.0        0.0 No        Scheduled
#32  20161101_05:03:35  2              1   2.0        0.0 No        2   2.0        0.0 No        Scheduled
#31  20161031_05:03:37  2              1   2.0        0.0 No        2   2.0        0.0 No        Scheduled
#30  20161030_05:03:34  2              0   0.0        0.0 No        2   2.0        0.0 No        Scheduled
#29  20161029_05:03:35  2              0   0.0        0.0 No        2   2.0        0.0 No        Scheduled
#28  20161028_05:03:35  2              0   0.0        0.0 No        2   2.0        0.0 No        Scheduled
#27  20161027_05:03:36  2              0   0.0        0.0 No        2   2.0        0.0 No        Scheduled
#26  20161026_05:01:14  2              1   2.0        0.0 No        2   2.0        0.0 No        Scheduled
#25  20161023_05:03:43  1              1   2.0        0.0 No        1   1.0        0.0 No        Scheduled
#24  20161022_05:03:43  1              1   2.0        0.0 No        1   1.0        0.0 No        Scheduled
#23  20161021_05:01:29  1              1   2.0        0.0 No        1   1.0        0.0 No        Scheduled
#22  20161019_08:26:55  1              1   2.0        0.0           1   1.0        0.0           Incremental
#21  20161019_07:56:53  1              1   2.0        0.0           1   1.0        0.0           Incremental
```

Not Deployed

Figure 8-7 View the AirMatch RF plan

When the Mobility Master first detects APs on the network, it enters its initial optimization phase, collects data from all APs, and generates an incremental solution every 30 minutes. The incremental solution highlighted in Figure 8-7 is in the new deployment phase. This may mean either an initial optimization or an ON-DEMAND OPTIMIZATION. In the Type column, Quick indicates on-demand optimization, Scheduled indicates periodic scheduling, and Incremental indicates initial optimization or on-demand optimization.

Cost

The lower the better. The metrics are susceptible to change and can be different between versions.

Conflict

The lower the better. The metrics are susceptible to change and can be different between versions.

Solution detail

```
(MM-1) *[mynode] #show airmatch solution 44

# 20161112_05:03:05      Scheduled
# 5GHz    capacity/network cost/solution cost/improvement: 2.0/2.0/2.0/0.0%
# 2.4GHz capacity/network cost/solution cost/improvement: 1.0/4.0/4.0/0.0%
# Band Radio             Mode Chan  CBW    EIRP(dBm)  APName
  ---- ----------------- ---- ----- ------ ---------  ----------
  5GHz 00:24:6c:2a:fe:f8 AP    153i    40i       22*  ap1

  2GHz 00:24:6c:2a:fe:f0 AP      1i    20*        9.  ap1

[*] regarded frozen | [.] no change |
[i] channel ignored because insufficient quality increase
```

Real Quality Increase Value

```
(MM-1) *[mynode] #show airmatch profile

AirMatch profile (Predefined (changed))
----------------------------------------
Parameter              Value
---------              -----
schedule               Enabled
deploy-hour            5 o'clock
eirp-offset            0 dB
quality-threshold      15 percent
```

Quality-threshold Setting

Figure 8-8 Solution detail

If the **Real Quality Increase** value does not exceed the quality-threshold (15% by default), the solution will be ignored (Figure 8-8). As per the previous example, solution 44 did not deploy because it did not exceed the 15% quality-threshold.

AirMatch, radar, and avoiding noise

AirMatch reacts to detrimental RF events such as radar and high noise levels.

When AirMatch responds to a dynamic high noise event, the AP considers channels that are not assigned to neighboring radios, as learned in the ARM interference indices. If it detects noise again, the AP will change channels and stay on that channel for another 12 hours, then return back. The AP will change channels after 30 minutes of detection of RF noise and radar event ends.

Events such as radar and high noise levels allow the network to manage sudden changes in the RF environment because although the channel change will cause clients to have to reconnect the high level of noise probably means there is no communication on the current channel.

Configuring AirMatch

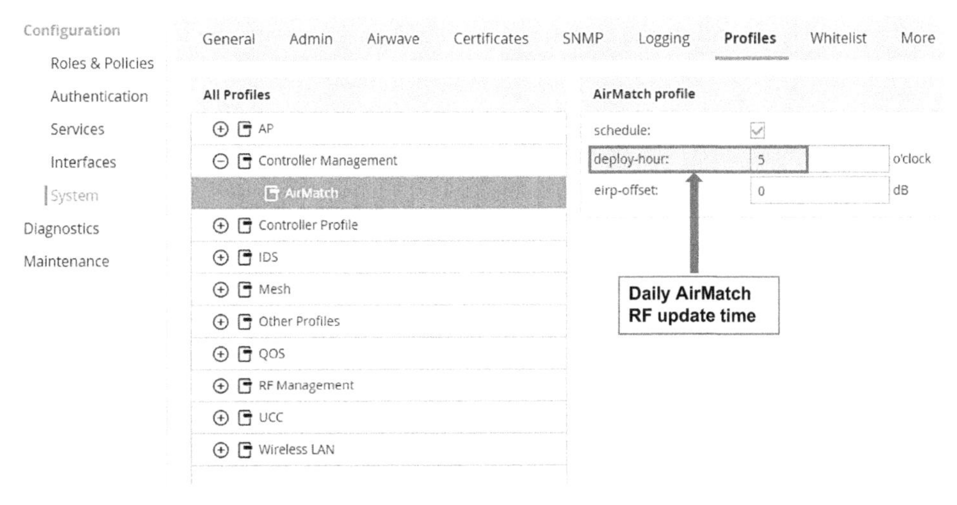

Figure 8-9 Configuring AirMatch

By default, the AirMatch deployment time is set for 5:00 AM. Figure 8-9 shows how to change the value under the AirMatch profile option. Also, keep in mind that the EIRP offset supports some power boosters. Typically AirMatch will assign an EIRP value(X)to the AP. So the AP will use "X" by default. However, if a non-zero offset (Y) is set, then the AP will get a new EIRP value (X+Y). Y can be (–6 to +6).

AirMatch LSM upgrade

Configuration
Diagnostics
Maintenance
Software Management
Configuration Management

Upgrade | Reboot | Boot Parameters | Service Module Packages | About

Service Module Packages

TYPE	NAME	AOS VERSION	BUILD #
AirGroup	default_airgroup_pkg	ArubaOS_MM_8.0.1.0	57204
airmatch	default_airmatch_pkg	ArubaOS_MM_8.0.1.0	57204
AppRF	default_appRF_pkg	ArubaOS_MM_8.0.1.0	57204
ARM CM	default_arm_cm_pkg	ArubaOS_MM_8.0.1.0	57204
NB API Helper	default_nbapi_helper_pkg	ArubaOS_MM_8.0.1.0	57204
UCM	default_ucm_pkg	ArubaOS_MM_8.0.1.0	57204

Load New Package

Access method: TFTP FTP SCP Local file
Host IP address:
Image file name:
Destination file name:

Figure 8-10 AirMatch LSM upgrade

AirMatch features are contained in Loadable Service Modules (LSM). This means it can be upgraded separately from the core AOS and other features. Figure 8-10 shows how you can do this from the GUI.

To upgrade from the CLI, import the new AirMatch LSM package with the following command:

```
(mobility-master) *[/] upgrade-pkg copy scp: <IP_ADDR> <LOGON_ID> <FILEPATH> flash: <ARBITRARY_NAME>

(mobility-master) *[/] upgrade-pkg activate <PACKAGE_NAME>
```

Example:

```
upgrade-pkg copy scp: 10.100.224.153 tftp

/tftpboot/ArubaOS_MM_8.0.0.0-svcs-ctrl_AirMatch_54937 flash: xyz

upgrade-pkg activate ArubaOS_MM_8.0.0.0-svcs-ctrl_AirMatch_54937
```

AirMatch FAQ

The following are some common questions related to AirMatch. Take a moment to review these questions and their answers.

Question 1—I just disabled the schedule. Will APs stop changing channels?

- **Answer**—No, you disabled the schedule, not the AP's channel change.

Question 2—Can APs change the channel even though the schedule is disabled?

- **Answer**—Yes, there are many triggers to change the channel:
  - Static configuration via “AirMatch ap” command
  - Radar detection and governments’ enforcement
  - High noise detection
  - Scheduled AirMatch optimization.

Note

When the schedule is disabled, only option (4) is disabled. The rest will continue to work.

Question 3—What happens if there is no radar, no high noise, and no static channel while the AirMatch schedule is disabled?

- **Answer**—No channel/bandwidth/EIRP change will be expected.

Question 4—If we disable ARM in the radio profile in AOS8, will AirMatch still work?

- **Answer**—Yes. ARM profile will not be used for AirMatch. Disabling assignment in ARM profile does not affect AirMatch.

Learning check

2. AirMatch will NOT be affected by high noise and radar signal
   a. True
   b. False

3. In AirMatch, who decides the channel and power of an AP?
   a. Mobility Master
   b. Managed Devices
   c. Wireless Client
   d. AP self

4. AirMatch and ARM cannot be used together. Standalone controllers and Master/Local deployment do not support AirMatch RF planning.
   a. True
   b. False

Answers to learning check

2. AirMatch will NOT be affected by high noise and radar signal
   a. True
   b. **False**

3. In AirMatch, who decides the channel and power of an AP?
   a. **Mobility Master**
   b. Managed Devices
   c. Wireless Client
   d. AP self

4. AirMatch and ARM cannot be used together. Standalone controllers and Master/Local deployment do not support AirMatch RF planning.
   a. **True**
   b. False

Legacy Client Match (CM)

What is CM?

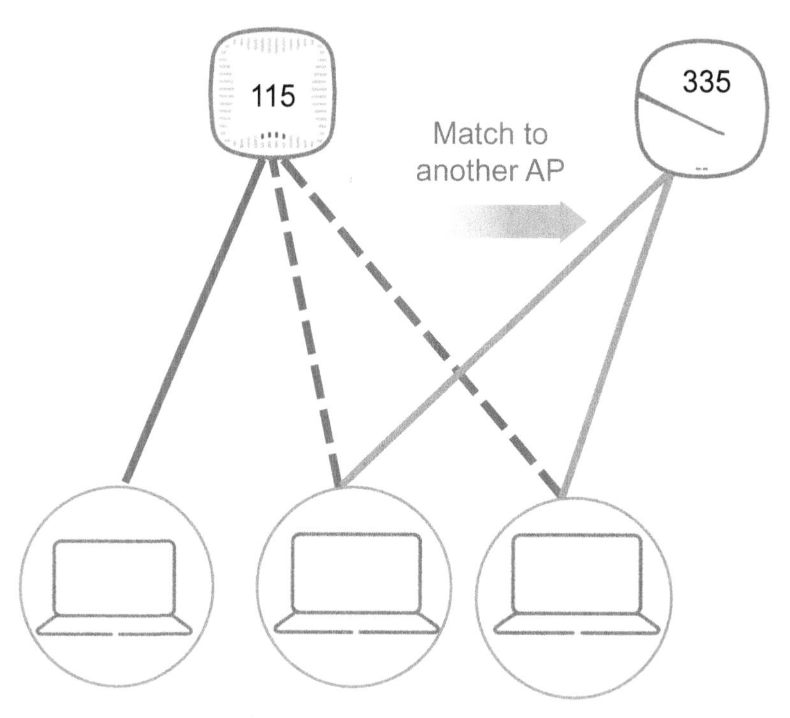

Figure 8-11 What is CM?

While ARM and AirMatch focus more on RF management at the AP level, Client Match (CM) focuses on the client level. Figure 8-11 shows that originally, all three clients were associated with AP 115, on the left. However, after the CM process, two of the clients were moved from AP 115 to AP 335. CM determined that better client services could be achieved with this change.
CM is an Aruba-patented technology, which provides three main functions:

- **Load balancing**—balances clients across APs on different channels, based on client load.
- **Sticky clients**—helps clients that tend to stay associated with an AP, despite low signal levels.
- **Band steering/Band balancing**—APs monitor RSSI for dual-band clients and moves them to the 5 GHz Radio

These features are implemented purely from within the HPE Aruba solution. No special client software or configuration is needed.

Legacy CM works on Master/Local architecture or standalone mode on both AOS 6.x and 8.x platforms.

How CM works

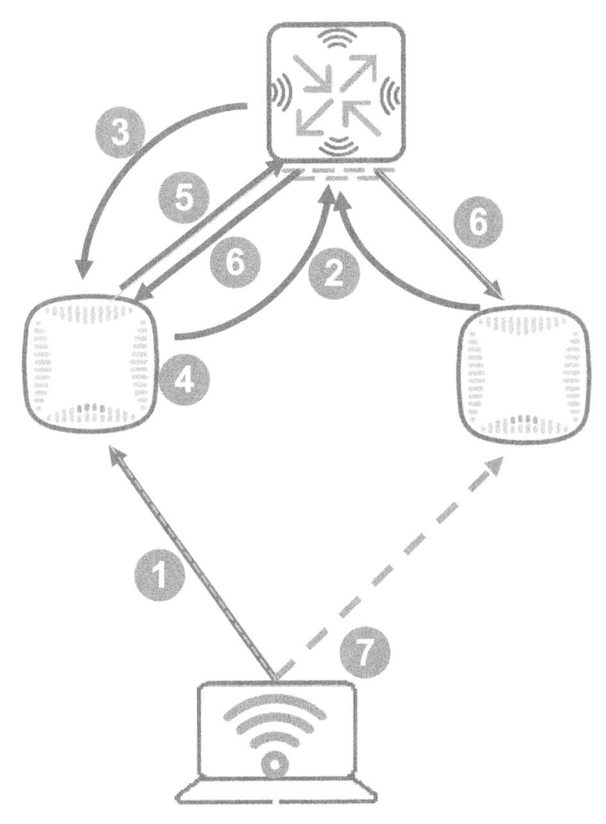

Figure 8-12 How CM works

Figure 8-12 shows how CM works, as described below:

1. Clients associate with a radio.
2. All APs report probe requests from clients (as long as it is a good signal). If the client is 802.11k compliant, then the client informs the AP of other radios.
3. The controller builds a virtual beacon report (VBR) for each client and shares this with the associated AP.
4. The AP now has visibility into all available AP radios for this client and determines the optimum radio for this client.
5. The AP notifies the controller if there is a better radio for the client.
6. The controller issues the command to steer the client.

7. The client is moved to another AP using a "DoS-like" mechanism, or 802.11v. The client is moved to the optimum radio. All management platforms are advised of the change (controller, GUI, and Airwave).

Default CM rule upgrade

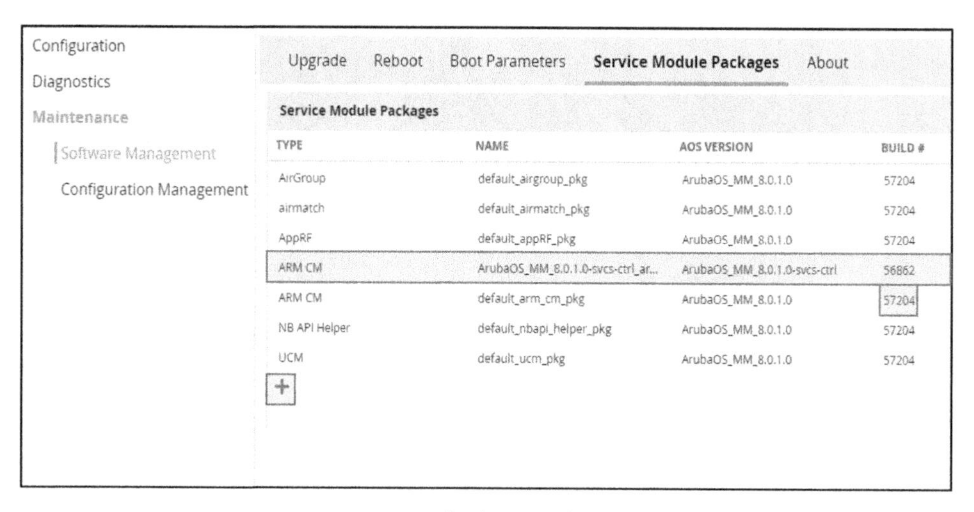

Figure 8-13 Default CM rule upgrade

Like AirMatch, CM can be upgraded independently of AOS version. Figure 8-13 shows two ARM_CM packages. One is the default, and the other has been newly uploaded. Only one can be active at a time.

Legacy CM caveat

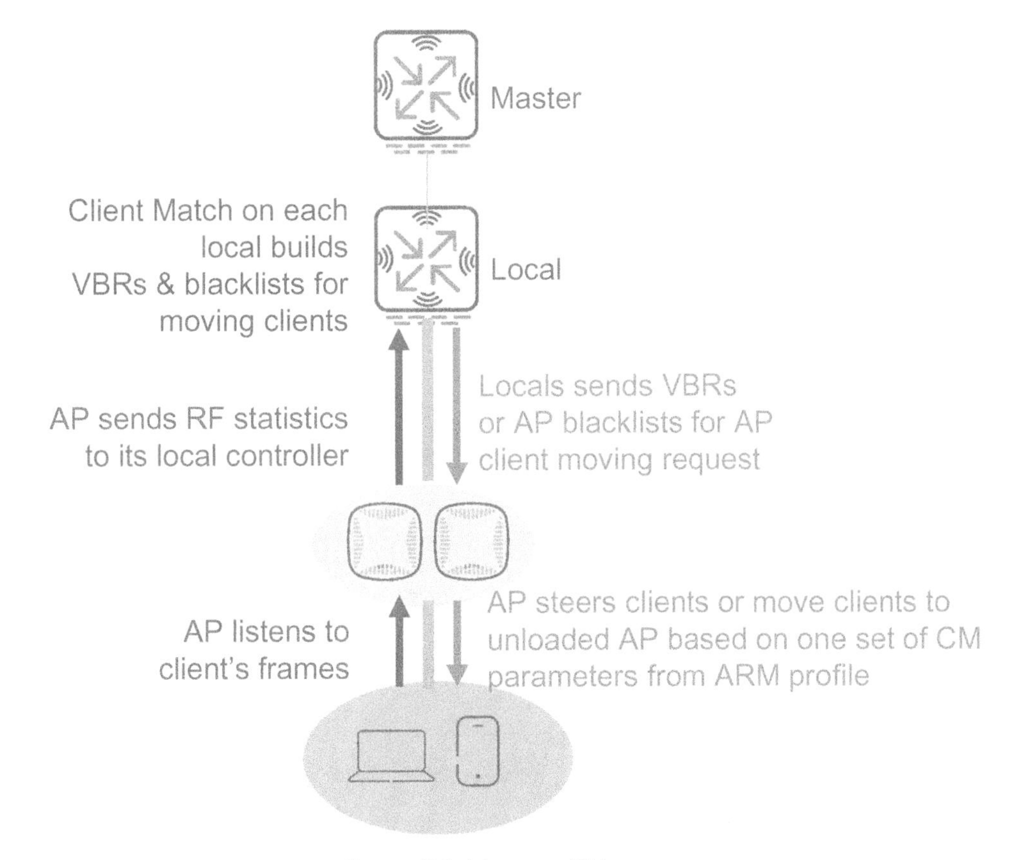

Figure 8-14 Legacy CM caveat

There are certain caveats associated with Legacy CM:

- Band-steering is done by the AP itself, since clients are steered from 2.4 GHz to 5 GHz radio on the same AP.
- In 6.x, only one set of CM parameters can be configured in the AP's ARM profile. Therefore, all types of client devices are treated alike.
- This one-rule-fits-all mechanism is not as effective for some types of clients with unique behaviors. These clients' connectivity may be impacted when the AP takes steering actions.
- There is no Rule-Based Client Match (RBCM) support—you are limited to using the default CM rule.

CM on MM/MD

How CM works on MM/MD deployment

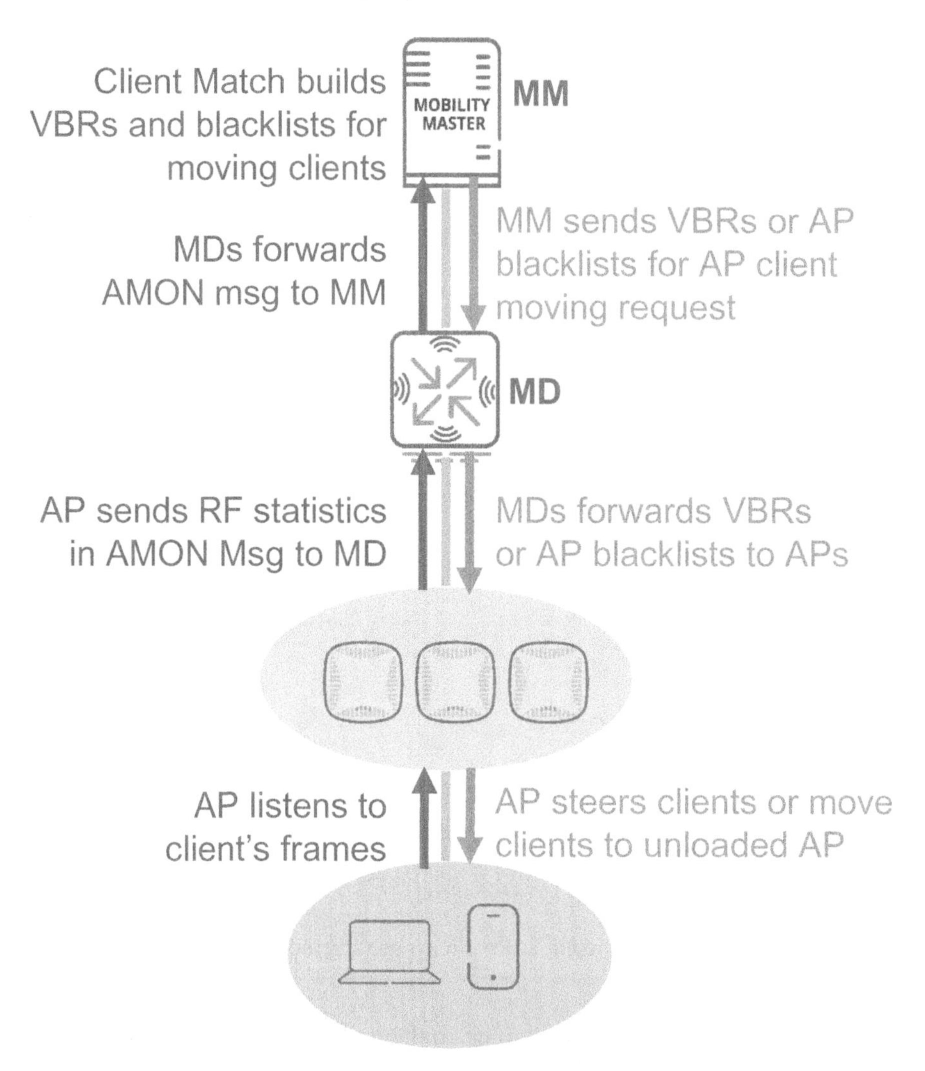

Figure 8-15 How CM works on MM/MD deployment

Let us compare AOS 8.x CM enhancements to legacy 6.x CM. First, recall that the 6.x environment is based on a Master/Local controller design.

Meanwhile, an 8.x Master/local deployment has a VM-based Mobility Master for more centralized control. In this scenario, APs listen to client frames, then send collected RF statistics to the MD in an AMON message.

The process is as follows:

1. The MD forwards AMON messages to the Mobility Master.
2. The MM's CM module builds a VBR table for each AP and sends it to the MD, which sends it on down to the AP.
3. Unlike v6.x CM, the 8.x MM builds the VBR, and so it coordinates client's steering and load balancing. Band-steering is still done right at the AP.
4. Similar to v6.x legacy CM, the AP does periodic health checks for the client and sends steering requests to the MM when anomalies are detected. Also like v6.x, the MM builds the AP blacklist and coordinates the steering.

Also, it is one of the Loadable Service Modules (LSM) in the Mobility Master, and so can be upgraded independently.

Rule-Based Client Match (RBCM)

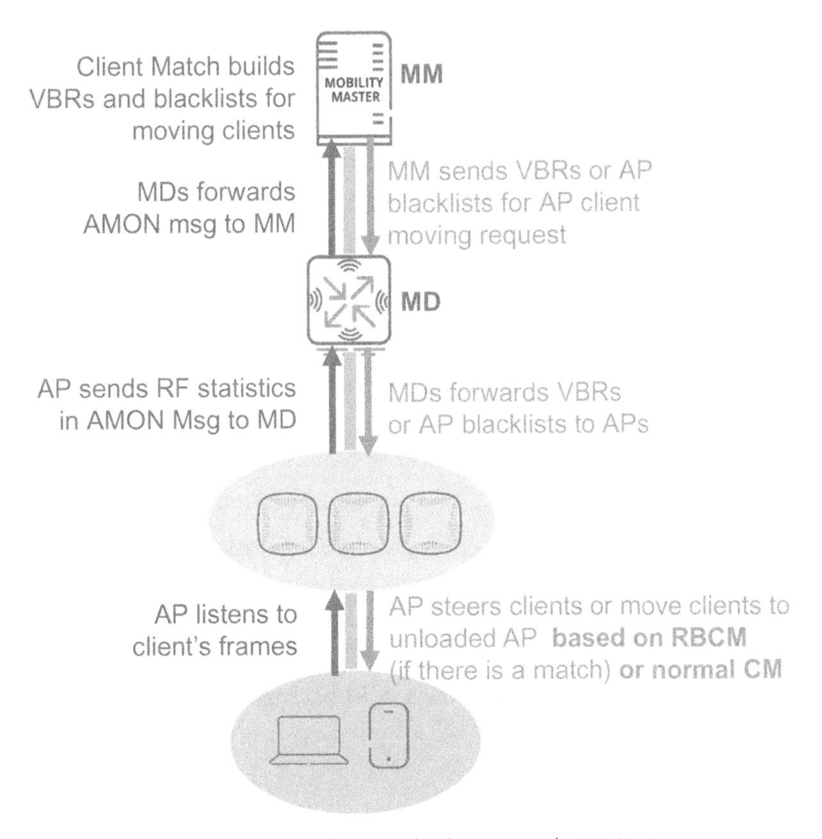

Figure 8-16 Rule-Based Client Match (RBCM)

As previously mentioned, CM functionality is similar between AOS 8.x and AOS 6.x The major difference is that steering coordination is moved to from mobility controllers to the Mobility Master. The big enhancement in AOS 8.x is the support of RBCM.

RBCM is described below:

- Runs only on MM. Standalone MDs revert to normal CM.
- Enables you to adapt CM behavior for specific devices or device types. Do this by configuring unique CM rules for certain client types.
- A CM rule can be applied based on a specific device type, specific MAC OUI, or a device MAC.
- You can define specific parameters for steering, sticky client, band-steering, device capability, 802.11v, and more.
- RBCM greatly improves client steering efficiency and stability.

CM Rule upgrade (Default ClientMatch rule)

Configuration
Diagnostics
Maintenance
Software Management
Configuration Management

Upgrade Reboot Boot Parameters **Service Module Packages** About

Service Module Packages

TYPE	NAME	AOS VERSION	BUILD #
AirGroup	default_airgroup_pkg	ArubaOS_MM_8.0.1.0	57204
airmatch	default_airmatch_pkg	ArubaOS_MM_8.0.1.0	57204
AppRF	default_appRF_pkg	ArubaOS_MM_8.0.1.0	57204
ARM CM	ArubaOS_MM_8.0.1.0-svcs-ctrl_ar...	ArubaOS_MM_8.0.1.0-svcs-ctrl	56862
ARM CM	default_arm_cm_pkg	ArubaOS_MM_8.0.1.0	57204
NB API Helper	default_nbapi_helper_pkg	ArubaOS_MM_8.0.1.0	57204
UCM	default_ucm_pkg	ArubaOS_MM_8.0.1.0	57204

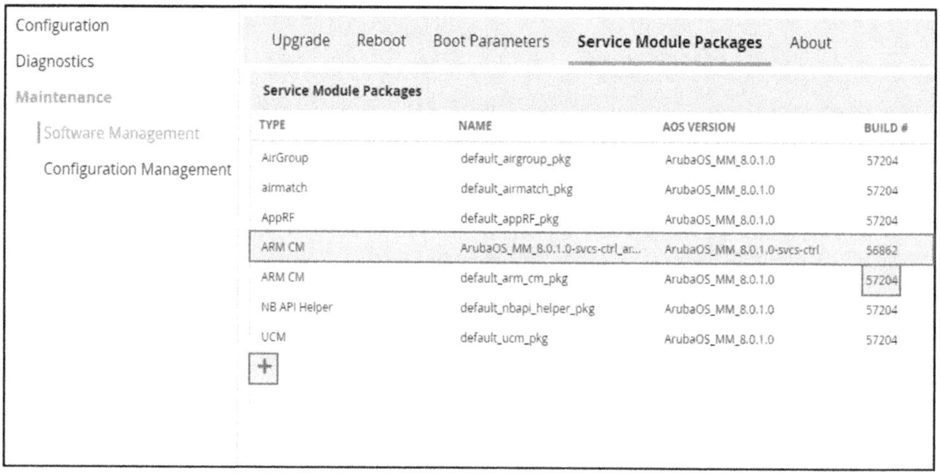

Figure 8-17 CM Rule upgrade (Default ClientMatch rule)

As stated before, CM can be independently upgraded. Figure 8-17 shows two ARM_CM packages. One is the default, while the other is a new one uploaded. However, only one can be active at a time.

By default, the RBCM rule file is part of the ARM_CM package.

RBCM Rule upgrade

Uploading a custom ClientMatch Rule update package

RBCM can be seen as an addendum to the existing CM. The user-defined rules in the RBCM table are checked as an additional lookup table along with the CM logic.

RBCM rules can be changed by installing a custom rule file. To upload a custom clientmatch rule, navigate to **Diagnostics > Technical Support >Client Match Rules**. From there you click **Upload File** to upload a file that contains clientmatch rules.

The following is a list of parameters, which can be defined in a rule:

- **Device type/MAC OUI/MAC address**—(the MAC address gets precedence over MAC OUI which gets precedence over the device type)
  - Device type —device type as seen in the "show user-table"
- **Steer Restrict**—can restrict sticky/band and steer/load-balance steers.
- **Steer Interval**—interval between steers.
- **Override dot11v**—overrides the default dot11v behavior for that device
- **Device capability**—"D" flag—Dual network (802.11 and cellular) capable.
- **Sticky params**—Sticky low SNR, Sticky delta SNR, Sticky min signal
- **Band steer params**—Band steer min 802.11a signal, Band steer max 802.11g signal
- **11v parameters**—Abridged bit for 11v BTM/ Steer restriction window /11v BTM attempts

Configuring ClientMatch

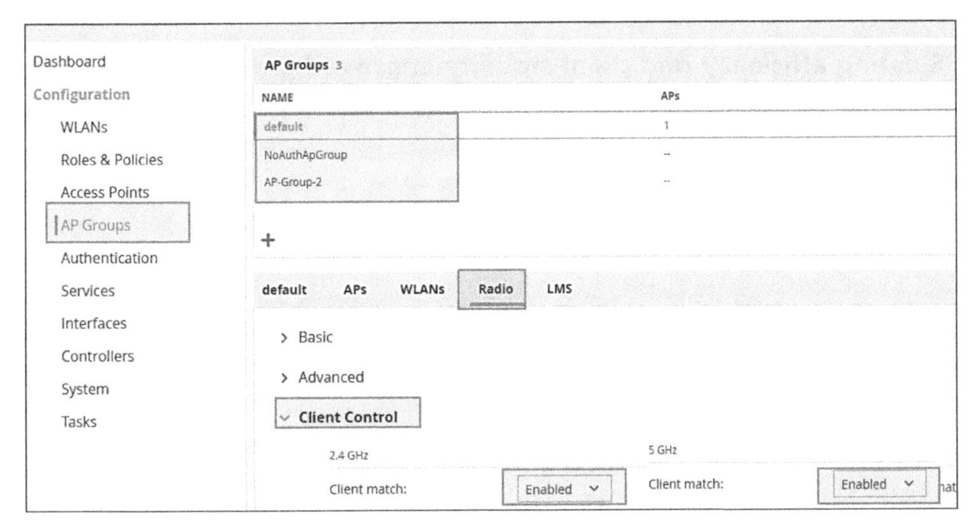

Figure 8-18 Configuring ClientMatch

For standalone controllers that have no associated managed devices, ClientMatch is enabled and disabled in the AP's ARM profile. To edit an ARM profile via the WebUI, navigate to **Configuration> AP-Group > Radio**, or to **Configuration > System > Profiles >RF Management >Adaptive Radio Management (ARM)**.

The ARM profile also includes advanced ClientMatch settings that can only be configured through via CLI. The default values for these settings are recommended for most users, and caution should be used when changing them. For complete details on all ClientMatch configuration settings, refer to the ArubaOS CLI Reference Guide.

Learning check

5. Which statements about RBCM are correct?
   a. RBCM runs only on MM, with standalone MD, reverts to normal CM
   b. Only a single set of CM configuration, and all clients are treated alike
   c. AP collects/sends RF info to local controller and local controller coordinates the client steering and load balancing
   d. Steering efficiency and client stability improved greatly

Answers to learning check

5. Which statements about RBCM are correct?
   a. **RBCM runs only on MM, with standalone MD, reverts to normal CM**
   b. Only a single set of CM configuration, and all clients are treated alike
   c. AP collects/sends RF info to local controller and local controller coordinates the client steering and load balancing
   d. **Steering efficiency and client stability improved greatly**

9 Guest Access

LEARNING OBJECTIVES

✓ This chapter begins with an overview of the HPE Aruba guest access solution. Then you will learn about the captive portal process and configuring a guest access solution, with a guest provisioning account. You will learn how to troubleshoot common guest access issues and get an introduction to guest access using the ClearPass product.

Aruba Guest Access solution

The HPE Aruba solution offers a single WLAN infrastructure for both internal employee and guest use. It provides differentiated access based on user and device characteristics. This is the basis for a number of guest access features. Guest access is often configured as a software option. No new hardware is required for basic guest access—just configure another WLAN on the existing Aruba MM and MC. Guests use the same APs that are currently in use for internal employee access.

With the Aruba guest solution, there is no need to configure a guest VLAN at every AP-connected LAN switch. Aruba's user-centric networks are added as an overlay on the existing wired LAN. Traffic from APs is directed to the MC via secure tunnels. The MC's stateful firewall maintains strict segregation between different traffic classes.

Internal employees are allowed access to corporate resources. Guest traffic remains within a secure tunnel, which is terminated to a Mobility Controller in the DeMilitarized Zone (DMZ). From there it travels to the Internet. Alternatively, traffic at the AP can be routed to either the controller or directly to the Internet. This is known as split tunnel mode and can be configured for an AP group.

The MC serves Captive Portal login screens and web forms for administration. For more sophisticated guest access solutions, Aruba's ClearPass server can be used for credit card processing, guest self-registration, access code authorization, and property management systems.

This gives you a general introduction to guest access. The solution varies based on the deployment model you choose. The next few pages describe various guest access deployment models. You should be able to pick one that is right for your particular needs.

Guest network with NAT

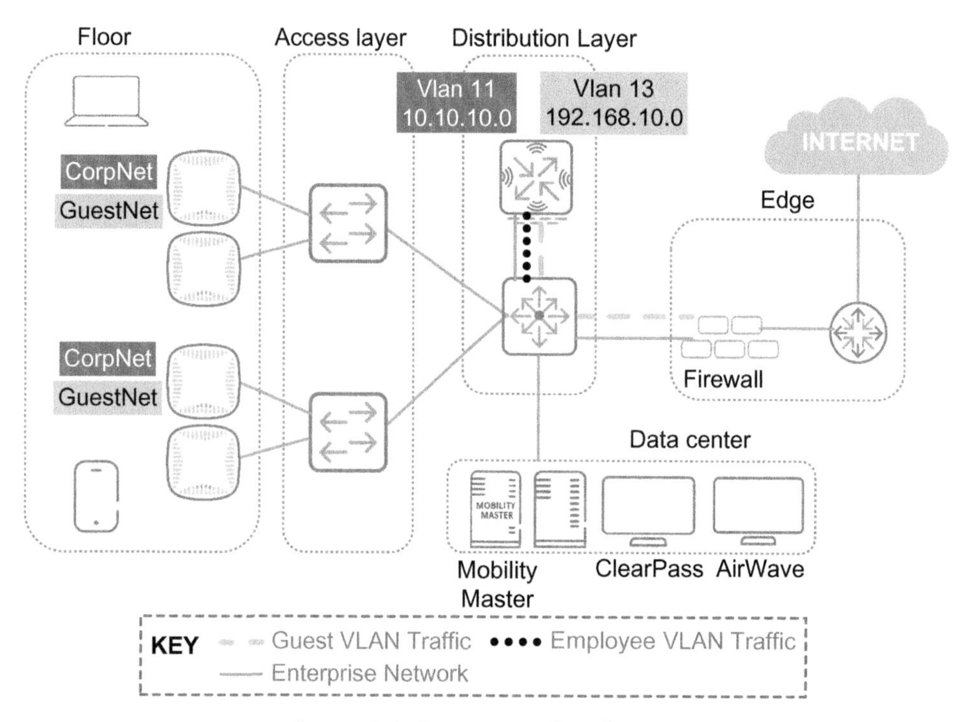

Figure 9-1 Guest network with NAT

As with most guest designs, guests are typically allocated distinct IP addresses, separate from internal corporate address space. You can configure the corporate DHCP server to assign these addresses, or use a separate DHCP server. The DHCP server assigns IP addressing information to the client, including the IP address of the DNS server. You must decide whether guests can leverage your internal DNS, or if they are to use external DNS.

Once guests have this IP addressing information they begin to send traffic, which is GRE tunneled to the controller. The controller removes the tunnel headers, revealing the native IP packet. The controller's NAT service can translate the guest's source address to an address of your choosing and forward this traffic into the corporate LAN. Or the controller can forward the native guest IP packets out an appropriate VLAN, toward a firewall, as shown in Figure 9-1. In this case, the firewall performs NAT translation, instead of the controller.

You will gain experience configuring guest access and firewall policies in the Lab exercise.

 Note

For this topology, it is highly recommended to maintain a logical separation of guest traffic at both Layers 2 and 3.

Guest network with dedicated WAN

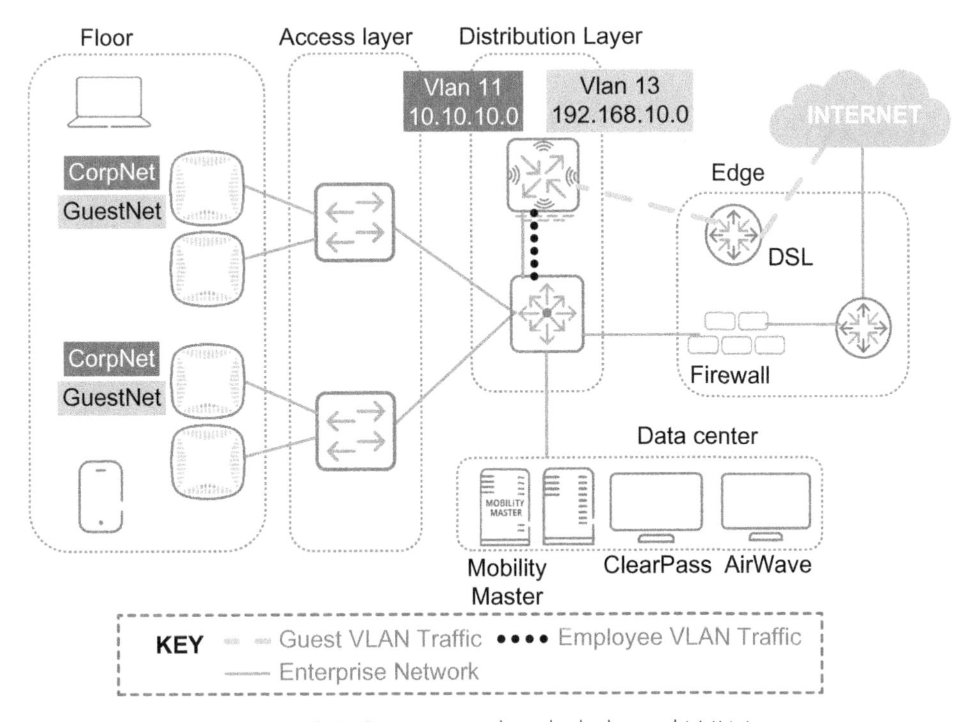

Figure 9-2 Guest network with dedicated WAN

This design features a dedicated WAN connection for guests. This physical isolation between guest and corporate WAN services may improve security. Compare this and the previous solution. You will notice the additional WAN connection to the Internet, via a DSL device in this case. Of course, these devices require a separate connection to the controller, as shown.

It is true that a bit more resources are needed, but this may help you meet certain compliance requirements, and it eases support and troubleshooting efforts. This solution also requires little to no focus on guest bandwidth utilization, since guest and corporate users do not share the same Internet connection.

Guest network tunnel to DMZ controller

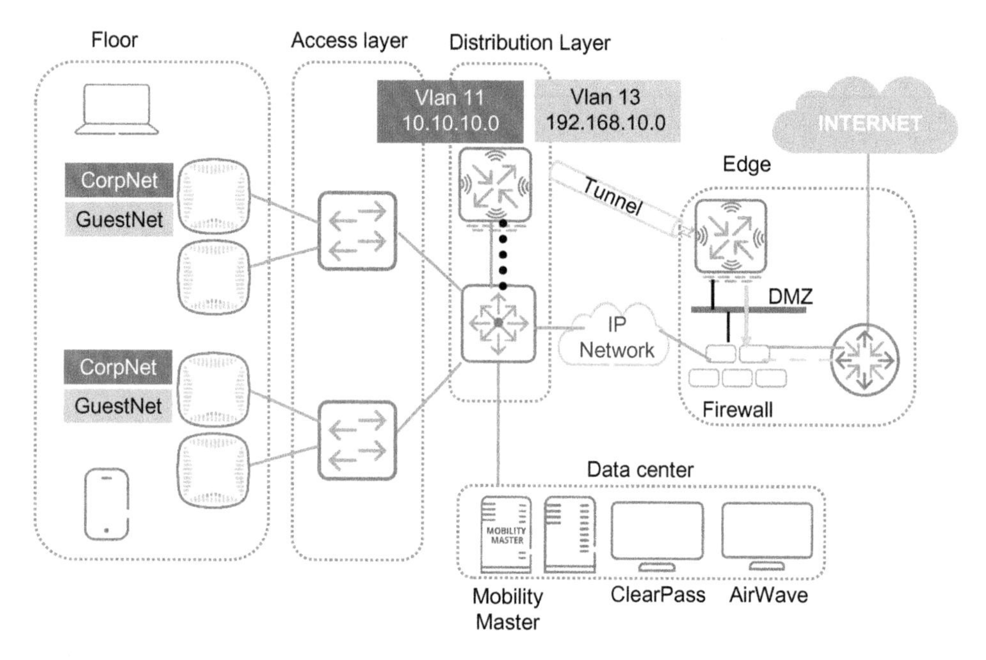

Figure 9-3 Guest network tunnel to DMZ controller

In this example, an additional MC has been added to the corporate DMZ. Large corporations can leverage this DMZ mobility controller for a guest solution.

Guest traffic is still tunneled from AP to the distribution layer MC, as in the previous two examples. This MC then tunnels the traffic to the DMZ MC. Yes, the physical data path for this tunnel is over the corporate IP network. Even so, you still have complete logical isolation of guest and corporate traffic because of the GRE encapsulation.

Guest Network using MultiZone AP

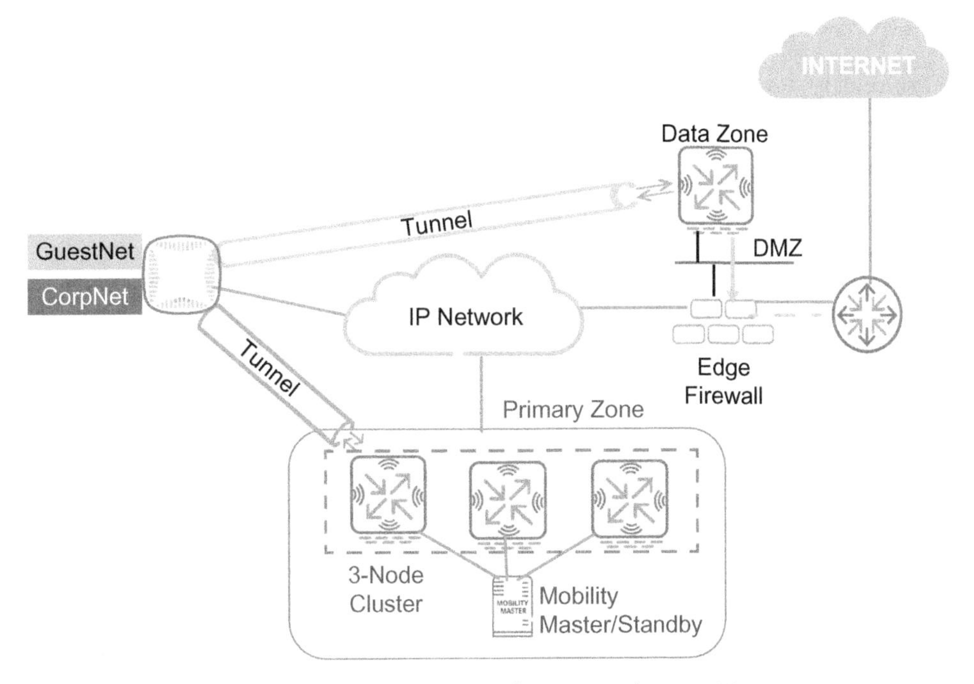

Figure 9-4 Guest Network using MultiZone AP

This example features the MultiZone AP feature of AOS8.x. This reduces the number of tunnels traversed by guest traffic before egress toward the Internet.

The previous DMZ example required two tunnels—one from AP to distribution-layer MC, and another from there to the DMZ MC. The MultiZone AP design tunnels guest traffic directly from AP to DMZ MC.

The MultiZone AP feature allows APs to terminate to different controllers that reside in different zones. A zone is a collection of controllers under a single administration domain. It can be as small as a single controller or as complex as a cluster of controllers. Zone provides a conceptual "Air Gap" between corporate and guest users, since there need not be a physical connection between zones.

The MultiZone feature uses two zone roles:

- Primary zone
  - The zone an AP first connects to upon initial boot up
  - In a single-zone architecture, full control over the AP—radio and channel configuration, and all features

- You can configure a MultiZone profile to enable the MultiZone feature. This would include the definition of one or more data zones

- Data zone
  - A secondary zone to which an AP connects, after getting MultiZone configuration from primary zone
  - Cannot reboot, provision, or upgrade image of AP
  - Only the tunnel mode virtual AP configuration is allowed

Each zone can have separate configuration and security classification, with separate cluster enabling if desired. You can only enable this feature on the Mobility Master. Each zone can have separate failover and rebootstrap. (Exception: the primary zone's rebootstrap disables all data zones). Zones can have separate ESSIDs and user info. If there is an ESSID duplicate, APs will reject the later one and generate a syslog error message.

Data zones must have the same AOS image version as the primary zone. AP names and AP group names must also match.

L3 deployment with Guest Services

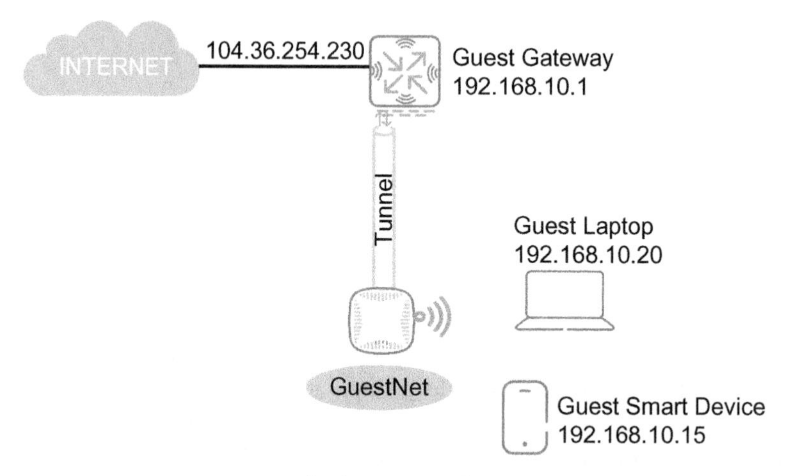

Figure 9-5 L3 deployment with Guest Services

The controller is typically the Layer 3 device when it exists as the default router for a nonroutable guest network. When a guest network is deployed in a private IP space and is not routable from the general network, the MC is normally configured to act as both the DHCP server and NAT device for the guests.

After guests get an IP address, they open a web browser and are redirected to the MC's Captive Web Portal. Depending on your needs, the guest may have to provide unique credentials, or they may simply have to agree to an Acceptable Use Policy (AUP).

Captive Portal process

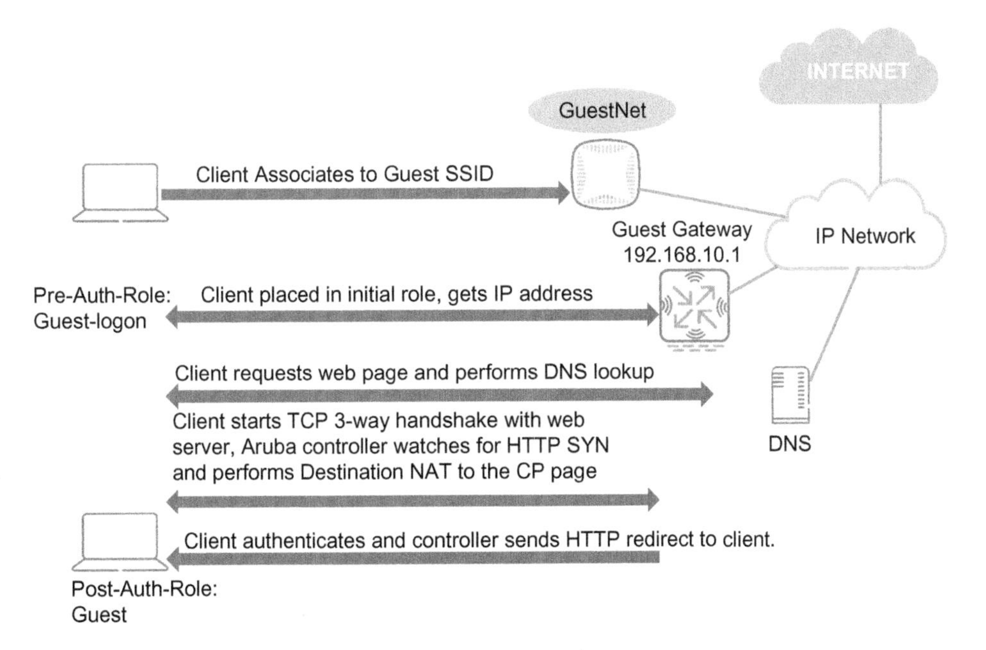

Figure 9-6 Captive Portal process

Captive portal is one authentication method supported by ArubaOS. A captive portal presents a web page, which requires user action before network access is granted. The required action can be simply viewing and agreeing to an AUP, or providing username/password credentials. The Captive Portal process is shown in Figure 9-6, and is described below.

1. Guest clients associate to the open guest SSID.
2. Guest clients are placed into a VLAN and assigned an IP address. The user has no access to network resources beyond that allowed in its initial, pre-authentication role. The initial role in Captive Portal allows for DHCP, DNS, and a few other optional protocols, soon discussed. The predefined role for preauthenticated guests is "Guest-logon." Of course, clients must be able to contact the DHCP server for IP addressing, and the DNS server for name resolution.
3. The client opens a web browser and enters some URL. This URL must be translated to an IP address, and so a request is sent to the DNS server.

4. The client knows the web site's IP address and initiates a TCP connection to the web server. The MC watches for the HTTPS SYN packet and redirects it to the captive portal page. It uses Destination NAT (d-NAT) to achieve this redirect.

The user then sees the captive portal web page in the browser windows, which prompts for username/password credentials. If the credentials are valid, the client role is updated from the Pre-auth-role to Post-auth-role. The predefined postauthentication role for guests is "Guest." Typically, this postauthentication role only grants Internet access. Access to internal networks and resources is strictly prohibited.

Learning check

1. What are the different type Guest Network deployment options Aruba recommends?
    a. Guest Network with NAT in the Core
    b. Guest Network with Dedicated WAN
    c. Guest Network using MultiZone AP
    d. Guest Network using Aruba controller
    e. Guest Network and Corporate User Traffic Mix

Answers to learning check

1. What are the different type Guest Network deployment options Aruba recommends?
    a. **Guest Network with NAT in the Core**
    b. **Guest Network with Dedicated WAN**
    c. **Guest Network using MultiZone AP**
    d. **Guest Network using Aruba controller**
    e. Guest Network and Corporate User Traffic Mix

Configuring guest WLAN using Aruba controller

Using the guest WLAN wizard

This subsection covers the configuration of guest WLANs using the wizard. Before or during the wizard process you should do the following:

- Design the IP network
- Choose a VLAN for Guest Access
- Designate a DHCP server and scope for Guest Access

During the wizard process, you will accomplish the following:

- Create a WLAN. This can be in the same AP group as existing employee WLANs
- Specify forwarding mode, VLAN, and radio type
- Designate the WLAN for guest access
- Specify Captive Portal Authentication. This includes whether to require user credentials, and/or a valid email address.
- Customize the look and feel of the captive portal Web Page
- Choose or create server to check guest credentials. This can be a controller's internal database, or external system.
- Create or use the suggested preauthenticated and authenticated roles and firewall policies

Create a guest WLAN

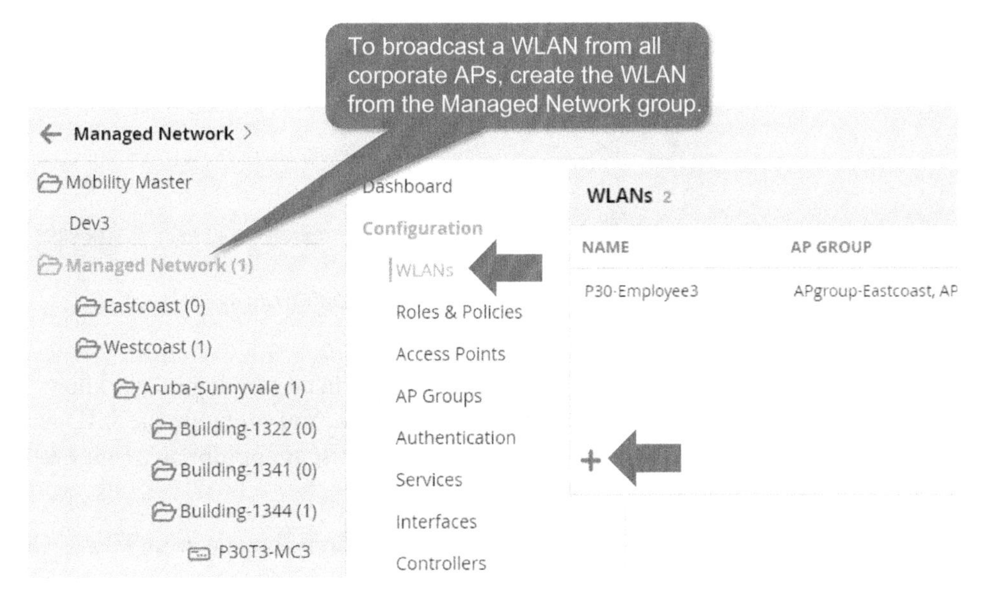

Figure 9-7 Create a guest WLAN

The recommended method for creating a new WLAN configuration is through the new WLAN wizard. Advanced users may also configure a WLAN manually via GUI or CLI.

If the guest WLAN is to be broadcast from all corporate APs, regardless of regions and locations, create the WLAN from the Managed Network group, at the top of the hierarchy, as shown in Figure 9-7. This ensures that the configuration flows down to all groups and subgroups.

To use the WLAN wizard for new WLAN creation, navigate to **Configuration> WLANs** under either Managed Networks or a specific group or subgroup, then in the WLANs section, click the "+" icon to start the wizard.

Guest WLAN general information

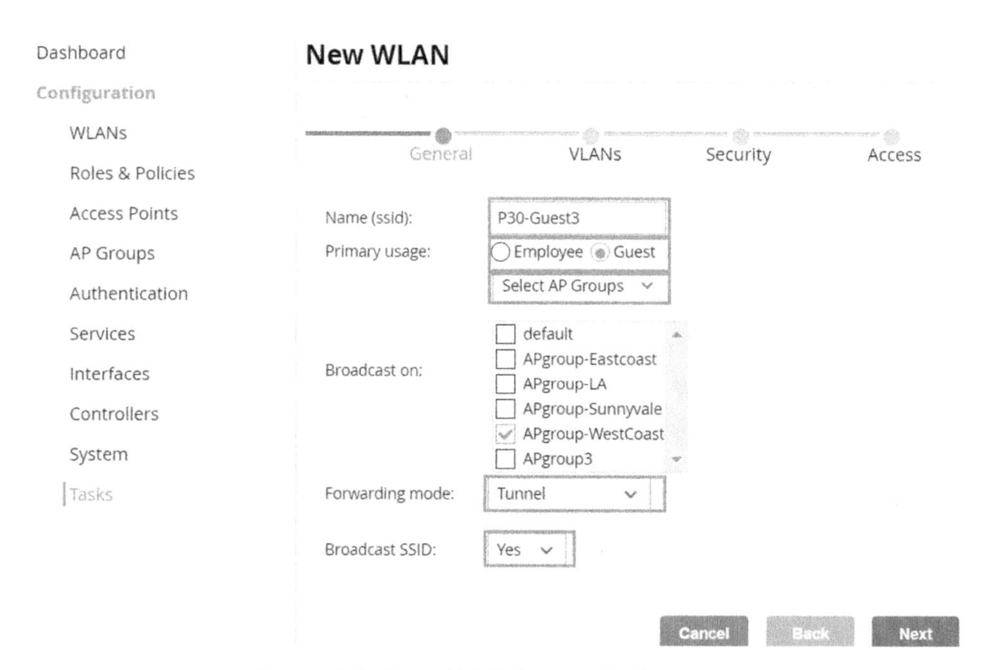

Figure 9-8 Guest WLAN general information

The WLAN wizard prompts you for many parameters, arranged in an easy, logical workflow. The first set of parameters are grouped within the general information area. You configure the SSID name, and then choose other setting options, or leave them at their defaults, as appropriate. Figure 9-8 shows these options, which are described below:

- **Name (SSID)**—Assign a WLAN name. This SSID name can be broadcast to WLAN clients.
- **Primary Usage**—Select whether the WLAN will support Employees or Guests.
  - Employee—You are presented with enterprise options for authentication and encryption, as previously covered.
  - Guest—You are presented with captive portal options.

- **Broadcast on**—Decide whether to broadcast the SSID on all APs associated with the MC or MM, or only for select AP Groups. If you choose the **Select AP Groups** option, you are prompted to select one or more AP groups.
- **Forwarding mode**—Select how the APs are to forward this traffic onto the LAN.
  - **Tunnel mode**—Tunnels *data traffic* to the Mobility Controller using GRE. PAPI *control traffic* is not tunneled—it is transported via UDP port 8211.
  - **Decrypt-Tunnel mode**—The AP decrypts and decapsulates all client 802.11 frames. It then translates them to 802.3 frames and sends them via a GRE tunnel to the MC, where firewall policies are applied. For data traveling from wired network to client, the MC sends 802.3 frames via GRE tunnel to the AP. The AP translates these to 802.11 frames, performs encryption and hashing functions, and sends them to the client. PAPI control frames arrive at the controller via a secure IPSec tunnel, such as CPSec.
  - **Other forwarding modes**—This includes Bridge mode in which the AP performs encryption/decryption and forwards traffic directly toward its destination, without tunneling traffic to the controller. Another forwarding mode is split-tunnel, where some traffic is tunneled, and other traffic is bridged.
- **Broadcast SSID**—By default, APs broadcast the new WLAN SSID upon creation. To avoid this, use the drop-down list and select No**.** Our focus here is on Guest SSIDs, which are typically broadcast.

Note

Options with the * symbol the default.

Note

Bridge Mode is not a recommended deployment mode, due to its limitations. Numerous features do not work in bridge mode. Bridge mode configuration can be done through the profiles or CLI only.

Guest WLAN VLAN

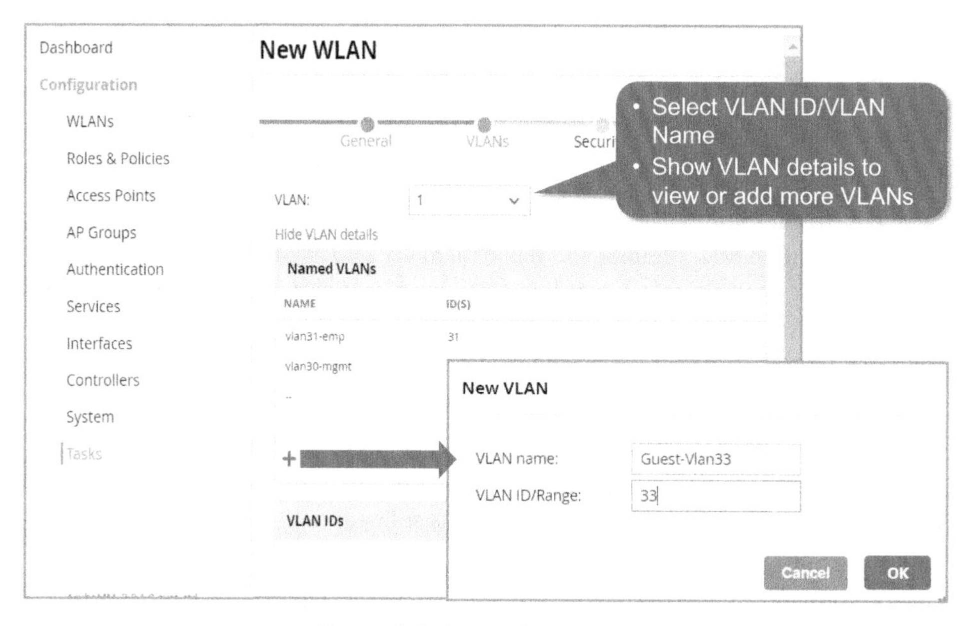

Figure 9-9 Guest WLAN VLAN

The next section of the WLAN wizard configures VLANs. The VAP needs a VLAN assignment. All users connecting to the guest SSID are placed in this VLAN. From the drop-down, select a VLAN ID/VLAN Name. You can click on **Show VLAN details** to view or add more VLANs. To add a VLAN, click the "+" symbol, as shown in Figure 9-9.

Guest WLAN VLAN IPv4 options

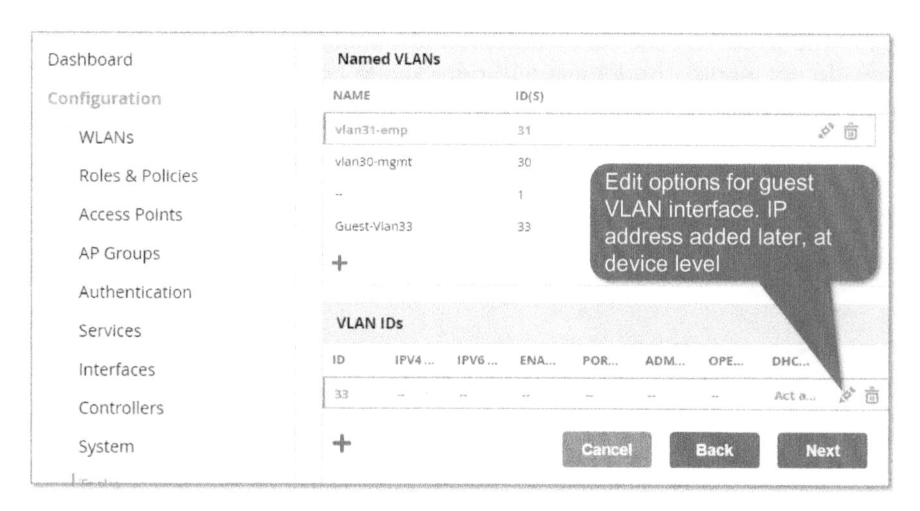

Figure 9-10 Edit a VLAN

In the scenario depicted starting with Figure 9-7, the WLAN was configured at the top of the hierarchy in the Managed Network group. Therefore, you simply need to enable the admin state of the VLAN. Figure 9-10 shows you how to enable the admin state of the VLAN—simply click the Edit icon (pencil) for the VLAN you want to enable. The Edit VLAN window appears, as shown below, in Figure 9-11.

Edit VLAN 33

VLAN ID:	33
Admin state:	Enabled
IPv4	
IPv4 address:	
Enable NAT:	Enabled
IPv6	
Local link address:	
Global unicast address:	

Cancel OK

Figure 9-11 Edit VLAN Window

In the Edit VLAN window's Admin State drop-down list, choose **Enabled** to enable the VLAN. You can also enable NAT for the guest WLAN from this window.

Notice that in Figure 9-11, the option to configure an IP address is not available. This is configured later, at the device level.

Guest WLAN security

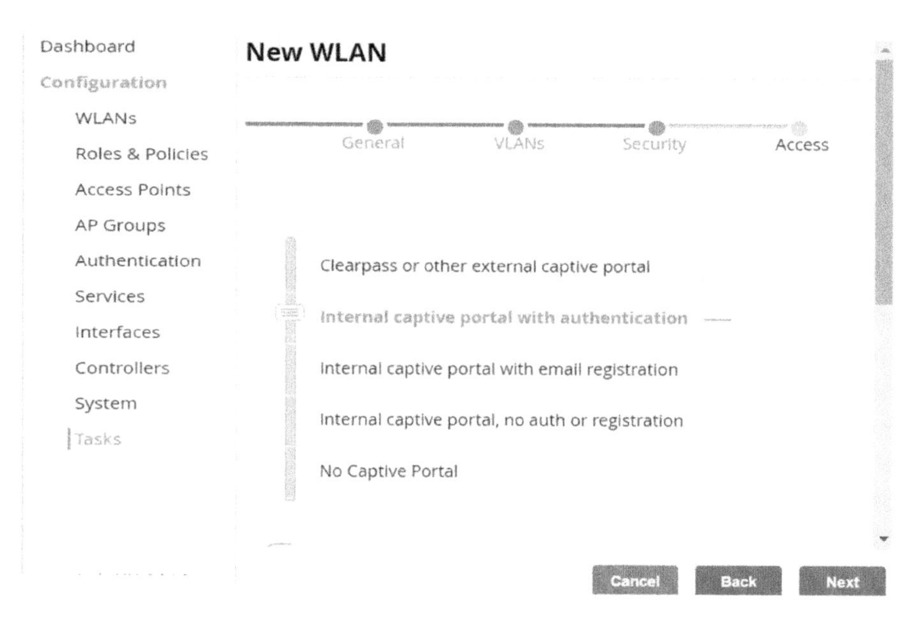

Figure 9-12 Guest WLAN security

The next section of the WLAN wizard is the Security section, where you can configure Captive Portal options for your guest VLAN. You can use captive portal with or without the PEFNG license installed in the Mobility Master. The PEFNG license provides identity-based security to wired and wireless clients, through user roles and firewall rules. You must purchase and install the PEFNG license on the Mobility Master to use identity-based security features. Use of the PEFNG license creates differences in captive portal configuration and functionality.

The Guest WLAN wizard allows for five different options for guest network setup, with or without captive portal web page:

- ClearPass or other external captive portal
- Internal captive portal with authentication
- Internal captive portal with email registration
- Internal captive portal, no authentication or registration
- No Captive Portal

The scenario in Figure 9-12 uses the internal captive portal with authentication. Later in this chapter, you will explore the use of ClearPass as an external captive portal.

ClearPass is an extensive and highly capable access control system for wired, wireless, and VPN infrastructures. Policy Manager comprises the core of the ClearPass system. Policy manager processes RADIUS and TACACS+ authentication requests for endpoints and Network Access Devices (NAD).

You can extend ClearPass capabilities beyond simple authentication by using expandable applications. For example:

- **OnGuard** helps gather health information from endpoints such as the state of their firewall or their antivirus software.
- **Guest** provides powerful visitor access management for organizations that need to support non-employee usage. Guest controls access for visitors, contractors, or other user types.
- **Onboard** helps employees use their personal devices on secure corporate networks. It can automate the process of provisioning these mobile devices with secure credentials like a certificate.

Guest WLAN security captive portal options

You can use captive portal for guest and registered users at the same time. The default captive portal web page displays login prompts for both registered users and guests. There are two options you can use to customize the default captive portal page:

- **Template:** Allows for template-based customization of the Login and Welcome pages (Figure 9-13).

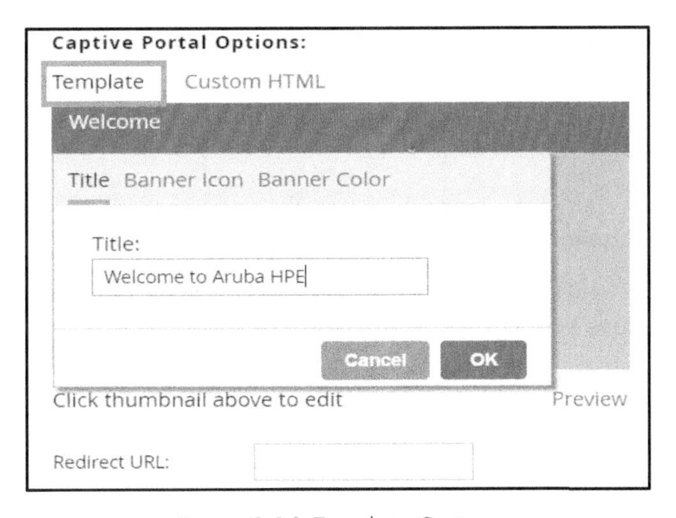

Figure 9-13 Template Options

- **Custom HTML**—Enables you to upload custom Login and Welcome pages (Figure 9-14).

Captive Portal Options:
Template Custom HTML
File for Login page: Browse Preview
File for Welcome page: Browse

Figure 9-14 Custom HTML Options

You can also load up to 16 different customized login pages into the Mobility Master. The login page displayed is based on client SSID.

Note
Some Captive Portal profile configuration can only be done via GUI. There are no CLI commands to perform some of these actions. This includes uploading a custom login or Welcome page, background images, logos, AUP texts, and so on.

Captive Portal Template customizations

From the Template page you can personalize the following:

- Banner title
- Banner icon
- Banner color
- Welcome text
- Acceptance Use Policy
- Captive portal background

To make changes to the Welcome Text or Acceptance Use Policy, click the thumbnail of the item you would like to change. Enter your changes in the pop-up box that appears, as shown in Figure 9-15.

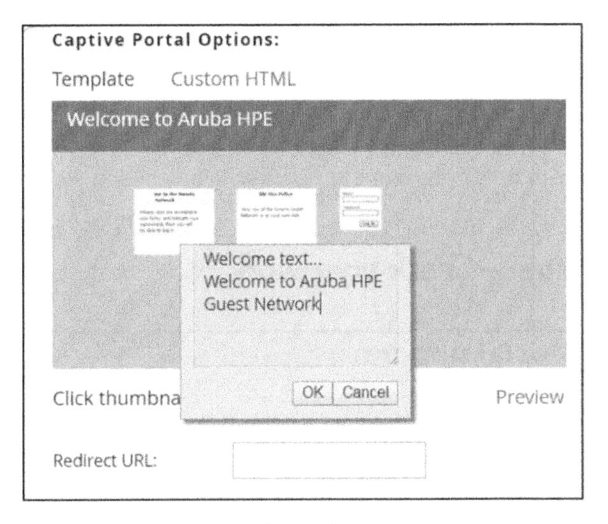

Figure 9-15 Edit Welcome Message

To make changes to the body color of the page, click anywhere in the body of the page, as show in Figure 9-16.

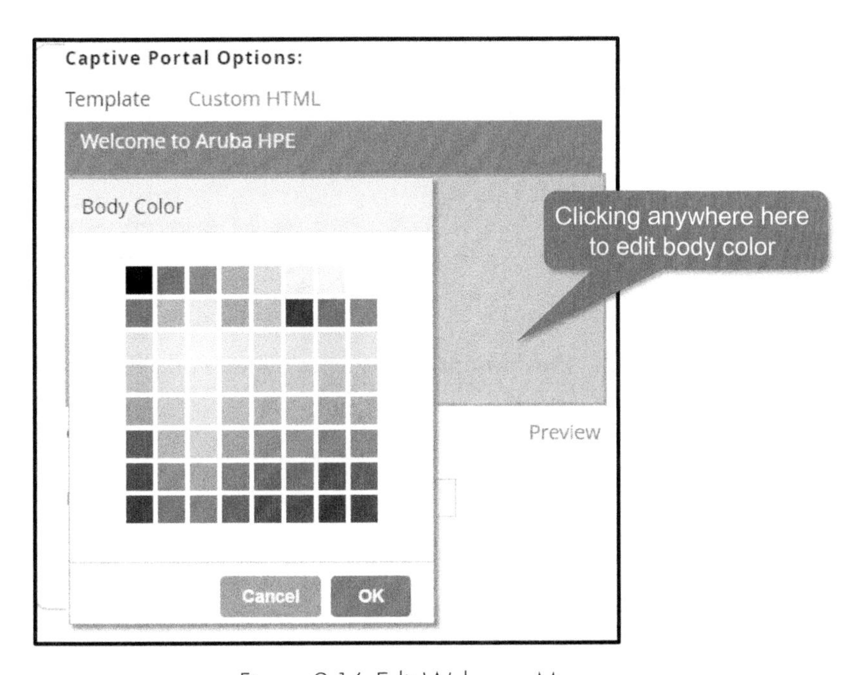

Figure 9-16 Edit Welcome Message

Starting with ArubaOS 8.0, the Reply-Message returned by the RADIUS server for a Captive Portal Authentication can be customized using the Standard Radius attribute **reply-Message** VSA.

The background image and text should be visible to users with a browser window on a 1024 by 768 pixel screen. The background should not clash if viewed on a much larger monitor. A good option is to have the background image at 800 by 600 pixels and set the background color to be compatible. The maximum image size for the background can be around 960 by 720 pixels, as long as the image can be cropped at the bottom and right edges. Leave space on the left side for the login box.

Captive Portal Template—Check your work

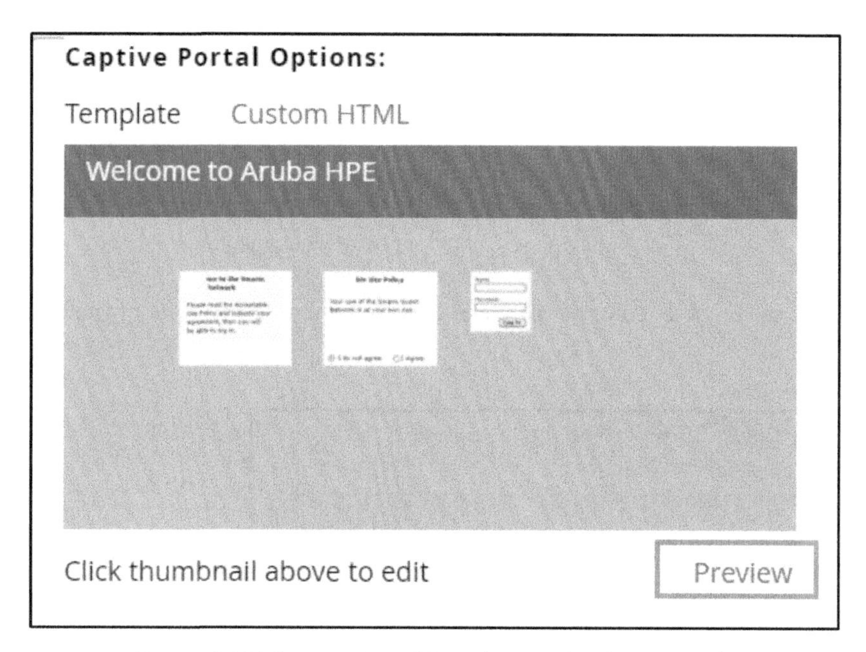

Figure 9-17 Captive Portal Template—Check your work

Before submitting changes, click the **Preview** option and check your work. Preview the customized values to be sure the page appears as desired.

Guest WLAN access

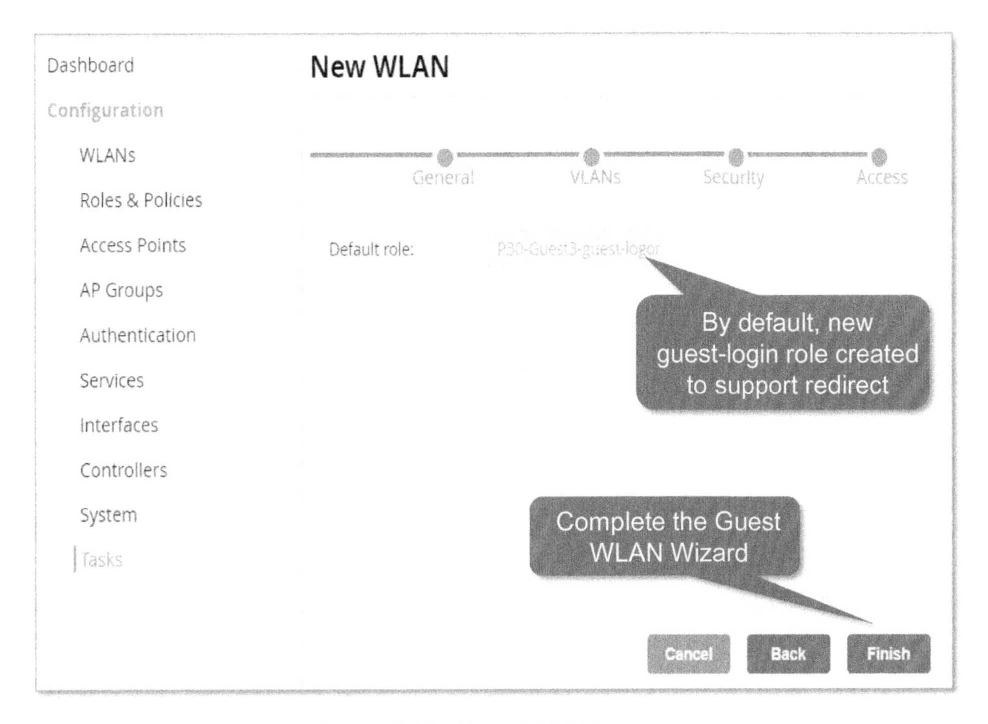

Figure 9-18 Guest WLAN access

Captive portal without PEFNG licensing

This discussion applies to the base ArubaOS, with no added licenses. This scenario allows full network access to all connected users—both guests and employees. However, you cannot configure or customize user roles. This is only available with PEFNG licensing. Captive portal allows you to control or identify who has access to network resources.

When you create a captive portal profile, an implicit user role is automatically created with the same name as the captive portal profile. This implicit user role allows only DNS and DHCP traffic between the client and network. This role also directs all HTTP or HTTPS requests to the captive portal. You cannot directly modify the implicit user role or its rules. Upon authentication, captive portal clients are allowed full access to their assigned VLAN.

Using captive portal with a PEFNG license

The PEFNG license provides identity-based security for wired and wireless users. There are two user roles that are important for captive portal:

- **Initial user role**—Specified in the AAA profile, it directs client web requests to the captive portal. This can be the predefined **guest-logon** system role. The captive portal authentication profile specifies the login page and other parameters. Initial user role configuration must include the applicable captive portal authentication profile instance.

Default user role—specified in the captive portal authentication profile. This role is granted to clients upon captive portal authentication. This can be the predefined **guest** system role.

Device level IP address and DHCP for Guest Network

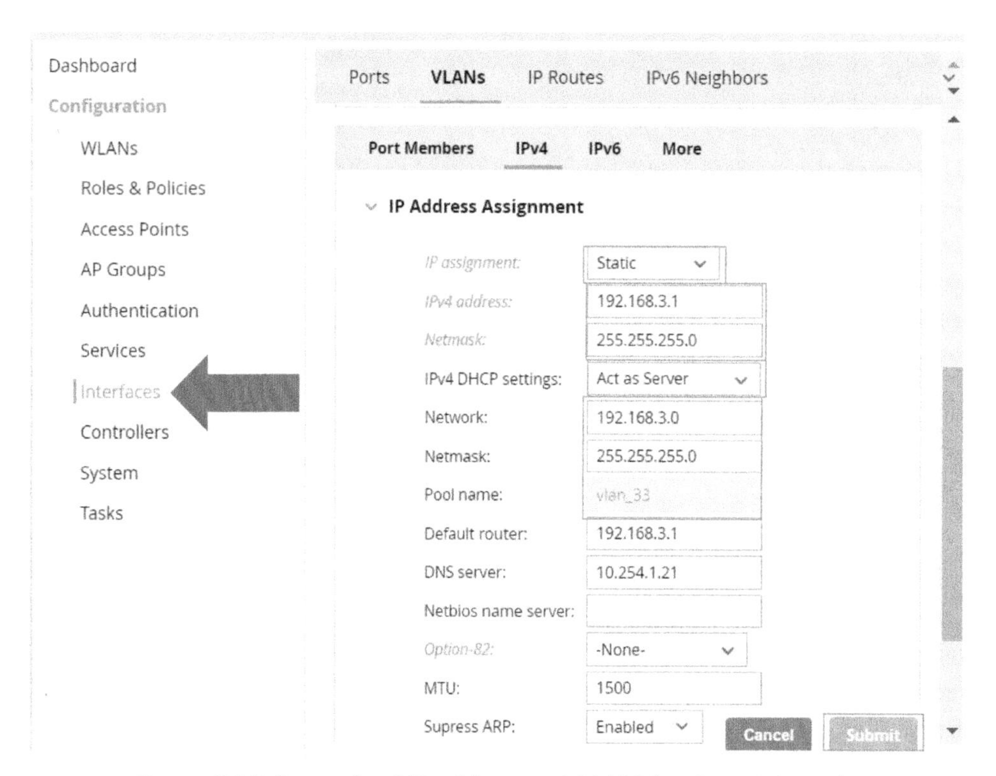

Figure 9-19 Device level IP address and DHCP for Guest Network

Figure 9-19 shows the configuration of IP addressing and DHCP services. In this scenario, the controller's guest VLAN interface is assigned a static address of 192.168.3.1, with a/24 mask.

A DHCP server is also configured, to serve IP address information to guest endpoints. Various options are configured, including the network, mask, default router, and DNS server.

Click Submit to save your changes.

Learning check

2. Which of the following guest roles are the predefined defaults for preauthentication and Postauthentication?
   a. Guest-logon
   b. Guest
   c. Logon
   d. Authenticated

3. Which two options can you use to customize the default captive portal page?
   a. Template
   b. Custom HTML
   c. CLI commands
   d. Remote console configuration

Answers to learning check

2. Which of the following guest roles are the predefined defaults for preauthentication and postauthentication?
   a. **Guest-logon**
   b. **Guest**
   c. Logon
   d. Authenticated

3. Which two options can you use to customize the default captive portal page?
   a. **Template**
   b. **Custom HTML**
   c. CLI commands
   d. Remote console configuration

Guest provisioning account

Create Guest provisioning account

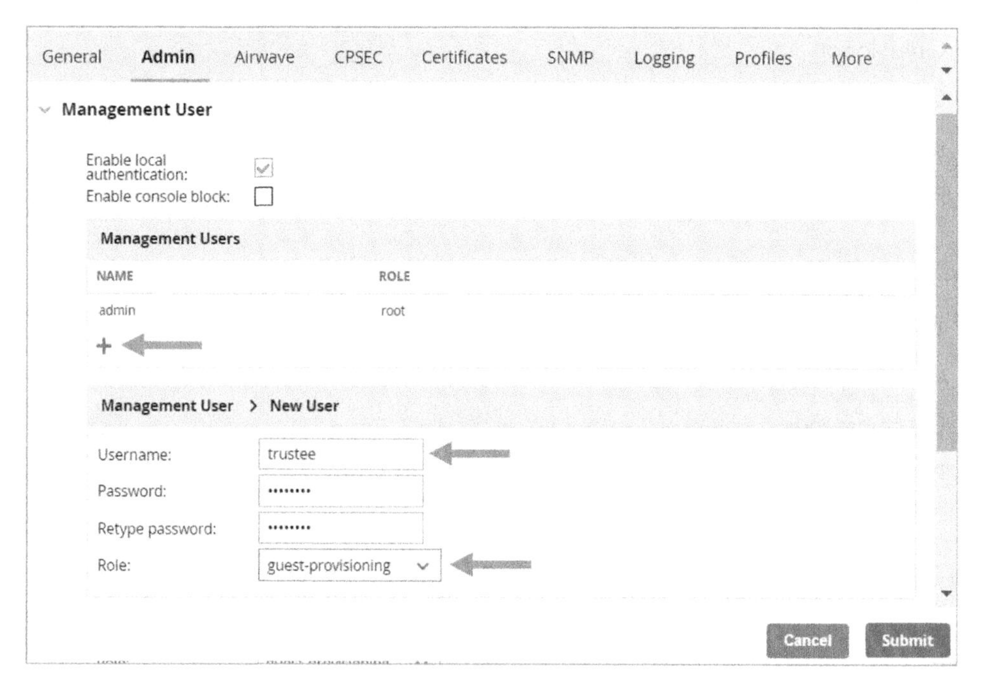

Figure 9-20 Create Guest provisioning account

Temporary user accounts are created in the internal database on the Mobility Master/Mobility Controller. You can create a user role to allow receptionists to create temporary user accounts. Guests can use the accounts to log into a captive portal login page to gain Internet access.

Figure 9-20 shows the creation of a user named Trustee, which is assigned to the administrative Role guest-provisioning. You can login as Trustee when you need to create user Guest accounts.

Create User Guest accounts

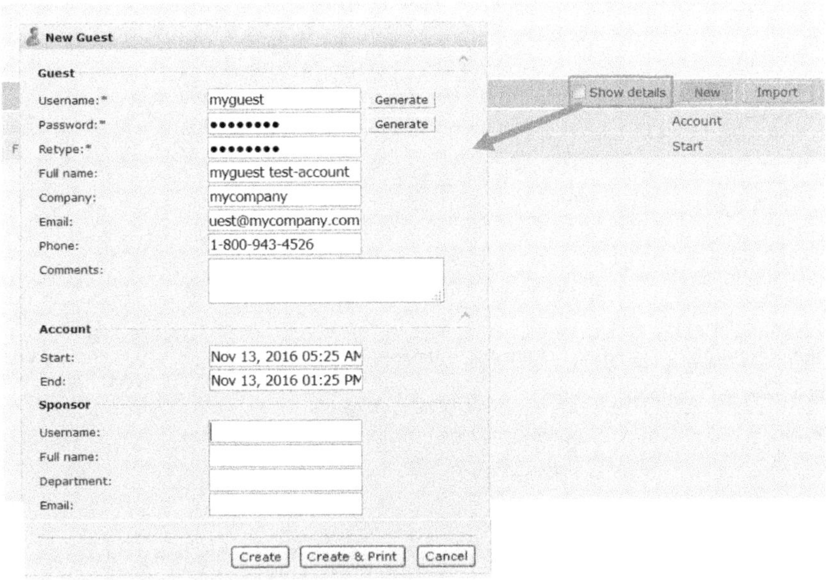

Figure 9-21 Create User Guest accounts

The receptionist logs into the controller interface using the designated user id and password. This account is restricted and can only create guest user accounts. To create a new user, click New, as shown, then enter the details and click Create. Note that the guest user has an expiration time that you can set.

Maintenance of the internal database and Guest Users

All Servers 2

NAME	TYPE	IP ADDRESS / HOSTNAME	SERVER GROUP
Dev-CPPM	Radius	10.254.5.100	--
Internal	Internal	--	default internal

Server >Internal Users Options Import Export Guest user page

USER NAME	ROLE	EMAIL	EXPIRATION	STATIC IP FOR RAPS	ENABLED
myguest	guest	myguest@mycomp...	11/19/2016 13:25	0.0.0.0	Yes
myguest2	guest	myguest2@mycom...	11/19/2016 14:03	0.0.0.0	Yes

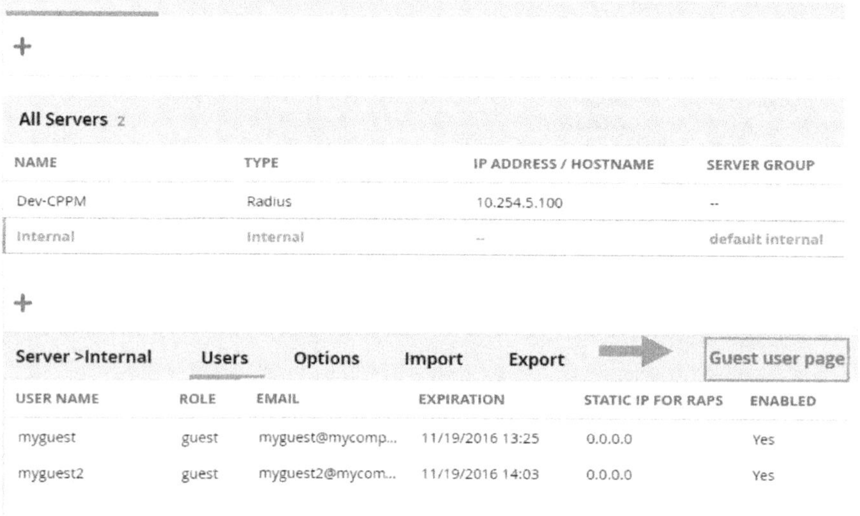

Figure 9-22 Maintenance of the internal database and Guest Users

Figure 9-22 shows the root administrator's path to manage the Internal-DB. The "Guest User Page" button brings up the same user administration menu that appeared when the new user was created.

From the top hierarchy, navigate as follows:

- **Mobility master > Configuration > Authentication > Auth Servers**. Click internal. This will push the username account to all mobility controllers, including the MM.
- **Managed Network > Westcoast > Aruba-Sunnyvale > Building-1344 > P1T3MC > Configuration > Authentication > Auth Servers**. Click internal. This is at the device-level.

At the CLI, use:

- `show local-userdb-guest` view a summary of the local user database
- `show local-userdb-guest verbose` see all user fields

Learning check

4. Which are the available options to set up a guest network?
   a. ClearPass or other external captive portal
   b. Internal captive portal with authentication
   c. Internal captive portal with email registration
   d. Internal captive portal with internal self-registration
   e. Internal captive portal with branding option

5. Guest Provisioning Account allows the receptionist to create, delete, and import guest user accounts, without having administrative privileges to mobility controller.
   a. True
   b. False

Answers to learning check

4. Which are the available options to set up a guest network?
   a. ClearPass or other external captive portal
   b. Internal captive portal with authentication
   c. Internal captive portal with email registration
   d. Internal captive portal with internal self-registration
   e. Internal captive portal with branding option

5. Guest Provisioning Account allows the receptionist to create, delete, and import guest user accounts, without having administrative privileges to mobility controller.
   a. True
   b. False

Troubleshoot captive portal

Captive portal troubleshooting

Many captive portal issues stem from failure to launch the Portal page. If this occurs, troubleshoot with the following steps:

- Determine if the client has an IP address.
- The portal page will not typically launch unless an http session was established. This could be because the DNS server was not configured in the client or is unreachable. Check DNS server configuration and reachability.
- Check the firewall settings; DNS and DHCP should be allowed. A DNS server address MUST be configured on guest clients' DHCP pool of addresses for captive portal to work properly.
- Check the client pre-auth role; it should have a Captive Portal policy and Captive Portal profile assigned.
- Check for certificate errors

The next few pages will give you details on each of these diagnostic steps.

Guest with preauthenticated role and Firewall Policy

To verify whether guests are assigned the correct preauthenticated role, navigate to **Dashboard > Clients**. In the list of clients, check the Role column. To check whether that role has the correct firewall policies.

To check the rules in those firewall polices to see if the appropriate traffic is permitted, navigate to **Configuration > Roles & Polices > Roles**. In the Roles table, click the role that was assigned to the user you are checking, and the policies for that role will show in the bottom table.

A **guest-logon** user role is assigned to a client that associate to a guest SSID. This role is more restrictive than the **logon** role, which is typically assigned to clients that associate to employee SSIDs.
The guest-logon user role consists of the following ordered policies:

- **Captiveportal**—A predefined policy that allows for redirection and captive portal authentication
- **Logon-control**—A policy that you create. If you use the WLAN wizard, then the system creates this pre-auth-role. Either way, this policy contains the following rules
  - Allows DHCP exchanges between a user and a DHCP server during business hours. Blocks other users from responding to DHCP requests
  - Allows ICMP exchanges between a user and MM during business hours
  - Allows DNS exchanges between a user and a public or internal DNS server during business hours

- Traffic is source-NATed using the MM's IP interface for the VLAN
- Denies a few blocks of reserved IP addresses 169.254.0.0/16 and 240.0.0.0/8
- Denies user access to internal networks and Internet

Guest with postauthenticated role

To verify whether guests are assigned the correct preauthenticated role, with its corresponding firewall policies, navigate to **Dashboard > Clients**. Then in the list of clients, check for the role listed in the Role column. Verify that each firewall policy permits the appropriate protocols.

You can also verify the role that is assigned to the user by entering the command **show user** in the CLI of the MC, as shown in Figure 9-23.

```
(P30T3-MC3) #show user
This operation can take a while depending on number of users. Please be patient

Users
-----
    IP              MAC              Name     Role     Age(d:h:m)  Auth  VPN link
ame
----------  ------------  ------  ----     ----------  ----  --------
---
192.168.3.2  ac:22:0b:94:97:ed  guest33  guest    00:00:11    Web

User Entries: 1/1
 Curr/Cum Alloc:2/1149 Free:0/1147 Dyn:2 AllocErr:0 FreeErr:0
(P30T3-MC3) #
```

Figure 9-23 Guest with postauthenticated role

The authenticated **Guest** user role consists of the following ordered policies:

- Allows DHCP exchanges between a user and a DHCP server during business hours. Blocks other users from responding to DHCP requests
- Allows ICMP exchanges between a user and MM during business hours.
- Allows DNS exchanges between a user and a public or an internal DNS server during business hours.
- Traffic is source-NATed using the MM's IP interface for the VLAN.
- Allows HTTP(S) traffic from the user during business hours. Traffic is source-NATed using the MM's IP interface for the VLAN.
- **drop-and-log** is a policy that you create. It denies all other traffic and logs the attempted network access.

WebUI Certificate

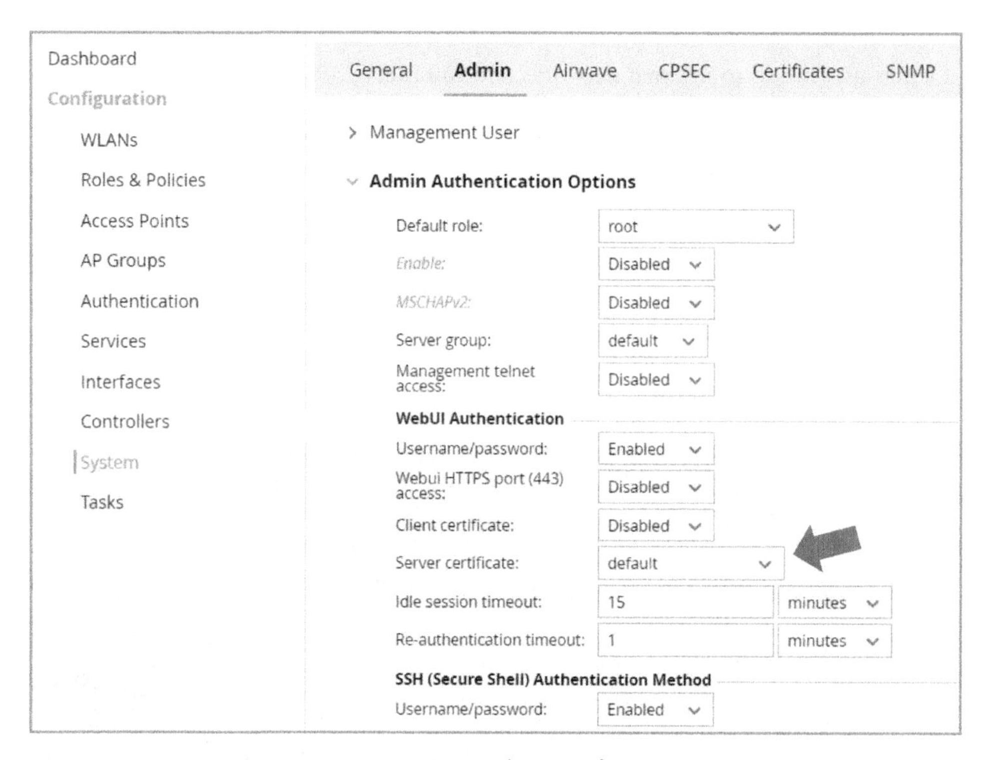

Figure 9-24 WebUI certificate

It is strongly recommended that you replace the default certificate with a custom certificate, issued for your site or domain by a trusted Certificate Authority (CA).

To replace the default certificate, first import a certificate or generate a CSR. To import a certificate or generate a CSR, navigate to **Configuration > System > Certificates**. Next change the default certificate. To change the default certificate, navigate to **Configuration > System > Admin > Admin Authentication Options**. In the WebUI Authentication section, choose the desired certificate from the Server certificate drop-down list, as shown in Figure 9-24.

For more information, please refer to Aruba VRD and support.arubanetworks.com for more tech notes.

Certificate error

You may see a server certificate error while using the WebUI or 802.1X termination. This is because the default certificate is self-signed and is not associated with a well-known root certificate. Such certificates are not trusted by endpoints. From your web browser, you can view and verify the certificate and add an exception to proceed.

WebUI certificate

```
(P30T3-MC3) #show web-server profile

Web Server Configuration
------------------------
Parameter                                              Value
---------                                              -----
Cipher Suite Strength                                  high
SSL/TLS Protocol Config                                tlsv1 tlsv1.1 tlsv1.2
Switch Certificate                                     default
Captive Portal Certificate                             default
IDP Certificate                                        default
Management user's WebUI access method                  username/password
User absolute session timeout <30-3600> (seconds)      60
User session timeout <30-3600> (seconds)               3600
Maximum supported concurrent clients <25-320>          75
Enable WebUI access on HTTPS port (443)                false
Enable bypass captive portal landing page              false
Exclude Security Headers from HTTP Response            false
(P30T3-MC3) #
```

Figure 9-25 WebUI certificate

The default server certificate on mobility controller is tied to securelogin.arubanetworks.com. You can use the CLI command show web-server profile to see what certificate names are in use. You can see certificates used by the switch and for the captive portal, as shown in Figure 9-25.

Guest Access with ClearPass

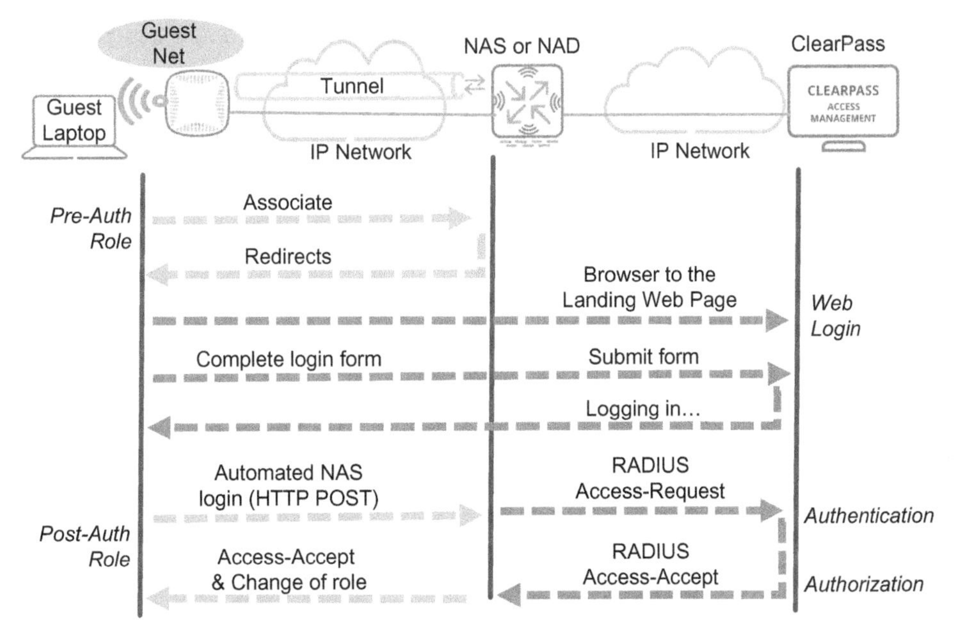

Figure 9-26 Guest Access with ClearPass

When guests connect to the Network Access Device (NAD), they are redirected to a login or a captive portal page. You can setup this login page so that user access is very limited, until they authenticate successfully. In this scenario, the MC serves the function of a NAD. In other scenarios, the NAD could be a wired switch, firewall, or IAP.

1. A user associates with the guest SSID and is assigned a Pre-auth-role by the MC, which limits their network access.
2. The user opens a browser, and requests some URL. The MC redirects this to the ClearPass Web Login page. It is important to remember that it is the NAD that redirects the user, and not ClearPass. Once redirected to ClearPass, the user is presented with web login landing page.
3. The user enters their credentials and submits the form, triggering an Automated NAS login. The guest device sends an HTTP POST message to the NAD/MC. This triggers a RADIUS request from MC to ClearPass, which contains user credentials.
4. ClearPass validates the user credentials. If they are OK, a RADIUS-Accept message is sent back to the MC. Thus, the user is logged in, and so is assigned their Post-Auth-Role. This role controls their access privileges.

Add ClearPass to the Controller Server list

In this scenario, the ClearPass server has already been configured. You need to configure the mobility controller to redirect the client web traffic to ClearPass. You must also ensure that the guest pre-auth role allows this traffic.

To add a ClearPass server to the MC's server list, navigate to **Configuration > Authentication > Auth Servers** and under the All Servers section click the "+" symbol to add new server. The New Server dialogue box appears, as shown in Figure 9-27.

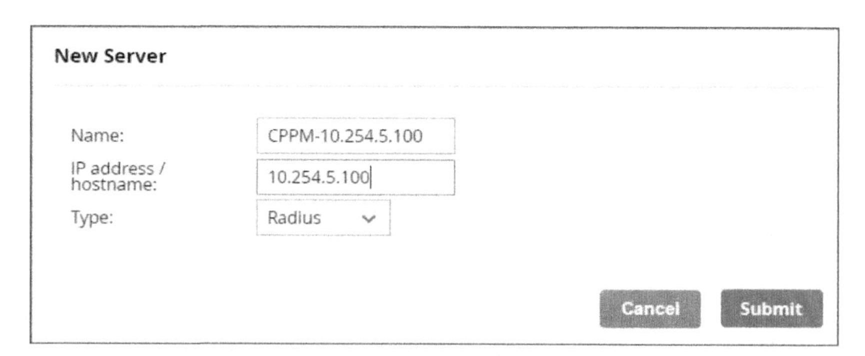

Figure 9-27 New Server Dialogue Box

To create a server group, navigate to **Configuration > Authentication > Auth Servers** and under the Server Group section, click the "+" symbol to create new server group. The Add Server Group dialogue box appears, as shown in Figure 9-28. The group you create in this window will be referenced in the captive portal profile. You cannot add servers directly to L3 or AAA profiles. You must define the server in a server group and then apply that group to a profile.

Figure 9-28 Add Server Group Dialogue Box

Add ClearPass Server to the server group

After you create the server group, you add the ClearPass server to this new group. This server group will be used with a L3 authentication profile (captive portal profile in this scenario).

To add the ClearPass server to the new server group, navigate to **Configuration > Authentication > Auth Servers** and select the server group to which you want to add the server. In the Servers detail section click the "+" symbol to add a server to the Server group. The New Server for "Group" dialog box appears. From this box, **choose Add Existing Server** and choose the ClearPass server you created, as shown in Figure 9-29.

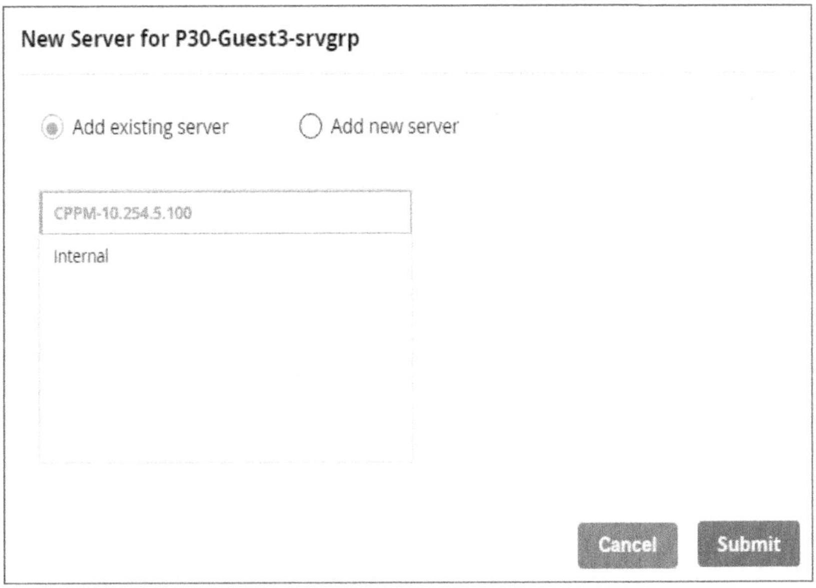

Figure 9-29 Add ClearPass Server to the server group

L3 authentication profile

To configure the L3 authentication profile, navigate to **Configuration > Authentication > L3 Authentication > Captive Portal Authentication > P30-Guest3 profile**.

You need to modify two things in this Captive Portal (CP) profile:

- Change the landing web page URL to point to the ClearPass captive portal web page (**http://***ClearPassIPaddress***/guest/***NameofWebPage*), as shown in Figure 9-30.

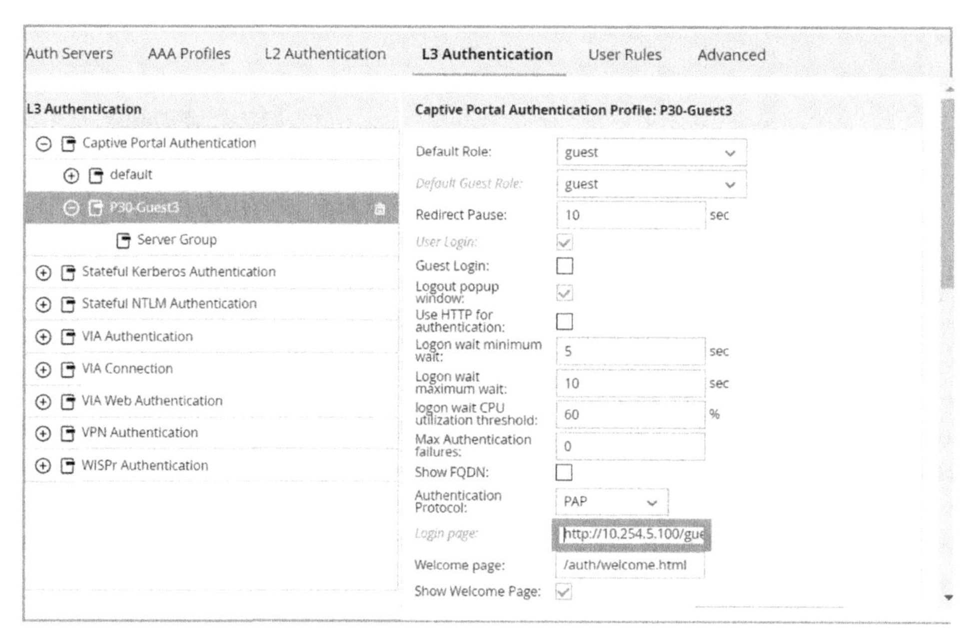

Figure 9-30 Change Landing Page

- Add the server group that contains the ClearPass server, as shown in Figure 9-31.

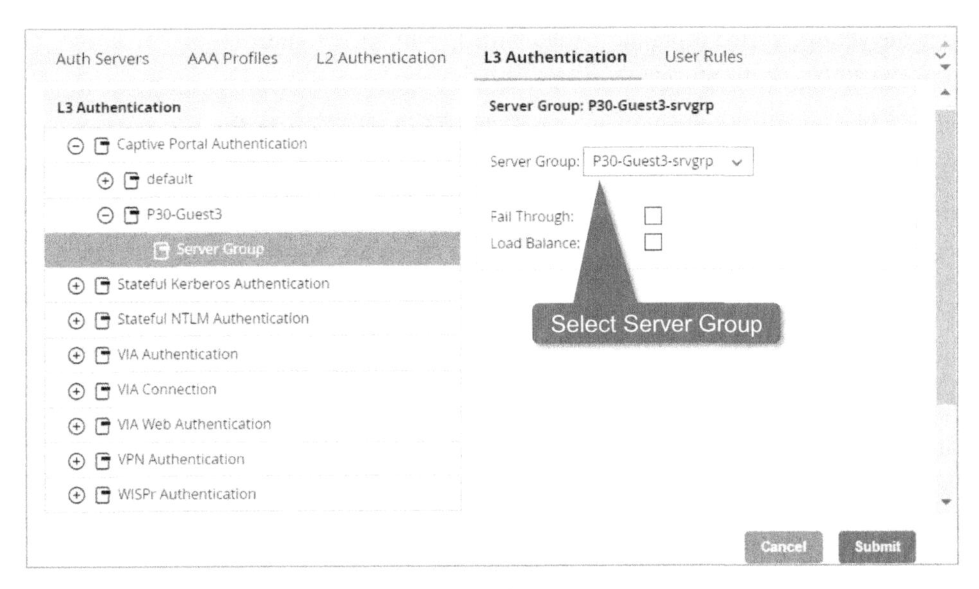

Figure 9-31 Add server group

Modify the pre-auth role

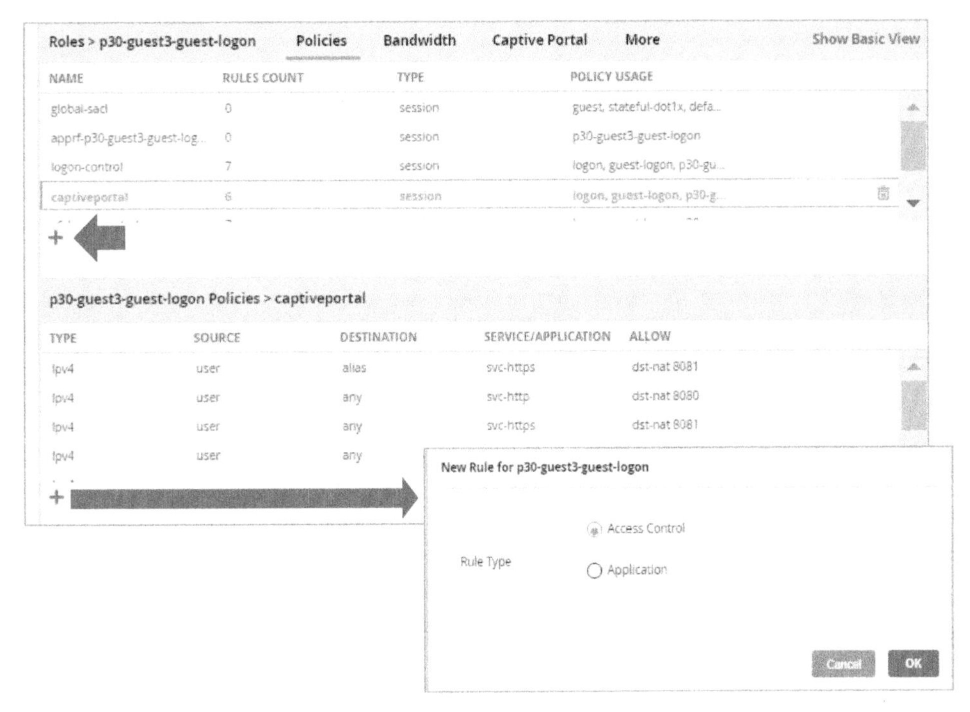

Figure 9-32 Modify the pre-auth role

Recall that the wizard created a pre-auth-role named "p30-guest3-guest-login." This role contains a firewall policy named "captiveportal."

You need to add two rules to this policy, so user HTTP & HTTPS traffic can reach the ClearPass server. Without these additions, the user cannot be successfully redirected to the ClearPass server.

To add the two rules to this policy, navigate to **Configuration > Roles & Policies > Roles** and choose **p30-guest3-guest-login** in the list of roles at the top. In the section that shows information for the p30-guest3-guest-login role, click **Show Advanced View** to see information about the policies assigned to this role. From the list of policies, choose **captiveportal**. This will open a section that shows the rules in the captiveportal policy.

To add a new rule to the captiveportal policy, click the "+" icon at the bottom of the list of rules. This will open the New Rule dialogue box, as shown in Figure 9-32.

Modify the policy captiveportal

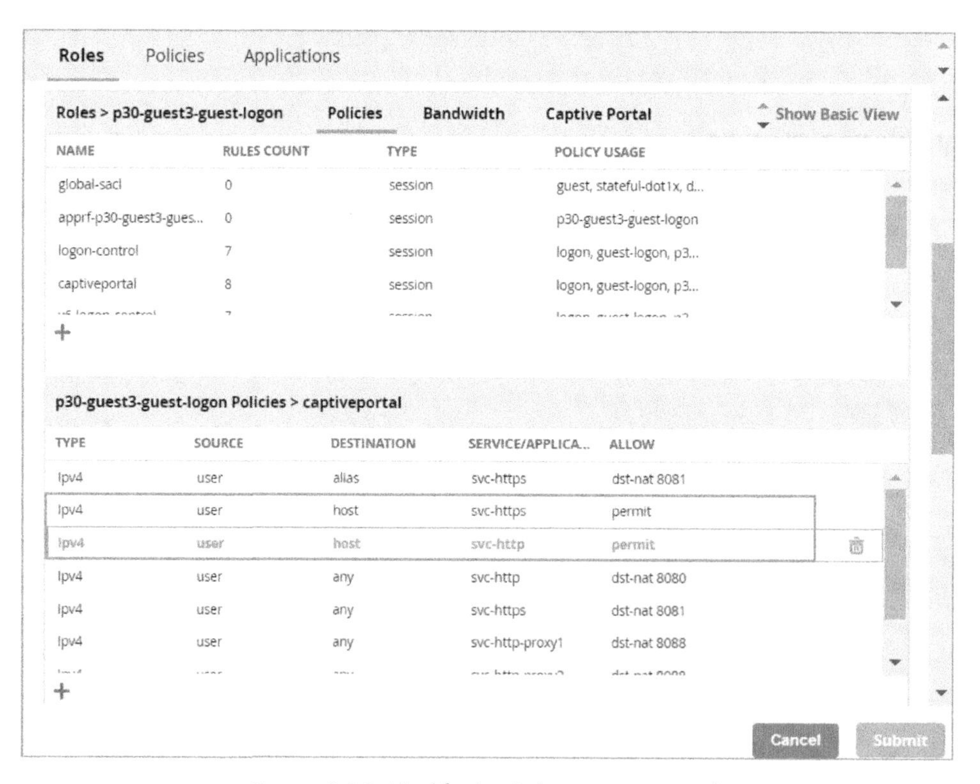

Figure 9-33 Modify the Policy captiveportal

You must add two rules to the captiveportal policy. This will allow guest HTTP/HTTPS traffic to reach the ClearPass server's IP address.

After you create the rules, do not forget to move them to the second and third lines of the session ACL, as shown in Figure 9-33. Remember, session ACLs are processed top-down. If you fail to move your new rules up above the fourth line, the redirect will not work. Notice the current fourth line redirects guests to the controller's *internal* captive portal. This means uses will not be redirected to the *external* ClearPass server.

Testing Guest access

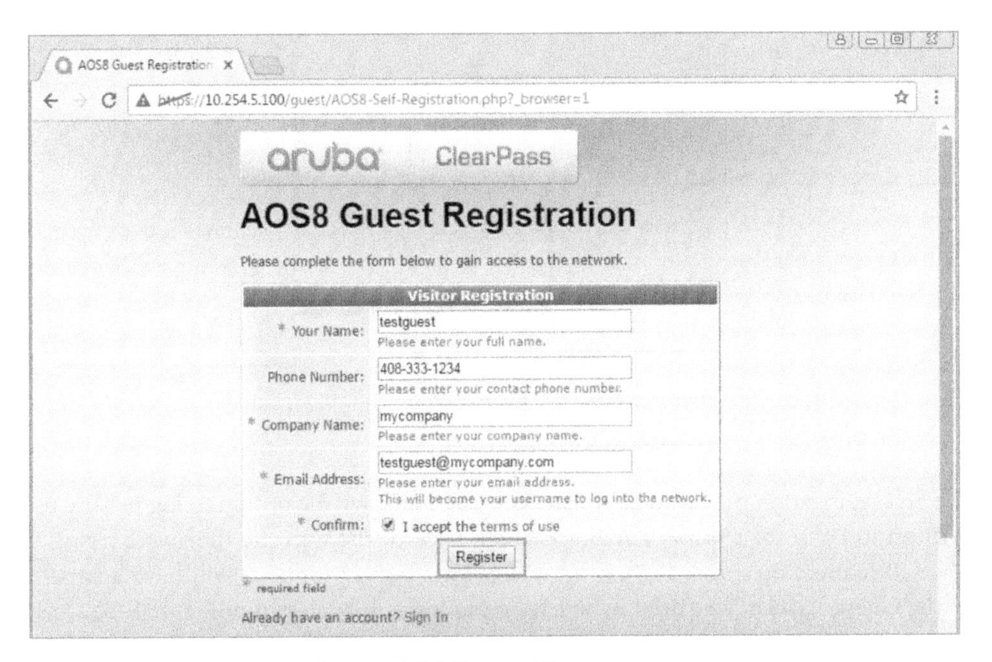

Figure 9-34 Testing Guest access

When associated users open a browser, they are redirected to the ClearPass Captive Portal (CP) web page. If the user has an account, they simply enter their credentials and log in. If the user has no guest account, they can click the Create Account option and use the self-registration process.

When guests click Create Account, they are redirected to a self-registration portal page, as shown in Figure 9-34. They fill out all required fields in the form and then click Register.

Self-Registration completed

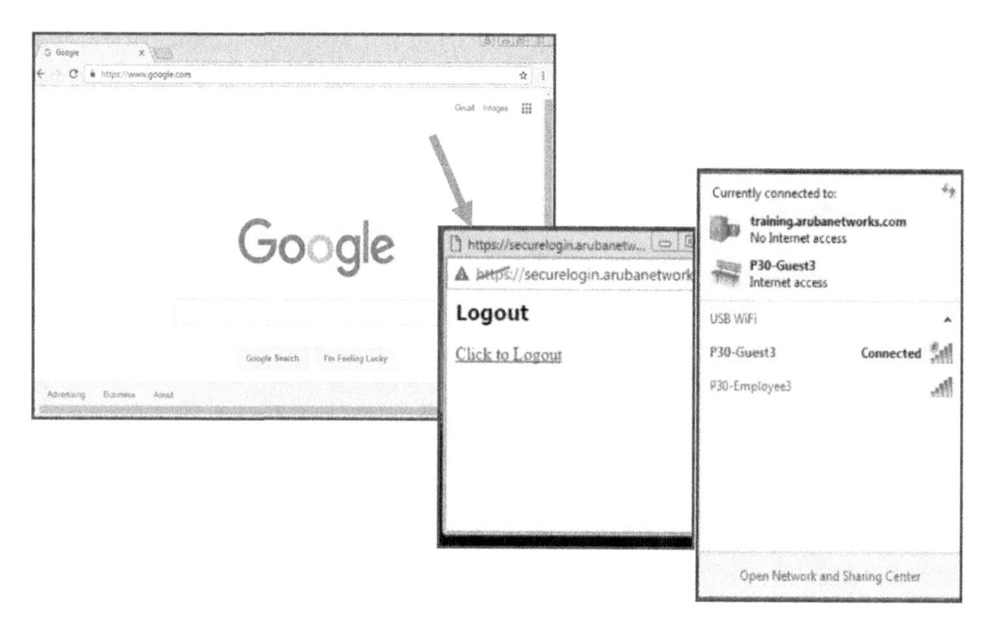

Figure 9-35 Self-Registration completed

When the Self-registration process is complete, the user will have the option to log in. The user clicks the Log In button and their browser redirects them back to the web site they were trying to access before the CP redirect occurred. If their browser allows pop-ups, guests will have a small window open for the Guest Self-Service Portal, where they can log out, as shown in Figure 9-35.

ClearPass Access Tracker Tool

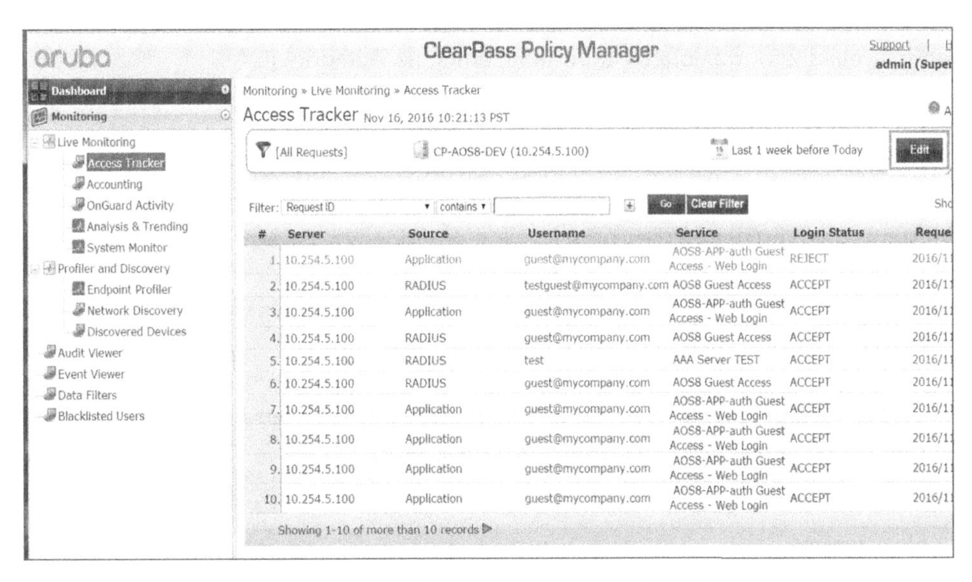

Figure 9-36 ClearPass Access tracker tool

Within the ClearPass product's Policy Manager, you can access various monitoring screens. Figure 9-36 shows the Access Tracker screen, which can be used to troubleshoot authentication issues.

This scenario shows several Accepted and Rejected requests. Click on an entry to get details about how the request was processed, why it was accepted or rejected, and what the ClearPass response was. You can adjust the columns that are displayed as well as the time frames by clicking on the Edit button.

Guest Access best practices

Here is a list of best practices when creating guest access:

- Use an open SSID, configured for open authentication
- For additional security, it is best to change the guest password on a periodic basis
- To keep the environment secure, guests get a separate IP scheme than corporate users
- You should write restrictive firewall policies for guest accounts
- Partition Guest IP access by using VLANs or separate physical links
- Use the Aruba internal database for user authentication (easy to configure expiration feature)
- Use the guest provisioning account on the Aruba controller (separate and limited administration)
- Should have a unique guest account for each guest user and then expires when not needed
- Ideally, integrate with ClearPass for guest Self-registration or sponsored guest access

Learning check

6. Aruba strongly recommends that you replace the default certificate with a custom certificate issued for your site or domain by a trusted Certificate Authority (CA).

    a. True

    b. False

7. ClearPass is an internal captive portal solution of mobility controller.

    a. True

    b. False

8. When using ClearPass to provide the external captive portal web page, which of the following captive portal profile settings must you modify?

    a. Login Page URL

    b. Server group

    c. Guest Login

    d. Welcome page

Answers to Learning check

6. Aruba strongly recommends that you replace the default certificate with a custom certificate issued for your site or domain by a trusted Certificate Authority (CA).

    a. **True**

    b. False

7. ClearPass is an internal captive portal solution of mobility controller.

    a. True

    b. False

8. When using ClearPass to provide the external captive portal web page, which of the following captive portal profile settings must you modify?

    a. Login Page URL

    b. Server group

    c. Guest Login

    d. Welcome page

10 Network Monitoring and Troubleshooting

LEARNING OBJECTIVES

✓ This chapter will explain how to use the dashboard for monitoring and troubleshooting controllers, APs, and client issues. You can also use the dashboard to analyze user traffic, potential issues, usage, and security. We will focus on several topics including:

- Use of the newly introduced banner for at-a-glance system health—controllers, APs, and clients
- Use AppRF and WebCC (deep packet inspection) to analyze user traffic
- Leverage alerts to understand and resolve basic wireless issues
- Introduction to AirWave

Banner

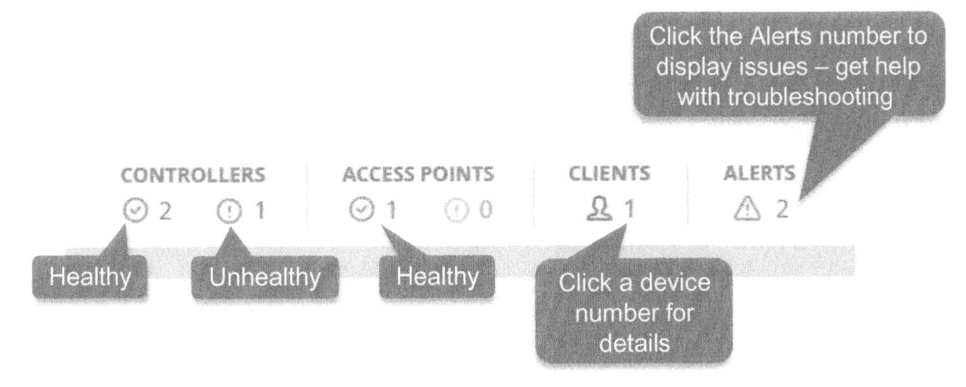

Figure 10-1 Banner

The banner feature was introduced in AOS 8 to provide health status for Controllers, APs, and clients. Banners also provide an alerting mechanism to potential wireless issues (Figure 10-1).

A green circle with a check mark indicates a healthy device and a red circle with an exclamation mark indicates an unhealthy device.

Click one of the numbers to learn more about a device. A new, more detailed page opens. This is an easy way to gain deeper insight into WLAN health and to troubleshoot Controller, AP, or client issues.

Controller dashboard

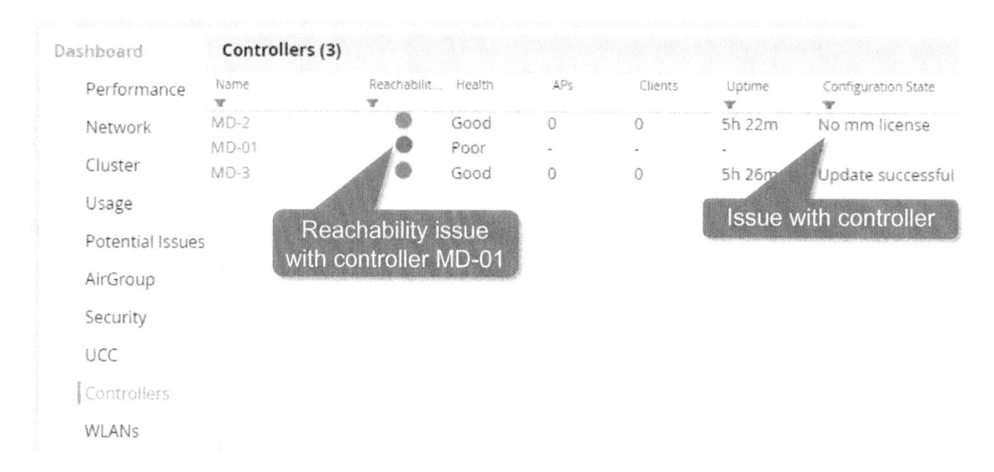

Figure 10-2 Controller dashboard

The **Controller dashboard** page lists all managed devices in the network and provides health-related information, as shown in Figure 10-2. To access the Controller dashboard navigate to **Managed Networks > Dashboard > Controllers**.

A green reachability icon means the device is reachable from the MM. Red means the device is not reachable from the MM. A health status of "good" means the MM can send and receive heartbeats to the device.

The Configuration state indicates whether the MM can successfully push configurations to the device. Click on any managed device to see more details about it.

AP dashboard

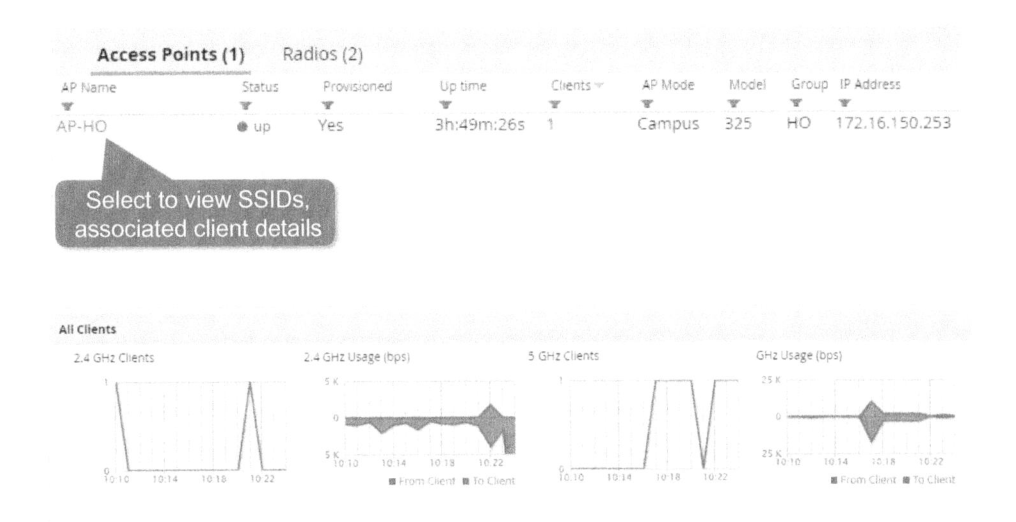

Figure 10-3 AP dashboard

The **Access Points dashboard** page displays details for all associated APs and AP radios (Figure 10-3). To access the Access Points dashboard, navigate to **Managed Networks > Dashboard > Access Points**.

Select a specific AP for trend analysis. You can see connected wireless clients and the client usage over the last 15 minutes, for both the 2.4 GHz and 5 GHz radio bands.

Client dashboard

The **Clients dashboard** page displays details for managed wireless clients. The dashboard also displays the trends for the connected clients and the client usage, under the 2.4 GHz and 5 GHz radio bands, over the last 15 minutes.

This dashboard is mainly used to view SNR, client Goodput, Retries, and dropped frames between AP and client.

Client dashboard (Cont.)

The Client dashboard shows the following key metrics:

- **SNR**—Compares the signal level to background noise level. SNR is defined as the comparison of the signal level to noise.

- **Goodput (bps)**—Displays the ratio of total bytes transmitted to the client or received from the client to the total air time required for transmitting or receiving. This should be as high as possible.
- **Dropped frames**—Displays the percentage of frames dropped when transmitted to client and received from client. This value should be low.
- **Retried frames**—Displays the percentage of frames requested for retransmission, this value will be higher when the SNR is very poor. This value should be low.

Trend analysis

Figure 10-4 Trend analysis—Client health trend

If you click the MAC address of a client, a graph displays that shows client's health trend over a period of time (Figure 10-4).

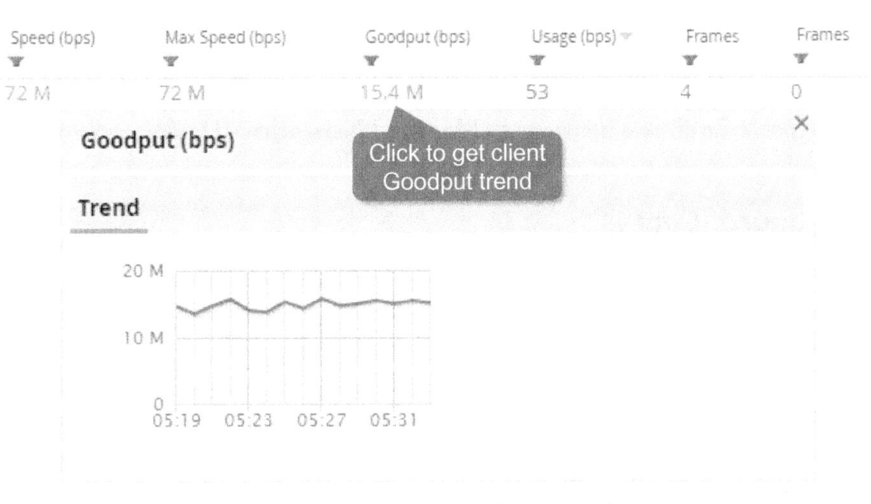

Figure 10-5 Trend analysis—Client Goodput trend

If you click the value of the Goodput column for one of the clients, a graph displays that shows client's Goodput trend over a period of time (Figure 10-5).

WLAN dashboard

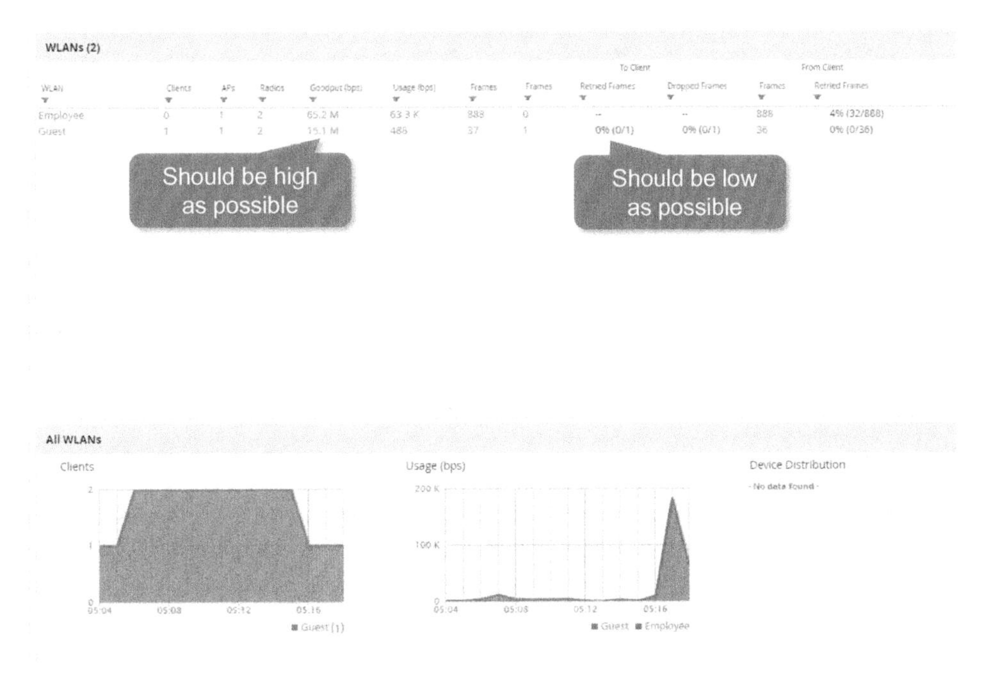

Figure 10-6 WLAN dashboard

The **WLANs dashboard** page displays WLAN details. This includes associated APs count, radios, wireless clients, and the WLAN usage in managed devices. You can also view the details of the associated APs and clients as tables (Figure 10-6).

To access the WLANs dashboard, navigate to **Managed Networks > Dashboard > WLANs**.

WLAN dashboard (Cont.)

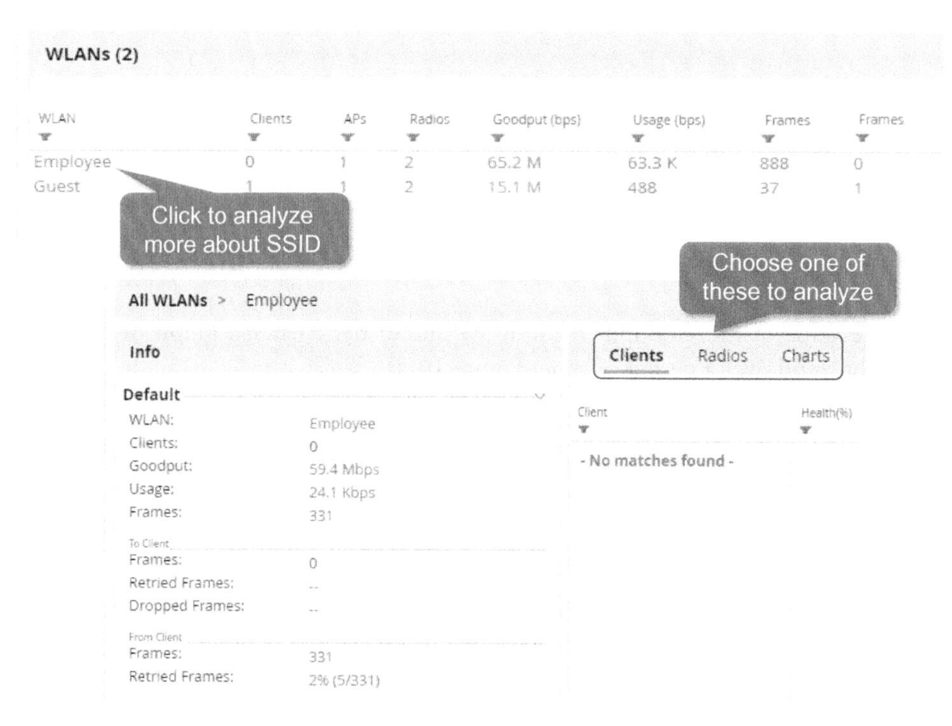

Figure 10-7 WLAN dashboard (Cont.)

When you click on a specific WLAN (SSID), a pop-up window reveals more information about clients and radios (Figure 10-7).

Potential issues

The **Potential Issues** page displays the total number of radios and wireless clients that may have potential issues in the network. You can click on the total number to view the trend of the clients and radios with potential issues in the last 15 minutes. You can also view the number of clients or radios that have a specific potential issue in each radio band.

To access the WLANs dashboard, navigate to **Managed Networks > Dashboard > Potential Issues**. The potential issues that a client may have are:

- **Low SNR**—Clients that have signal-to-noise ratio of 30 dBm or lower.
- **Low speed**—Clients that have a connection speed of 36 Mbps or lower.
- **Low goodput**—Clients that have an average data rate of 24 Mbps or lower.
- The potential issues that a radio may have are:
- **High noise floor**—Radios that have a noise floor of –85 dB or greater.
- **Busy channel**—Radios that have a channel utilization of 80% or greater.
- **High interference**—Radios that have an interference of 20% or greater.
- **High client association**—Radios that have 15 or more clients connected.

You can click on the hyperlinked number to view the details of the respective clients or radios in the bottom pane of the page in the **Radios with potential issues** section. This section has two views, **Listing** and **Trend**.

The **Listing** view displays clients with potential issues as a list in a table format. You can sort by clicking a column header of the table based on the entries on the active column.

The **Trend** view displays the radios with potential issues in a chart.

Monitoring performance

The **Performance dashboard** page displays the performance details. This includes overall client health, SNR, Goodput, noise, and interference. These statistics are tracked for all managed wireless clients and APs.

To access the WLANs dashboard, navigate to **Managed Networks > Dashboard > Performance**.

Performance dashboard

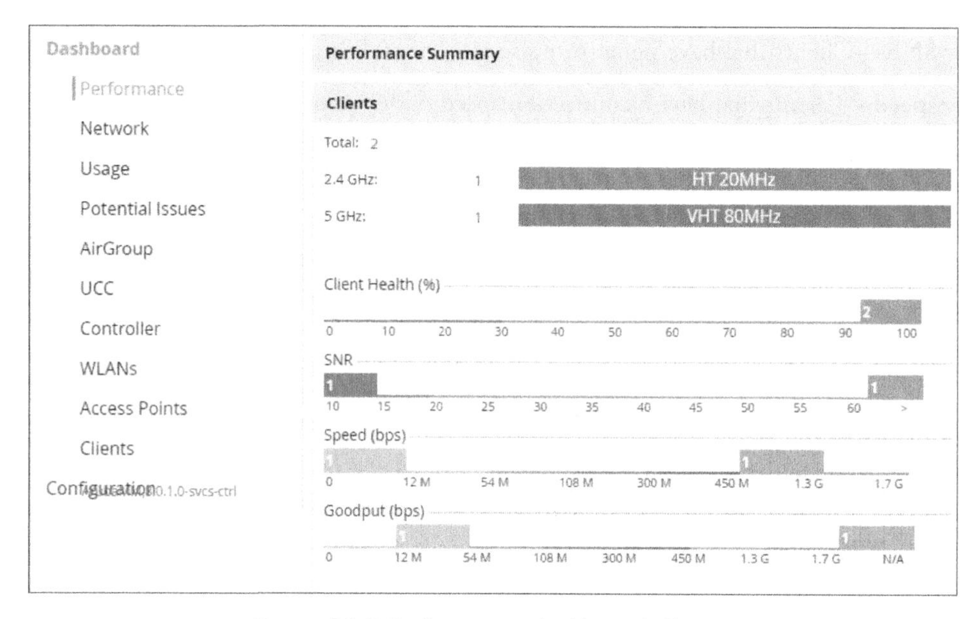

Figure 10-8 Performance dashboard Clients

The Performance dashboard Clients section displays the total number of wireless clients connected. You start with a top-level view of wireless clients. Then drill down into a specific managed device to view the wireless clients connected to it (Figure 10-8).

You can view the distribution of clients in different client health ranges, SNR ranges, associated data rate ranges, and data transfer speed ranges. This information is viewed in histograms and distributed charts.

You can click on the hyperlinked number to view the data in different screens with histograms.

A client health metric of 100% means the actual airtime the AP spends transmitting data is equal to the ideal amount of time required to send data to the client. A client health metric of 50% means the AP is taking twice as long as is ideal or is sending one extra transmission to that client for every packet. A metric of 25% means the AP is taking four times longer than the ideal transmission time or sending 3 extra transmissions to that client for every packet.

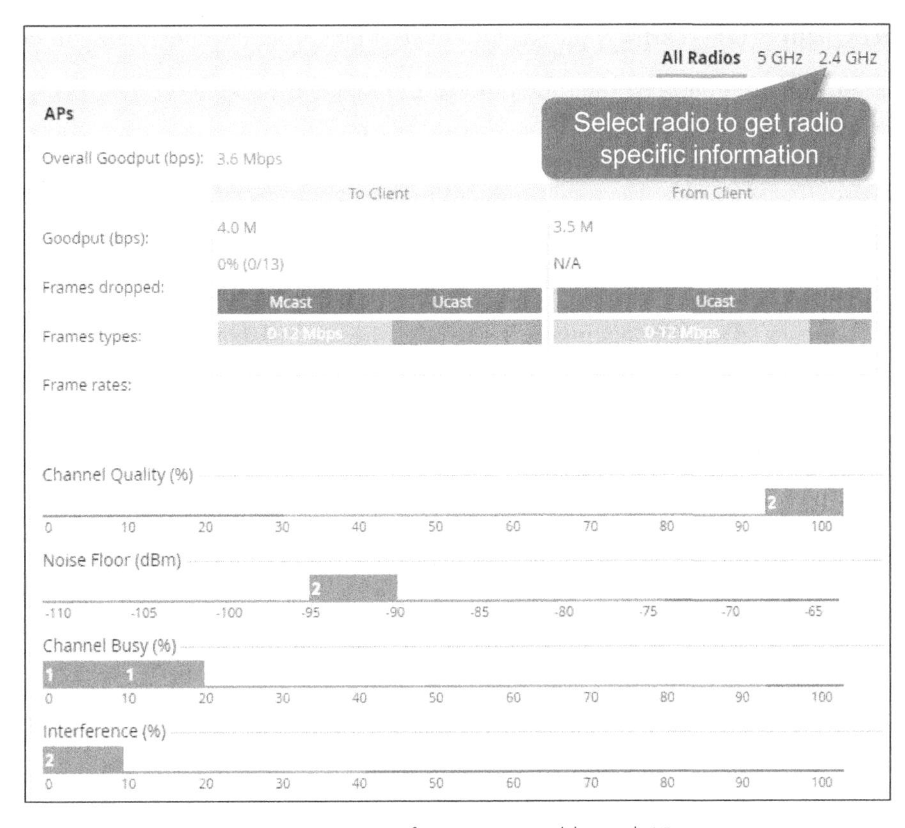

Figure 10-9 Performance Dashboard APs

The APs section of the Performance dashboard displays the following performance details of the APs on the Mobility Master. You can click the hyperlinked text and histograms to view the AP specific performance information as a trend chart (Figure 10-9).

- **Overall goodput (bps)**—Displays the ratio of the total bytes transmitted or received in the network to the total air time required for transmitting or receiving the bytes.
- **Goodput (bps)**—Displays the ratio of total bytes transmitted to a client or received from a client to the total air time required for transmitting or receiving.
- **Frames dropped**—Displays the percentage of frames dropped when transmitted to a client and received from a client.
- **Frame types**—Displays the type of frame (broadcast, multicast, and unicast) transmitted to a client or received from a client.
- **Frame rates**—Displays the range of frame rates transmitted to a client or received from clients in Mbps.

- **Channel Quality** (%)—Displays the total number of radios per channel quality in percentage.
- **Noise Floor (dBm)**—Displays the total number of radios per noise floor (dBm) range.
- **Channel Busy** (%)—Displays the total number of radios per channel busy in percentage.
- **Interference** (%)—Displays the total number of radios per Interference in percentage.
- **2.4 GHz Channels**—Displays the number of radios using each channel of the 2.4 GHz spectrum.
- **5 GHz Channels**—Displays the number of radios using each channel of the 5 GHz spectrum.
- **EIRP (dBm)**—Displays the number of radios per EIRP level.

Usage dashboard

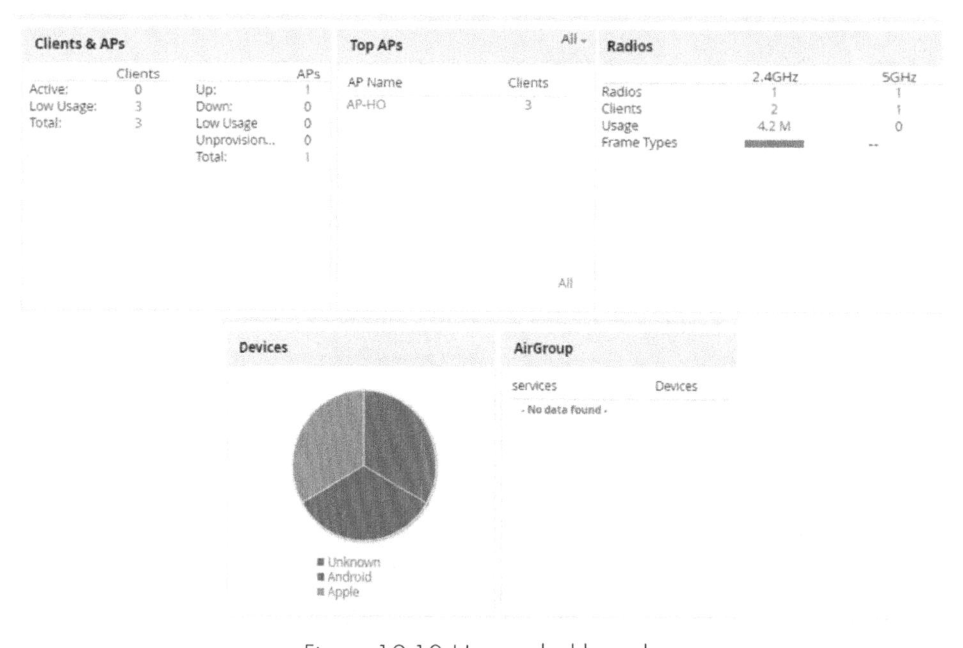

Figure 10-10 Usage dashboard

The **Usage dashboard** page displays usage summaries for all managed devices. Click the hyperlinked text in the sections on this dashboard to view 15-minute lists and trend charts. You can see AP and client summaries in new windows. Figure 10-10 shows an example of some of the information that is available from the Usage dashboard.

To access the WLANs dashboard, navigate to **Managed Networks > Dashboard > Usage**. The following information is available from the Usage dashboard:

- **Clients & APs**—The active wireless clients, status of APs, and its usage.
- **Top APs**—The list of APs with the number of clients on the managed device. The list of APs is in the descending order based on the number of clients associated with an AP. You can filter the APs for the 2.4 GHz and 5 GHz radio band options.
- **Radios**—The radios and clients connected to an AP, usage, and frame types transmitted and received by the radio.
- **Devices**—The pie chart of the clients based on the device type. Clicking on the pie chart segment opens the client details page filtered on the device type.
- **AirGroup**—All the AirGroup services available and number of servers offering the service. It is aggregated by the total number of AirGroup servers sorted by the services they advertise
- **Overall Usage**—The total number of clients and APs that have the low usage and throughput data in the last 15 minutes.
- **Usage by WLANs**—The total number of clients per WLAN and throughput data in the last 15 minutes. You can view only three WLANs in a graph, and the remaining WLANs are displayed in other graph. Click the graph to view the blown up chart and information on the **Clients** page.

Traffic analysis

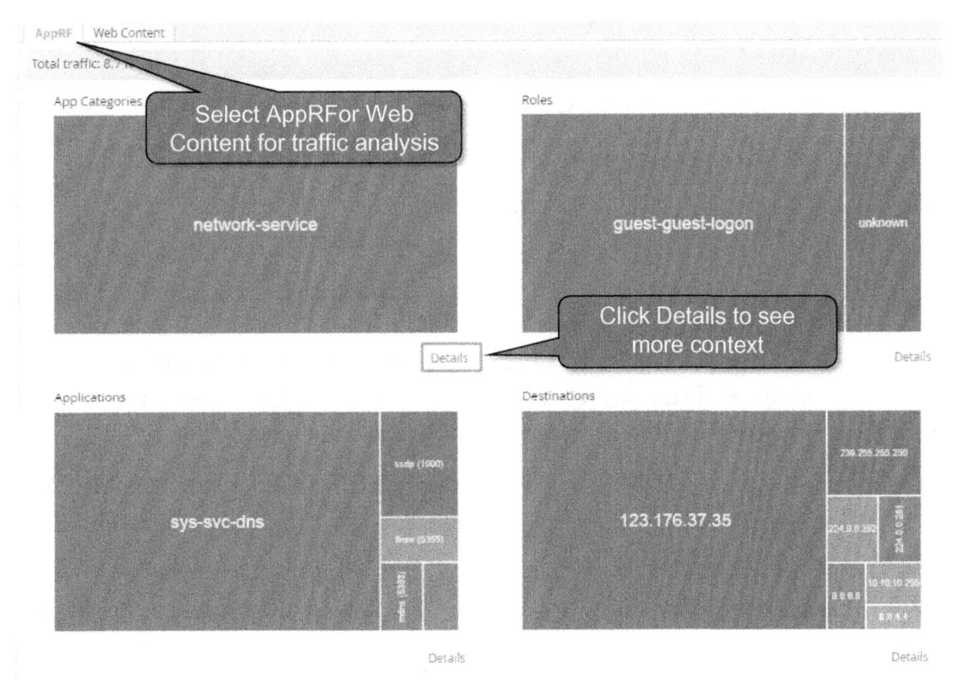

Figure 10-11 Traffic Analysis

The information in the **Traffic Analysis dashboard** page is gathered by AppRF, an application visibility and control feature. AppRF performs Deep Packet Inspection (DPI) for local traffic and detects over 1500 applications on the network. AppRF allows you to configure both application and application category policies within a given user role.

Note

The Traffic Analysis dashboard application visibility feature is supported only in 7000 Series and 7200 Series controllers and requires WebCC the PEF-NG license.

Figure 10-11 shows an example of some of the information that is available from the Traffic Analysis dashboard. To access the WLANs dashboard, navigate to **Managed Networks > Dashboard > Traffic Analysis**. From this dashboard you can view traffic analysis based on the following two options:

- **AppRF**—This page displays the PEF summary of all the sessions in the managed device aggregated by users, devices, destinations, applications, WLANs, and roles. The applications, application categories, and other containers are represented in box charts.
- **Web Content**—This tab displays the summary of only the web traffic in the managed device. When the WebCC feature is enabled, all web traffic is classified. This includes both HTTP and HTTPS traffic. The classification is done in data path as the traffic flows through the managed device and updates dynamically

Benefits of WebCC

- Prevention of malicious malware, spyware, or adware by blocking known dangerous websites
- Visibility into web content category-level
- Visibility into web sites accessed by the user

Filter view

On the Traffic Analysis dashboard page if you click on any rectangle tile in a container that filter is applied across all containers. Data is filtered by all selected rectangles. When a filter is applied, the **Details** link changes to **User filtered by filter option** in that container.

For example, if you click on the **Web** rectangle in the **Application Categories** container, Application Categories = = Web filter is applied to all other containers (Roles, WLANs, Application, Destination, and Devices).

You can apply multiple filters from different containers by clicking on multiple rectangle tiles in various containers.

Details view

User Destination Application WLAN Device Role

Application Categories (2)

Application Category	Bytes	Packets	Application	User	Destination	WLAN	Device	Role
network-service	13.0 K	105	3	1	3	1	0	1
unknown	0	0	1	1	1	1	0	0

Aggregated Sessions (4)

Source IP	Destination Name/IP	IP Protocol	Application	Application Category	Tx Bytes	Rx Bytes
10.10.10.253	239.255.255.250	udp	ssdp	network-service	7.6 K	0
10.10.10.253	10.10.10.255	udp	nbns	network-service	3.1 K	0
10.10.10.253	224.0.0.252	udp	llmnr (5355)	network-service	2.2 K	0
10.10.10.253	224.0.0.251	udp	mdns (5353)	unknown	129	0

Figure 10-12 Details view for App categories

Click on **Details** link in a container to navigate to the corresponding details page for that container. Figure 10-12 shows the Applications Categories and Aggregated Sessions section of the details page for the App Categories container. Figure 10-13 shows the Usage Breakdown section of the details page.

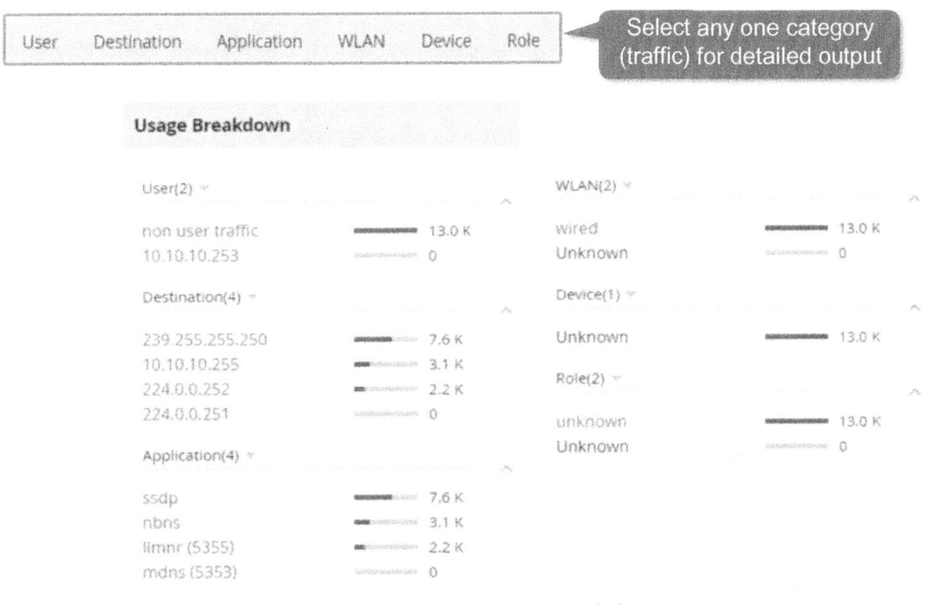

Figure 10-13 Details view Usage Breakdown section

The Usage Breakdown section of the Details view shows the summary of all categories. To view detailed output, select one of the categories. (Example—User specific traffic, WLAN specific traffic, and so forth.)

Security analysis

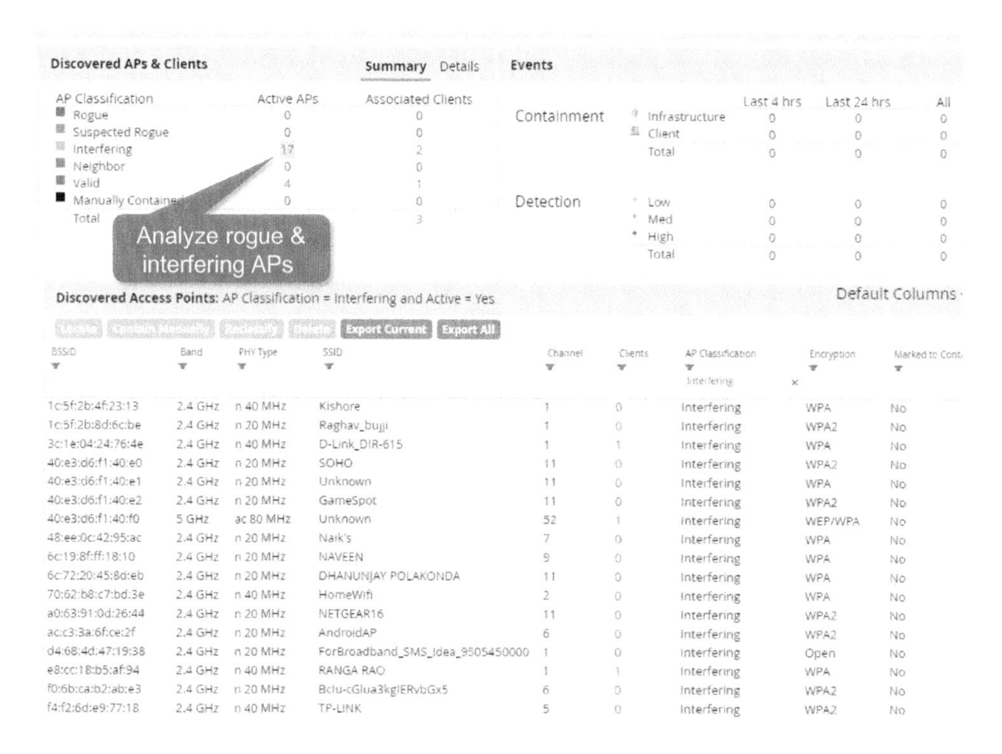

AP Classification	Active APs	Associated Clients
Rogue	0	0
Suspected Rogue	0	0
Interfering	17	2
Neighbor	0	0
Valid	4	1
Manually Contained	0	0
Total		3

		Last 4 hrs	Last 24 hrs	All
Containment	Infrastructure	0	0	0
	Client	0	0	0
	Total	0	0	0
Detection	Low	0	0	0
	Med	0	0	0
	High	0	0	0
	Total	0	0	0

BSSID	Band	PHY Type	SSID	Channel	Clients	AP Classification	Encryption	Marked to Cont.
1c:5f:2b:4f:23:13	2.4 GHz	n 40 MHz	Kishore	1	0	Interfering	WPA	No
1c:5f:2b:8d:6c:be	2.4 GHz	n 20 MHz	Raghav_bujji	1	0	Interfering	WPA2	No
3c:1e:04:24:76:4e	2.4 GHz	n 40 MHz	D-Link_DIR-615	1	1	Interfering	WPA	No
40:e3:d6:f1:40:e0	2.4 GHz	n 20 MHz	SOHO	11	0	Interfering	WPA2	No
40:e3:d6:f1:40:e1	2.4 GHz	n 20 MHz	Unknown	11	0	Interfering	WPA	No
40:e3:d6:f1:40:e2	2.4 GHz	n 20 MHz	GameSpot	11	0	Interfering	WPA2	No
40:e3:d6:f1:40:f0	5 GHz	ac 80 MHz	Unknown	52	1	Interfering	WEP/WPA	No
48:ee:0c:42:95:ac	2.4 GHz	n 20 MHz	Naik's	7	0	Interfering	WPA	No
6c:19:8f:ff:18:10	2.4 GHz	n 20 MHz	NAVEEN	9	0	Interfering	WPA	No
6c:72:20:45:8d:eb	2.4 GHz	n 20 MHz	DHANUNJAY POLAKONDA	11	0	Interfering	WPA	No
70:62:b8:c7:bd:3e	2.4 GHz	n 40 MHz	HomeWifi	2	0	Interfering	WPA	No
a0:63:91:0d:26:44	2.4 GHz	n 20 MHz	NETGEAR16	11	0	Interfering	WPA2	No
ac:c3:3a:6f:ce:2f	2.4 GHz	n 20 MHz	AndroidAP	6	0	Interfering	WPA2	No
d4:68:4d:47:19:38	2.4 GHz	n 20 MHz	ForBroadband_SMS_Idea_9505450000	1	0	Interfering	Open	No
e8:cc:18:b5:af:94	2.4 GHz	n 40 MHz	RANGA RAO	1	1	Interfering	WPA	No
f0:6b:ca:b2:ab:e3	2.4 GHz	n 20 MHz	Bclu-cGlua3kglERvbGx5	6	0	Interfering	WPA2	No
f4:f2:6d:e9:77:18	2.4 GHz	n 40 MHz	TP-LINK	5	0	Interfering	WPA2	No

Figure 10-14 Security analysis

The **Security dashboard** page allows you to monitor intrusion detection and protection events (Figure 10-14).

The two top tables—**Discovered APs & Clients** and **Events**—contain data as links. When these links are selected, they arrange, filter, and display the appropriate information in the lower table, **Discovered Access Points**.

To access the WLANs dashboard, navigate to **Managed Networks > Dashboard > Security.**

Using alerts

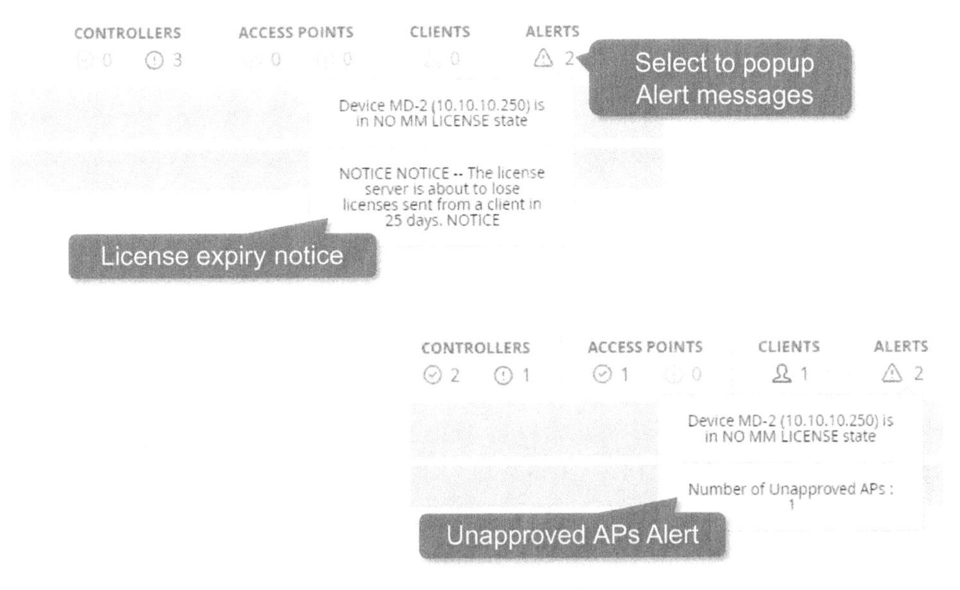

Figure 10-15 Using alerts

Alerts were introduced in AOS 8.x. Alerts are shown on the banner. You see a red triangle with an exclamation mark, along with a number to count new alerts. An alert description will pop-up when you click on the Alert (number) (Figure 10-15).

Navigate to the relevant page by clicking on the Alert message.

Learning check

1. Mobility controller alerts appear on the banner.
   a. True
   b. False

2. Network usage can be monitored through dashboard.
   a. True
   b. False

3. Which of the following is used to monitor application usage?
   a. Banner
   b. Dashboard
   c. AppRF
   d. WebCC

Answers to learning check

1. Mobility controller alerts appear on the banner.
   a. **True**
   b. False

2. Network usage can be monitored through dashboard.
   a. **True**
   b. False

3. Which of the following is used to monitor application usage?
   a. Banner
   b. **Dashboard**
   c. AppRF
   d. WebCC

Monitoring through Airwave

Airwave

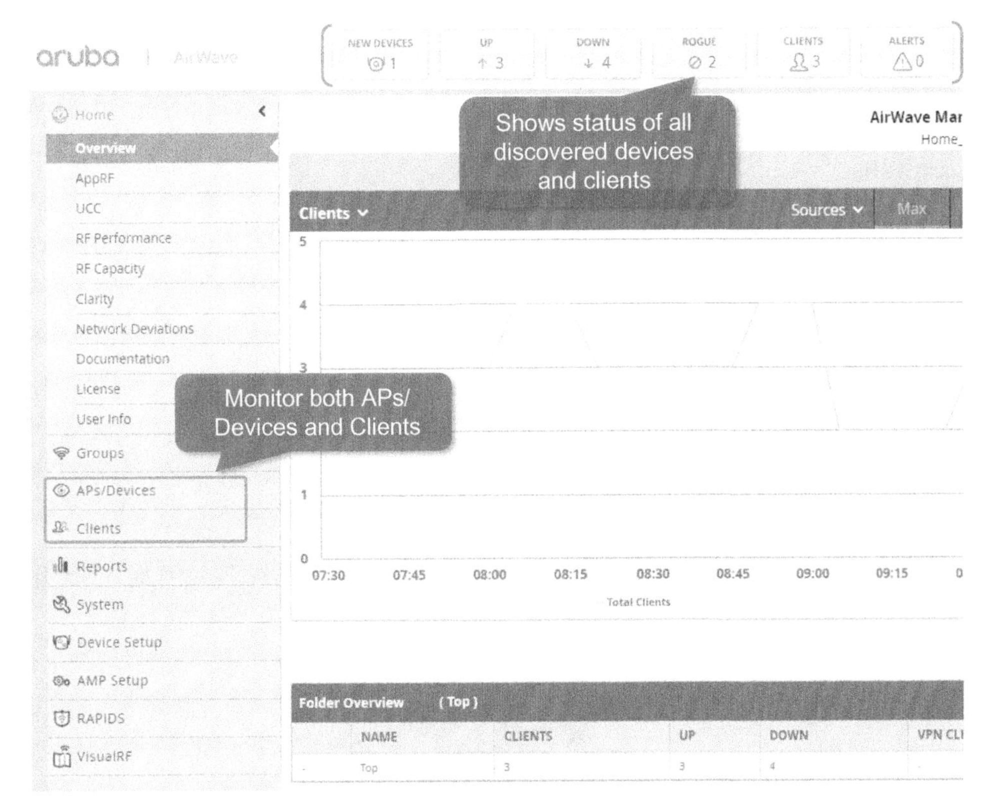

Figure 10-16 Airwave

Aruba AirWave is a powerful and easy-to-use network operations system that manages wired and wireless infrastructure from Aruba and third-party manufacturers. It also provides granular visibility into devices, users, and applications on the network. AirWave lets IT organizations proactively optimize network performance, strengthen wireless security, and improve the end-user experience.

AirWave provides real-time monitoring, proactive alerts, historical reporting, and fast, efficient troubleshooting. Dedicated dashboard views quickly help view potential RF coverage issues, unified communications and collaboration (UCC) traffic, application performance, and network services health.

In Figure 10-16, the banner shows the status of all discovered devices and clients, as described below.

Status

Down

- May appear until AWMS polls device.
- Click the Down Link to see more info

Configurations

- Unknown—Means the device configuration has not yet been fetched.
- Mismatched—Device config does not match the group config.
- Error—Communication issues

Network overview

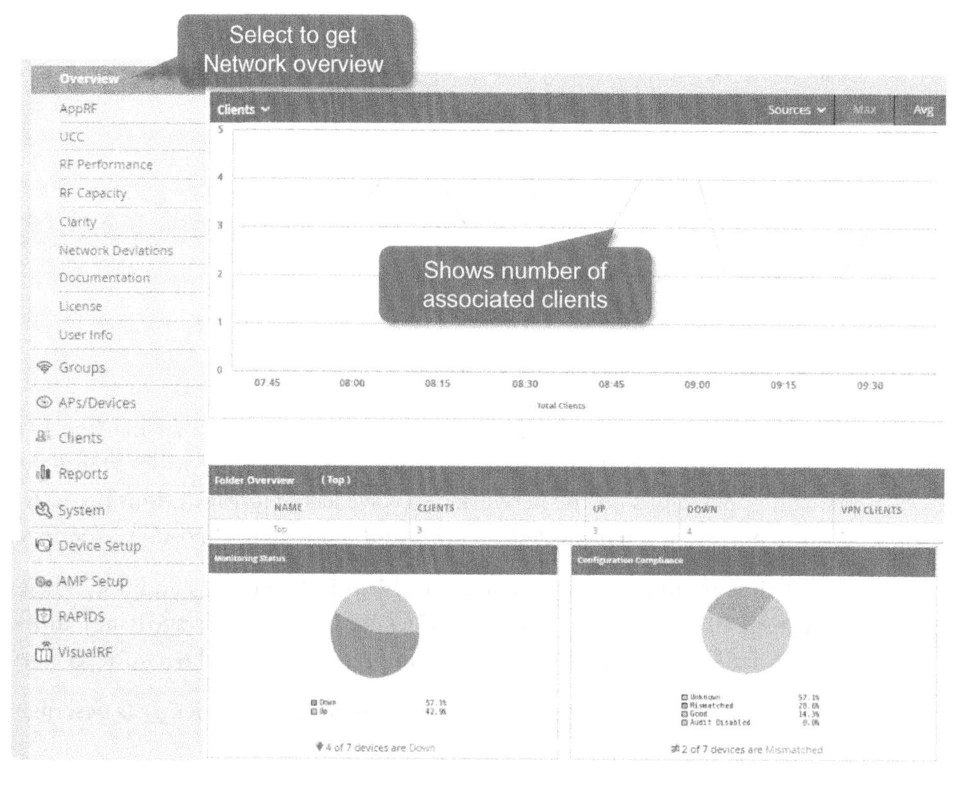

Figure 10-17 Network overview

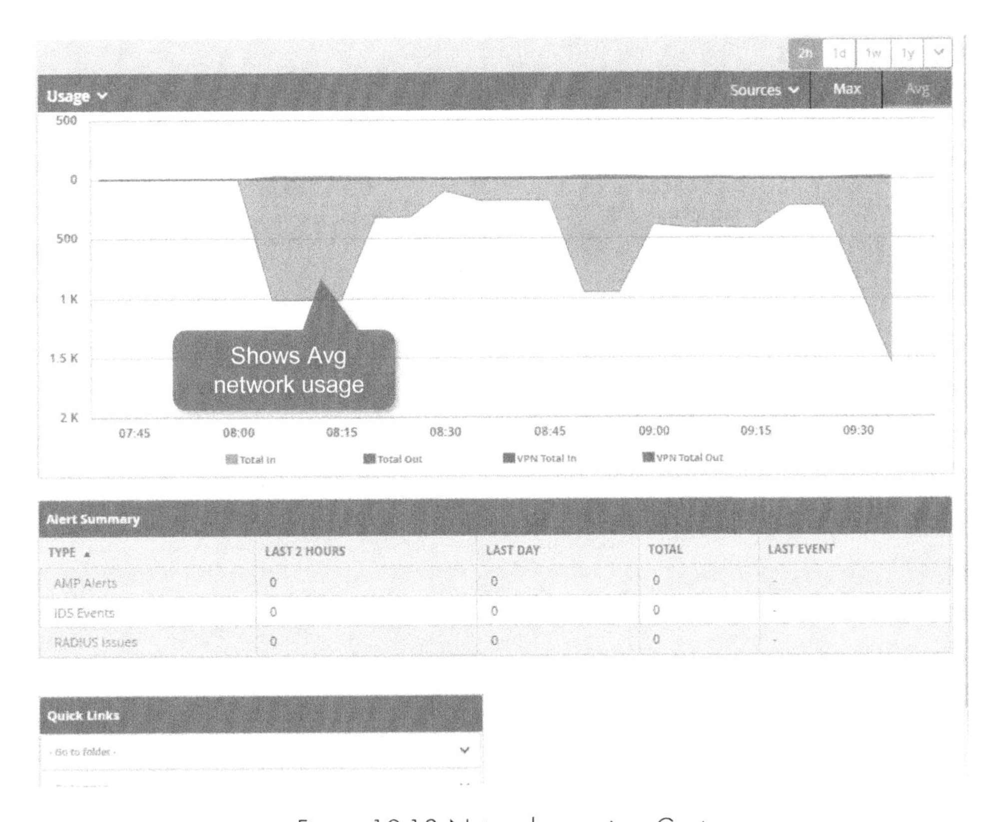

Figure 10-18 Network overview Cont.

Figure 10-17 and Figure 10-18 show the AirWave network overview. You see the number of associated clients, along with average network utilization.

Monitoring devices

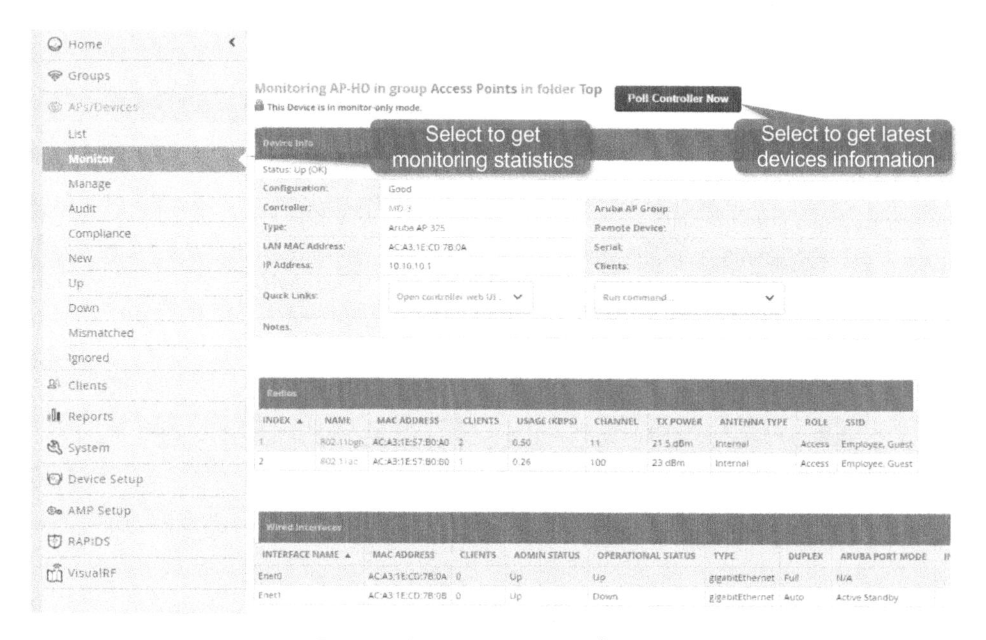

Figure 10-19 Monitoring devices

You can find much useful information in the **APs/Devices > Monitor** page. Click "Poll Controller Now" to ensure that you are looking at the very latest information.

The top section is labelled "Device Info." The example in Figure 10-19 shows the status of a 325 model AP controlled by a controller named MD-3. The status is OK, and the configuration is good. There are quick links to open the controller's Web UI and to run commands.

The next section is labelled "Radios." You can see that the 2.4 GHz radio (802.11bgn) is currently supporting two clients on channel 11. The 5 GHz radio is currently supporting one client, on channel 100. You can see the Transmit (Tx) power settings for each radio, and the SSIDs supported. You can click on the radio name to view more detailed information.

The bottom section is labelled "Wired Interfaces." You can see that this AP has two Ethernet interfaces. One has an operational status of "Up," and the other is currently "Down," but is configured to Active Standby mode.

Monitoring clients

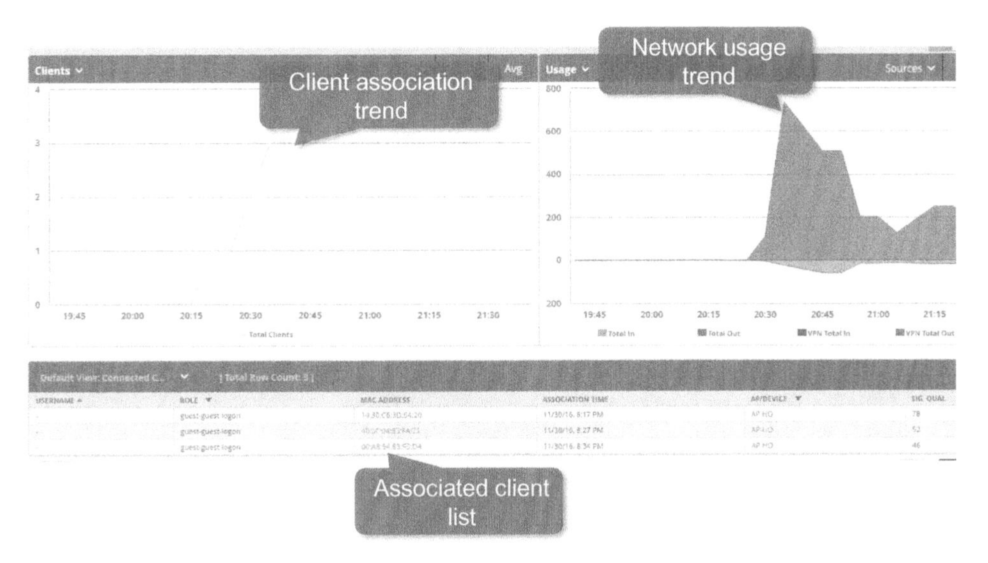

Figure 10-20 Monitoring devices

The Clients page displays multiple types of user data for existing WLAN clients and VPN users. The data comes from a number of locations, including AP data tables, RADIUS accounting server information, and AirWave-generated data.

Figure 10-20 shows the **Clients > Overview** page. You see how client association trends, revealing total number of clients connected over time. You also see a network usage trend. This can be a helpful baseline. You become aware that under certain traffic loads, clients are happy and do not complain of performance issues. Should clients one day begin to complain, you can tell if it is a WLAN performance-related issue.

The bottom of the output shows a listing of all clients, their role, and their signal quality. You can click on the MAC address link of any client to get more detail.

Client graphs

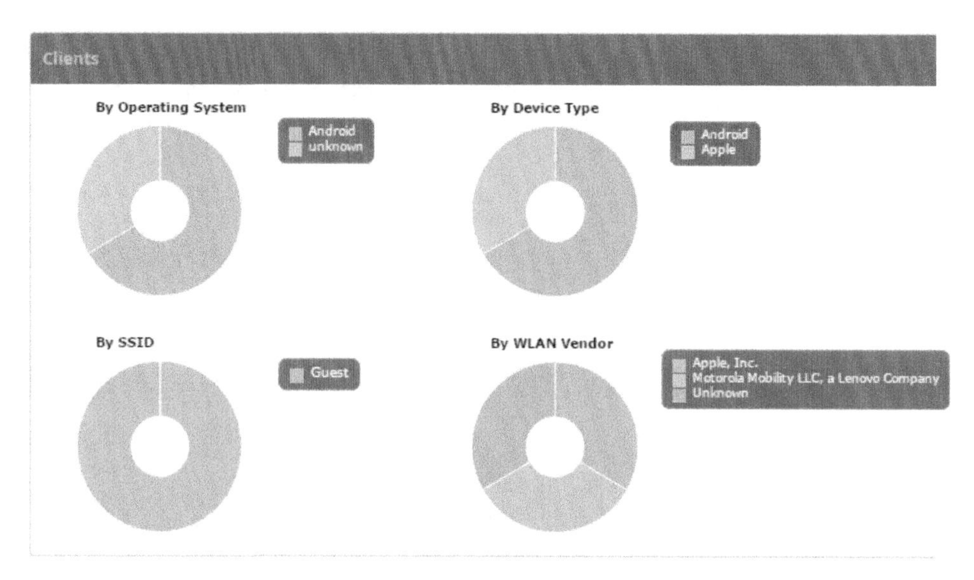

Figure 10-21 Client graphs

The **Clients > Overview** page provides pie charts, which summarize information about all clients across your network. This information is broken down by operating system, device type, SSID, and WLAN vendor. You can click on any section of these charts to see a table of clients in that category. If any clients on your network are specified as Watched Clients, then a Watched Clients table will appear in this page, showing the client's health, speed, SNR value, SNR trend, when the client was last heard, and whether the client is active (Figure 10-21).

Reports

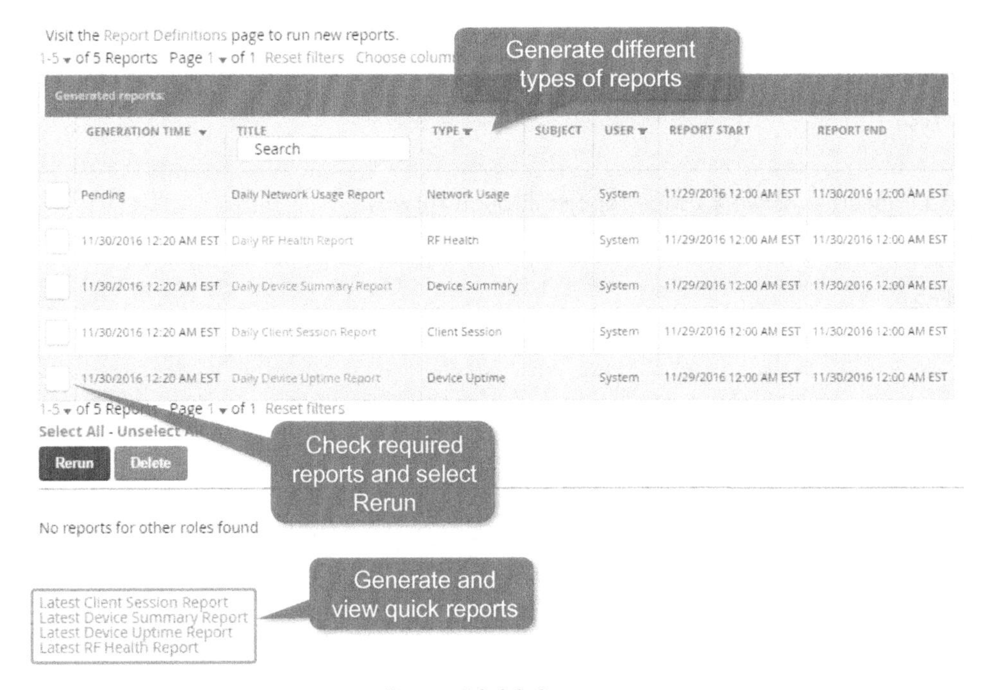

Figure 10-22 Reports

The **Reports > Generated** page supports several general viewing options (Figure 10-22). By default, reports are sorted by Generation Time. You can sort reports by any other column header, in sequential or reverse sequential order. You can also choose columns, export the Generated Reports list in CSV, and modify the pagination of this list.

The **Reports > Detail** page launches when you select any report title from this page. The Generated Reports page contains fewer columns and information than the Definitions page.

Learning check

4. Airwave can be used to monitor all rogue APs and clients.
    a. True
    b. False

5. We can generate network usage reports through Airwave.
    a. True
    b. False

Answers to learning check

4. Airwave can be used to monitor all rogue APs and clients.
   a. **True**
   b. False

5. We can generate network usage reports through Airwave.
   a. **True**
   b. False

11 Practice Test

1. Which organization develops Layer 2 standards for transmitting data over a wireless network?
    a. IETF
    b. FCC
    c. IEEE
    d. ETSI

2. Which channels are typically used in the 2.4GHz spectrum, and why?
    a. 1,6,11 are used, because the other channels can cause co-channel interference.
    b. 36, 40, 44, 48, 52, 56, 60, 64 are used, because they are available
    c. 100 through 140 are used, because they are in the so-called UNII-2 Extended range
    d. 1,6, 11 are used, because they do not interfere with aircraft radar systems
    e. 36, 40, 44, 48, 52, 56, 60, 64 are used, because the other channels can cause co-channel interference.

3. What is the advantage of using a "capacity-based" approach to WLAN design, as compared to a "coverage-based" design?
    a. Capacity-based designs are more economical
    b. A coverage-based design requires more APs
    c. Capacity-based designs tend to provide better performance and reliability for a typical office environment.
    d. With a capacity-based design, each AP covers more floor space

4. Which of the following APs do not support end user connectivity?
    a. Campus AP
    b. Mesh AP
    c. Air Monitor AP
    d. Remote AP
    e. Instant AP

5. Choose all options below that accurately describe the differences between a Mobility Master (MM) and a Master Controller (MC)?

    a. Unlike the MC, an MM can be deployed on both physical and virtual appliances

    b. The MM can push full configurations to all managed devices

    c. You can deploy an MM using either AOS 6.x, or AOS 8.x

    d. The MC cannot push L2 and L3 configuration to managed devices.

    e. An MM can directly terminate AP tunnels

6. Which HPE Aruba feature enables you to manage guest and corporate WLANs on the same APs, but with an "Air Wall" between them?

    a. AirMatch

    b. AirPlay

    c. MultiZone

    d. ClientMatch

    e. Zone filtering

7. Which statements below accurately describe Aruba OS 8.x hierarchical configuration?

    a. Licenses are centrally managed by the Mobility Master

    b. Configurations that you create in the mn group automatically flows into all groups below that level

    c. You can configure an individual managed device, if you need to override something in a group configuration

    d. You can create hierarchical management groups, but you cannot create subgroups

    e. The system does not automatically create groups. You must create them all manually.

8. Which two features help to automate controller configuration?

    a. Zero Touch Provisioning

    b. netconf

    c. ClearPass

    d. Activate

    e. MultiTouch

9. Controllers at remote locations can connect to corporate headquarters over an IPSec tunnel, which terminates on a VPN Concentrator (VPNC).

    a. True

    b. False

10. Which statements below accurately describe HPE Aruba's AP group structure?

    a. APs advertise SSIDs based on their AP-group membership

    b. APs can belong to a primary AP-group, and a secondary AP-group

    c. If you want all APs to advertise the same SSIDs, you only need one group

    d. AP groups contain all AP configuration, organized into profiles

    e. It is a best practice to modify the default profile.

11. The recommended method for creating WLANs is through the WLAN wizard.

    a. True

    b. False

12. During WLAN configuration, you have the option of defining authentication servers. Choose the options below that accurately describe this option.

    a. You typically define authentication servers for WLANs defined to use Enterprise and Personal security

    b. You can configure RADIUS and LDAP authentication server types

    c. You can configure TACACS and AD authentication types

    d. You need to define the IP address and the shared key of the authentication server

    e. You cannot define authentication servers via the WLAN wizard

13. Control Place Security (CPSec) includes which of the following features and capabilities?

    a. CPSec is used to encrypt management traffic between APs and controllers

    b. CPSec is used to encrypt end user wireless traffic between APs and controllers

    c. CPSec relies on the standards-based IPsec protocol suite

    d. You can disable auto-certificate provisioning to prevent rogue APs from connecting to your controller

    e. You must ensure that all APs are first connected to the controller before you enable CPSec

14. Which of the options below can be used for AP provisioning?
    a. Manual provisioning, via GUI
    b. Provision the AP directly, from its own CLI
    c. Use the wizard to provision APs
    d. Use Zero Touch Provisioning
    e. Use Activate

15. An Access Point in your network is named AP274. This AP supports three WLANs, for both the 2.4GHz and 5GHz bands. Which statement below accurately describes the tunnels between the AP and the MC?
    a. There will be three CPSec tunnels
    b. There will be six CPSec tunnels and one GRE tunnel
    c. A single tunnel is used for all WLANs
    d. A single tunnel is used for all WLANs for the 2.4GHz radio, and another tunnel is used for the 5GHz radio.
    e. There will be seven GRE tunnels

16. Which statement below are true, as relates to 802.11i authentication and guest web authentication.
    a. The 802.11 negotiation phase is identical for both methods
    b. 802.11i authentication is a layer 2 process, while guest web authentication is a layer 3 process
    c. With both methods, the client IP address is assigned after the authentication phase
    d. Guest web authentication is a layer 3 process that relies on the Layer 2 security protocols
    e. Guest web relies on role assignment for its security policy, while 802.11i authentication relies on 802.1x

17. Extensible Authentication Protocol over LAN (EAPOL) is typically used for which of the following authentication methods?
    a. PEAP
    b. WEP
    c. Web authentication
    d. PSK
    e. EAP-TLS

18. Which of the options below are Layer 2 Denial of Service (DoS) attacks?
    a. RF Jamming
    b. Deauthentication attacks
    c. Association/disassociation attacks
    d. Asymmetric key encryption
    e. Packet snooping

19. Which statements below accurately describe user roles?
    a. User roles change as users roam from subnet to subnet
    b. User roles can be assigned directly from a Mobility Controller (MC)
    c. User roles can be assigned from a RADIUS server
    d. Role assignment does not occur until after the user is fully authenticated
    e. Each role is associated with a single policy

20. A policy is merely an ordered set of rules, processed in alphabetical order?
    a. True
    b. False

21. What is the typical method used to override a locally derived role?
    a. You cannot override a locally derived role.
    b. Use an AP derived role
    c. Configure a new role via the CLI
    d. Configure role derivation for that user on an external RADIUS server
    e. Configure a role override at any place at or above the same level in the hierarchy

22. Which statements below accurately describe Adaptive Radio Management (ARM)?
    a. It automatically responds to a changing RF environment
    b. It automatically adjusts AP channel and power settings
    c. Aruba OS version 8.x and higher must use ARM
    d. Each AP scans its current operating channel, and relies on other APs to scan the other channels
    e. ARM uses the Interference Index and Coverage Index to help determine optimal AP settings

23. Which feature ensures that APs will not change channels while clients are attached?
    a. ClientMatch
    b. Over-the-Air (OTA) updates
    c. Client Aware
    d. Mode Aware
    e. AirWatch

24. Rule-Based Client Match (RBCM) runs on an AOS 8.x Mobility Master (MM)
    a. True
    b. False

25. As relates to guest WLANs, what are the advantages of the MultiZone feature?
    a. Reduces the number of required controllers
    b. Simplifies the configuration
    c. Eliminates the need for an external RADIUS server
    d. Reduces the number of tunnels traversed by guest traffic
    e. Allows each AP to terminate guest and corporate traffic to different controllers

26. Which authentication option is most appropriate for guest WLAN access?
    a. 802.1X authentication
    b. PSK authentication
    c. Certificate-based authentication
    d. Captive portal authentication

27. What are two basic network services must be available for pre-authenticated guests, to ensure that they can authenticate?
    a. FTP
    b. SQL
    c. HTTP/HTTPS
    d. TFTP
    e. DNS

28. You are analyzing the Aruba controller GUI interface's Client dashboard. Which value indicates a ratio of total bytes transmitted to/from client to total airtime required.

    a. Retried frames

    b. Signal to Noise Ratio (SNR)

    c. Goodput

29. You using the Dashboard to analyze APs. Which of the options would you most hope to see, as an indication of a healthy RF environment?

    a. –95dBm Noise Floor

    b. –65dBm Noise Floor

    c. 65% Channel Busy

    d. 40% Channel Quality

    e. 40% Interference

30. Which of the following are benefits of WebCC?

    a. Prevent malicious malware

    b. Visibility into firewall access-list usage

    c. Ability to visualize the distribution of trusted vs. untrusted users on the network

    d. The ability to use a web browser for more advanced guest options

    e. Visibility into web sites accessed by the user

Answers to Practice Test

1. Which organization develops Layer 2 standards for transmitting data over a wireless network?

    a. IETF

    b. FCC

    c. IEEE

    d. ETSI

2. Which channels are typically used in the 2.4GHz spectrum, and why?
   a. **1,6,11 are used, because the other channels can cause co-channel interference**.
   b. 36, 40, 44, 48, 52, 56, 60, 64 are used, because they are available
   c. 100 through 140 are used, because they are in the so-called UNII-2 Extended range
   d. 1,6, 11 are used, because they do not interfere with aircraft radar systems
   e. 36, 40, 44, 48, 52, 56, 60, 64 are used, because the other channels can cause co-channel interference.

3. What is the advantage of using a "capacity-based" approach to WLAN design, as compared to a "coverage-based" design?
   a. Capacity-based designs are more economical
   b. A coverage-based design requires more APs
   c. **Capacity-based designs tend to provide better performance and reliability for a typical office environment**.
   d. With a capacity-based design, each AP covers more floor space

4. Which of the following APs do not support end user connectivity?
   a. Campus AP
   b. Mesh AP
   c. **Air Monitor AP**
   d. Remote AP
   e. Instant AP

5. Choose all options below that accurately describe the differences between a Mobility Master (MM) and a Master Controller (MC)?
   a. **Unlike the MC, an MM can be deployed on both physical and virtual appliances**
   b. **The MM can push full configurations to all managed devices**
   c. You can deploy an MM using either AOS 6.x, or AOS 8.x
   d. **The MC cannot push L2 and L3 configuration to managed devices.**
   e. An MM can directly terminate AP tunnels

6. Which HPE Aruba feature enables you to manage guest and corporate WLANs on the same APs, but with an "Air Wall" between them?
   a. AirMatch
   b. AirPlay
   c. **MultiZone**
   d. ClientMatch
   e. Zone filtering

7. Which statements below accurately describe Aruba OS 8.x hierarchical configuration?
   a. **Licenses are centrally managed by the Mobility Master**
   b. **Configurations that you create in the mn group automatically flows into all groups below that level**
   c. **You can configure an individual managed device, if you need to override something in a group configuration**
   d. You can create hierarchical management groups, but you cannot create subgroups
   e. The system does not automatically create groups. You must create them all manually.

8. Which two features help to automate controller configuration?
   a. **Zero Touch Provisioning**
   b. netconf
   c. ClearPass
   d. **Activate**
   e. MultiTouch

9. Controllers at remote locations can connect to corporate headquarters over an IPSec tunnel, which terminates on a VPN Concentrator (VPNC).
   a. **True**
   b. False

10. Which statements below accurately describe HPE Aruba's AP group structure?
    a. **APs advertise SSIDs based on their AP-group membership**
    b. APs can belong to a primary AP-group, and a secondary AP-group
    c. **If you want all APs to advertise the same SSIDs, you only need one group**

d. **AP groups contain all AP configuration, organized into profiles**

e. It is a best practice to modify the default profile.

11. The recommended method for creating WLANs is through the WLAN wizard.

a. **True**

b. False

12. During WLAN configuration, you have the option of defining authentication servers. Choose the options below that accurately describe this option.

a. You typically define authentication servers for WLANs defined to use Enterprise and Personal security

b. **You can configure RADIUS and LDAP authentication server types**

c. You can configure TACACS and AD authentication types

d. **You need to define the IP address and the shared key of the authentication server**

e. You cannot define authentication servers via the WLAN wizard

13. Control Place Security (CPSec) includes which of the following features and capabilities?

a. **CPSec is used to encrypt management traffic between APs and controllers**

b. CPSec is used to encrypt end user wireless traffic between APs and controllers

c. **CPSec relies on the standards-based IPsec protocol suite**

d. **You can disable auto-certificate provisioning to prevent rogue APs from connecting to your controller**

e. You must ensure that all APs are first connected to the controller before you enable CPSec

14. Which of the options below can be used for AP provisioning?

a. **Manual provisioning, via GUI**

b. **Provision the AP directly, from its own CLI**

c. **Use the wizard to provision APs**

d. Use Zero Touch Provisioning

e. Use Activate

15. An Access Point in your network is named AP274. This AP supports three WLANs, for both the 2.4GHz and 5GHz bands. Which statement below accurately describes the tunnels between the AP and the MC?

    a. There will be three CPSec tunnels

    b. There will be six CPSec tunnels and one GRE tunnel

    c. A single tunnel is used for all WLANs

    d. A single tunnel is used for all WLANs for the 2.4GHz radio, and another tunnel is used for the 5GHz radio.

    e. **There will be seven GRE tunnels**

16. Which statement below are true, as relates to 802.11i authentication and guest web authentication.

    a. **The 802.11 negotiation phase is identical for both methods**

    b. **802.11i authentication is a layer 2 process, while guest web authentication is a layer 3 process**

    c. With both methods, the client IP address is assigned after the authentication phase

    d. Guest web authentication is a layer 3 process that relies on the Layer 2 security protocols

    e. Guest web relies on role assignment for its security policy, while 802.11i authentication relies on 802.1x

17. Extensible Authentication Protocol over LAN (EAPOL) is typically used for which of the following authentication methods?

    a. **PEAP**

    b. WEP

    c. Web authentication

    d. **PSK**

    e. **EAP-TLS**

18. Which of the options below are Layer 2 Denial of Service (DoS) attacks?

    a. RF Jamming

    b. **Deauthentication attacks**

    c. **Association/disassociation attacks**

    d. Asymmetric key encryption

    e. Packet snooping

19. Which statements below accurately describe user roles?
    a. User roles change as users roam from subnet to subnet
    b. **User roles can be assigned directly from a Mobility Controller (MC)**
    c. **User roles can be assigned from a RADIUS server**
    d. Role assignment does not occur until after the user is fully authenticated
    e. Each role is associated with a single policy

20. A policy is merely an ordered set of rules, processed in alphabetical order?
    a. True
    b. **False**

21. What is the typical method used to override a locally derived role?
    a. You cannot override a locally derived role.
    b. Use an AP derived role
    c. Configure a new role via the CLI
    d. **Configure role derivation for that user on an external RADIUS server**
    e. Configure a role override at any place at or above the same level in the hierarchy

22. Which statements below accurately describe Adaptive Radio Management (ARM)?
    a. **It automatically responds to a changing RF environment**
    b. **It automatically adjusts AP channel and power settings**
    c. Aruba OS version 8.x and higher must use ARM
    d. Each AP scans its current operating channel, and relies on other APs to scan the other channels
    e. **ARM uses the Interference Index and Coverage Index to help determine optimal AP settings**

23. Which feature ensures that APs will not change channels while clients are attached?
    a. ClientMatch
    b. Over-the-Air (OTA) updates
    c. **Client Aware**
    d. Mode Aware
    e. AirWatch

24. Rule-Based Client Match (RBCM) runs on an AOS 8.x Mobility Master (MM)

    a. **True**

    b. False

25. As relates to guest WLANs, what are the advantages of the MultiZone feature?

    a. Reduces the number of required controllers

    b. Simplifies the configuration

    c. Eliminates the need for an external RADIUS server

    d. **Reduces the number of tunnels traversed by guest traffic**

    e. **Allows each AP to terminate guest and corporate traffic to different controllers**

26. Which authentication option is most appropriate for guest WLAN access?

    a. 802.1X authentication

    b. PSK authentication

    c. Certificate-based authentication

    d. **Captive portal authentication**

27. What are two basic network services must be available for pre-authenticated guests, to ensure that they can authenticate?

    a. FTP

    b. SQL

    c. **HTTP/HTTPS**

    d. TFTP

    e. **DNS**

28. You are analyzing the Aruba controller GUI interface's Client dashboard. Which value indicates a ratio of total bytes transmitted to/from client to total airtime required.

    a. Retried frames

    b. Signal to Noise Ratio (SNR)

    c. **Goodput**

29. You using the Dashboard to analyze APs. Which of the options would you most hope to see, as an indication of a healthy RF environment?
    a. **–95dBm Noise Floor**
    b. –65dBm Noise Floor
    c. 65% Channel Busy
    d. 40% Channel Quality
    e. 40% Interference

30. Which of the following are benefits of WebCC?
    a. **Prevent malicious malware**
    b. Visibility into firewall access-list usage
    c. Ability to visualize the distribution of trusted vs. untrusted users on the network
    d. The ability to use a web browser for more advanced guest options
    e. **Visibility into web sites accessed by the user**

Index

A

B

C

E

F

X

Z